Palgrave Studies in Agricultural Economics and Food Policy

Series Editor
Christopher Barrett
Cornell University
Ithaca, NY, USA

Agricultural and food policy lies at the heart of many pressing societal issues today and economic analysis occupies a privileged place in contemporary policy debates. The global food price crises of 2008 and 2010 underscored the mounting challenge of meeting rapidly increasing food demand in the face of increasingly scarce land and water resources. The twin scourges of poverty and hunger quickly resurfaced as high-level policy concerns, partly because of food price riots and mounting insurgencies fomented by contestation over rural resources. Meanwhile, agriculture's heavy footprint on natural resources motivates heated environmental debates about climate change, water and land use, biodiversity conservation, and chemical pollution. Agricultural technological change, especially associated with the introduction of genetically modified organisms, also introduces unprecedented questions surrounding intellectual property rights and consumer preferences regarding credence (i.e., unobservable by consumers) characteristics. Similar new agricultural commodity consumer behavior issues have emerged around issues such as local foods, organic agriculture, and fair trade, even motivating broader social movements. Public health issues related to obesity, food safety, and zoonotic diseases such as avian or swine flu also have roots deep in agricultural and food policy. And agriculture has become inextricably linked to energy policy through biofuels production. Meanwhile, the agricultural and food economy is changing rapidly throughout the world, marked by continued consolidation at both farm production and retail distribution levels, elongating value chains, expanding international trade, and growing reliance on immigrant labor and information and communications technologies. In summary, a vast range of topics of widespread popular and scholarly interest revolve around agricultural and food policy and economics. The extensive list of prospective authors, titles, and topics offers a partial, illustrative listing. Thus a series of topical volumes, featuring cutting-edge economic analysis by leading scholars, has considerable prospect for both attracting attention and garnering sales. This series will feature leading global experts writing accessible summaries of the best current economics and related research on topics of widespread interest to both scholarly and lay audiences.

More information about this series at
http://www.palgrave.com/gp/series/14651

James Shortle • Markku Ollikainen
Antti Iho

Water Quality and Agriculture

Economics and Policy for Nonpoint Source Water Pollution

palgrave
macmillan

James Shortle
Distinguished Professor Emeritus
Agricultural and Environmental
Economics, College of Agricultural
Sciences, Pennsylvania State University
University Park, PA, USA

Markku Ollikainen
Professor Emeritus, Environmental
Economics, Department of Economics
and Management
University of Helsinki
Helsinki, Finland

Antti Iho
Senior Scientist Natural Resources
Institute Finland
Helsinki, Finland

ISSN 2662-3889 ISSN 2662-3897 (electronic)
Palgrave Studies in Agricultural Economics and Food Policy
ISBN 978-3-030-47086-9 ISBN 978-3-030-47087-6 (eBook)
https://doi.org/10.1007/978-3-030-47087-6

This Palgrave Macmillan imprint is published by the registered company Springer Nature Switzerland AG.
The registered company address is: Gewerbestrasse 11, 6330 Cham, Switzerland

This book was largely written during the first year of the Covid-19 pandemic which added much to the usual stresses on family of book writing. We are deeply appreciative of the support and patience of our families and especially our spouses, Sanna Lindahl, Heta Pyrhönen, and Lou Anne Shortle.

Foreword

Water is perhaps the most overlooked and complex element of agri-food systems. People concerned about the impacts that agri-food systems have on human or planetary health routinely focus on the land or on the foods, fuels, and fibers produced on it. Far too often agricultural water inputs and pollutants are an afterthought. Perhaps agricultural water gets little attention because much of the action is unobservable, occurring belowground or off the downslope edges of fields. But it is a potentially dangerous oversight, given the central role water plays in regulating all life on the planet. Because agriculture accounts for roughly 70% of the world's freshwater use currently, farming practices inevitably deeply affect the water quality on which all life depends.

Regrettably, chemicals, pathogens, sediments, and other pollutants originating in agricultural production and on-farm processing indeed compromise water quality all over the planet, albeit to varying degrees. Coastal regions and freshwater bodies increasingly experience hazardous algal blooms and even resulting dead zones, and agricultural water withdrawals and pollution degrade drinking water quality from aquifers, sometimes even to the point of making the water toxic.

These are not new problems. But as the world has gotten better control over water pollution from point-based industrial and municipal sources, the complex challenges concerning the management and regulation of nonpoint-sourced agricultural water pollution have become increasingly apparent. Moreover, as food production continues expanding to meet rising demand due to population and income growth, the urgency of addressing the water quality impacts of agriculture only grows.

Thankfully, serious scholars like Jim Shortle, Markku Ollikainen, and Antti Iho have been tackling these issues for years. They and other scholars have built a helpful analytical framework built on the insights of environmental and welfare economics. This approach provides a coherent, intuitive way of conceptualizing the problems and the mechanisms that cause them. Even more important, this same apparatus can be used to build valuable empirical evidence and to identify and assess candidate solutions so as to guide policy makers and instruct students. In this terrific volume they share valuable lessons they have learned about the economics of improving the efficiency and effectiveness of water quality protection.

The approach Shortle et al. advocate turns on recognizing that water quality is a matter of choices. Lots of choices. Made by huge numbers of people over space and time. The interactions among so many independent choices make agricultural water pollution control a wicked problem, meaning its complexity makes it difficult to solve. Yet, as this book makes clear, the extent and nature of water pollution is largely the predictable result of many layers of individual and collective choices that ultimately determine how water gets used and impacted. The water quality problems we face arise because too many people make choices—over farmers' use of land, irrigation, or fertilizer; over the tariff rates regulators set for water or electricity; over tax rates levied on land, income, or imported products; etc.—that ultimately lead to adverse environmental outcomes. Of course, if the problems have anthropogenic origins, then so too must solutions likewise turn on human behavior. Technological fixes—whether via prevention or treatment—offer tempting shortcuts, but ultimately only work if we get the right choice architecture in place to guide billions of people to make choices that prove less harmful to our increasingly scarce water resources. Economics, the study of choices subject to scarcity and other constraints, thus provides a crucial lens for understanding both the origins of and solutions to water pollution problems related to agri-food systems.

A brief foreword cannot do justice to the clarity and power of the insights Drs. Shortle, Ollikainen, and Iho deliver in the chapters that follow. They methodically develop an intuitive yet rigorous analytical apparatus in a way that will empower nonspecialists. And they clearly discuss the concepts behind, and the empirical evidence on, key policy instruments, like auctions, credits, subsidies, taxes, and trading mechanisms. If I could recommend only a single source to someone to learn all the key issues concerning agricultural water pollution policy, this book would be it, hands down.

Shortle et al. assume little or no prior familiarity with key economic concepts, nor with the science of water quality, making their monograph an easy entry point to the topic. Indeed, the clarity of their prose will make this an especially valuable resource for nonexperts who seek to learn more about the issues surrounding water pollution from agriculture and the policy instruments available to address those challenges. I could easily imagine a legislative staffer with no prior background in the subject but newly assigned responsibility for these issues getting up to speed quickly by reading this volume. Instructors of courses in environmental policy or sustainability sciences might fruitfully use this volume to help guide noneconomics students through the economics of water quality issues. At the same time, the book's encyclopedic coverage will also make this a tremendous resource for specialists in agricultural, environmental, and resource economics or in water quality policy, as its coverage far exceeds that of anything else currently available.

It is a pleasure to include Jim Shortle, Markku Ollikainen, and Antti Iho's excellent new volume in the Palgrave Studies in Agricultural Economics and Food Policy series. This insightful contribution to the literature couldn't be more timely.

Ithaca, NY, USA Christopher B. Barrett

Preface

Water pollution control has been a top environmental policy priority of the world's most developed countries for decades, and the focus of significant regulation and public and private spending. Yet, serious water quality problems remain, and trends for some pollutants are in the wrong direction. A key premise of this book is that economics is crucial to understanding why water quality problems remain in countries and places which have developed extensive policy architectures and devoted enormous resources to water quality protection. Another key premise is that economics is crucial to finding policies that are effective and efficient. The book focuses on the economics of water pollution control and water pollution control policy in agriculture as it is the leading cause of water quality problems in rich countries and an emerging problem in developing countries. Our goals are to provide students, environmental policy analysts, and other environmental professionals with economic concepts and tools essential to under standing the problem and crafting solutions that can be effective and efficient.

University Park, PA, USA
Helsinki, Finland

James Shortle
Markku Ollikainen
Antti Iho

Acknowledgments

We have benefited greatly from review and comments on various sections of this book by Patrick Fleming, Richard Horan, Richard Ready, Matthew Royer, and Tamie Veith. We have also benefited from discussions while writing the book with Lassi Ahlvik, Mazdak Arabi, Zach Easton, Paul Ferraro, Ronald Griffin, Daniel Hellerstein, Armen Kemanian, Jussi Lankoski, Sanna Lötjönen, Marc Ribaudo, Leonard Shabman, Gary Shenk, Kurt Stephenson, Hanna Tuomisto, Lisa Wainger, and David Zilberman. We are grateful to them all. The ideas we present in this book are also shaped by collaborations with many colleagues. These include David Abler, Robert Brooks, Richard Horan, Peter Kleinman, Marc Ribaudo, Matthew Royer, Tamie Veith, and Lisa Wainger. Environmental economics students at the University of Helsinki kindly helped in producing figures and checking many details. We thank Ia Ahl, Linda Pesu, Oula Rinne, and Ursula Rinta Jouppi. Antti Iho gratefully acknowledges funding from the Shared Waters project of the Finnish Cultural Foundation. Markku Ollikainen thanks research consortium "Enhancing Adaptive Capacity for Sustainable Blue Growth (BlueAdapt)," funded by the Strategic Research Council of Academy of Finland (Contract No. 312650 BlueAdapt).

Praise for *Water Quality and Agriculture*

"If I could recommend only a single source to someone to learn all the key issues concerning agricultural water pollution policy, this book would be it, hands down."
—Christopher B. Barrett, *Cornell University*

"Helping farmers minimize their losses of valuable nutrients and soil from their agricultural production facilities and lands has posed a seemingly insurmountable change to watershed and water quality restoration efforts around the world. Even after decades of partnership-based collaborative efforts within the Chesapeake Bay watershed yielding implementation of conservation best practices on millions of acres of agricultural lands and tens of thousands of agricultural facilities, stream, river and estuarine monitoring networks spanning the 64,000 square mile watershed attribute only a very small portion of the positive trends observed to date to reductions in agricultural sources of pollutants. In this one volume, Shortle, Ollinkainen and Iho deliver what is missing—wholistic market and performance-driven approaches leading to investments which yield sustainable agriculture, profits to producers, clean water and expanded ecosystem services. Their book effectively weaves together their in-depth understanding of the many external forces which drive decision-making within agricultural production with the economic market forces and private capital investment opportunities which have yet to be employed at a scale and depth needed to truly turn the corner on this formidable challenge and reach sustainability on all fronts."
—Richard Batiuk, retired, Associate Director for Science, Analysis and Implementation, *United States Environmental Protection Agency Chesapeake Bay Program Office*

"Shortle, Ollikainen and Iho have provided a clear, timely and comprehensive presentation of the economics of effectively and efficiently managing water quality, an issue that is steadily growing in importance globally. Students, researchers and policy makers seeking a solid introduction to water quality economics will find this an excellent starting place. In a logical systematic way it guides the reader through the issues that arise when attempting to alter a complex economic and biophysical system to increase efficiency, and provides rich examples from international experience. By introducing the fundamental science that drives water quality it alerts those designing or evaluating policy to the science they will need to access and account for if their efforts to protect and enhance it are to be effective."
—Suzi Kerr, Chief Economist at the Environmental Defense Fund

"This book is a must-read for students and professionals interested in the economic perspective on the challenges and approaches to managing one of the most critical environmental problems of our times, water pollution from agriculture. The book identifies the limitations of current voluntary approaches to achieving water quality goals and the policy innovations needed to achieve them efficiently. It distills insights from a vast body of research in an easy-to-ready text book that should be required reading for those seeking an in-depth and technically accessible understanding of the applications of economics for designing better policies."

—Madhu Khanna, Distinguished Professor of Agricultural and Consumer Economics, *University of Illinois, Urbana-Champaign*

"This book provides a very up-to-date and comprehensive treatment of the complex issues of controlling agricultural nonpoint source pollution in terms of both theory and empirical modeling. It covers the design of several policy instruments recommended by the OECD and promotes their implementation."

—Jussi Lankoski, Economist, *Organization for Economic Co-operation and Development*

"Agricultural pollution is the largest source of water quality impairment in many industrialized countries, but current policies fall short of solving this problem. In addition to damaging human health and the environment, farms feed the world, so progress involves difficult tradeoffs. This terrific new book shows persuasively and in accessible language how market forces can be harnessed to reduce agricultural water pollution, and do so cost-effectively. Weaving together basic economic concepts and rigorous discussion of water pollution as a problem of incentives, the authors provide a roadmap to better aligning farmers' choices with human and environmental well-being."

—Sheila Olmstead, Professor of Public Affairs, *University of Texas at Austin*

"This important book should change the way that agricultural water pollution is thought about around the world. It combines cutting edge economic thinking with a wealth of practical knowledge and delivers its information in a highly accessible way.

Shortle, Ollikainen, and Iho make a powerful case for putting economics at the center of efforts to reduce water pollution from agriculture. The authors' focus is on the role of government policies and programs and how they can improve water quality for the benefit of the community as a whole. One of their most important observations is that the persistence of water-quality problems arising from agriculture is not due to a lack of policy effort. Rather, it is because the policies we have are not working. The authors lay out the logic of economics and demonstrate its power as a tool for integrating information of diverse types to inform better decision making about these policies. Their economic concepts and insights are not just about making money. They are about how policy makers can deliver the greatest improvements in water quality within the constraints that they face. This requires recognition of the science of water pollution as well as the science of human behaviour, including the influence of economics on farmer behaviour.

The book contains a wealth of practical advice on how to bring economic thinking to the design and implementation of water-pollution policy. Readers of the book will clearly understand that there is no simple fix to the problem of agricultural water pollution, but will come away with a new awareness of practical policy tools, underpinned by economic thinking, that can be applied to improve policy performance. The authors argue persuasively that compelling economic evidence and arguments can help to address one of the key problems that holds back reform of water-pollution policy: a lack of political motivation to reduce the problem.

Water Quality and Agriculture is written in a highly accessible way that will make sense to non-economists and will enlighten economists who are not already working in this area. The authors make frequent use of boxes to provide illustrations and real-world examples. They demonstrate a wealth of practical knowledge of real-work policies."

—David Pannell, Professor of Agricultural and Resource conomics, *University of Western Australia*

"Water Quality and Agriculture is a comprehensive and authoritative source on the economics of water pollution and will inspire fresh thinking on solutions. This impressive, internationally-renowned author team addresses the persistent problem of how to control pollution from agricultural sources with a thorough set of economic incentives and analysis tools.

The book's compelling anecdotes clarify why some common policies can fail to deliver water quality benefits and how economic theory can be applied to improve results."

—Lisa Wainger, Professor of Environmental Economics, *University of Maryland Center for Environmental Science*

Contents

List of Figures

List of Tables

CHAPTER 1

Introduction

1.1 Introduction

The importance of water to agriculture is common knowledge. Agricultural systems cannot thrive without adequate and reliable water supplies. In much of the world, water is supplied directly to crops by rainfall. Irrigation is utilized where rainfall is insufficient or unreliable. About 69% of the water withdrawn from developed surface and groundwater supplies globally is allocated to agriculture, mostly for irrigation (FAO 2020). Law and policies governing water use and supply in arid regions of the world often prioritize agricultural use. Agricultural conservation policies in irrigated regions are often intended to improve the efficiency and sustainability of agricultural water use.

Less well known is the importance of agriculture to water. There are multiple dimensions to this importance. Historical agricultural development has had enormous and enduring impacts on water resources and aquatic ecosystems as forests, prairies, and other natural landscapes were converted to agricultural use (Rosenberg et al. 2000). Disturbing natural landscapes to create crop and pasture lands affects flow regimes, temperature, and a host of other variables that affect the structure and health of

J. Shortle et al., *Water Quality and Agriculture*, Palgrave Studies in Agricultural Economics and Food Policy,
https://doi.org/10.1007/978-3-030-47087-6_1

aquatic ecosystems, water quality, and the utility of water for various uses. The environmental consequences of land conversion are an issue in places where agricultural expansion continues. But, as most arable land is already developed, attention to water quality impacts of agriculture is largely focused on the impacts of water pollutants resulting from agricultural production (Tyagi and Shortle 2019).

Contemporary agricultural systems excel in their capacity to meet the ever-increasing food and fiber demands of growing populations and higher-income societies. The contributions to human welfare of modern agricultural systems are tremendous. But these systems also cause environmental degradation, one of the most important being water pollution. In intensively farmed humid regions, runoff from fields, pastures, and barnyards following rainfall events or snowmelt moves fertilizers, pesticides, sediments, pathogens, and other pollutants into surface waters. Pollutants also leach from agricultural lands into groundwater supplies. Leaching and return flows[1] in irrigated regions cause contamination of surface and groundwater. Whereas water scarcity is primarily a concern for agriculture in arid regions, water quality damage is a problem in humid and arid regions.

Recognition that agriculture has adverse impacts on water quality is not new. Pesticides, fertilizers, salts, and sediments from agricultural lands have been known to damage water quality for decades. The significance of agriculture as a source of water pollution has become more apparent over the past four decades in many high-income countries as water quality problems have persisted or in some cases become worse after strong regulation of industrial and municipal wastewater. Public policy to address water pollution from agriculture is also not new. Many high-income countries implemented policies to address water pollution from agriculture beginning in the 1980s. Assessments indicate these policies have generally fallen short and that policy reforms are essential. Some key messages from a 2012 report on agriculture and water quality in countries that are members of the Organisation for Economic Co-operation and Development (OECD 2012, p. 9):

- While the current situation varies within and across OECD countries, agriculture is often the main source of water pollution. Achieving further reductions is a challenge for policy makers, espe-

[1] Return flows are water that leaves fields in surface or subsurface flows after irrigation.

cially as a major part of agricultural water pollution is from diffuse sources.

- Growth and intensification of agricultural production could further heighten regional pressures on water systems in some countries. Moreover, the task of achieving water quality objectives in agriculture will become more difficult as a result of climate change.
- Policies have generally fallen short of requirements to meet water quality policy goals in agriculture.

A 2017 report from the Food and Agriculture Organization (Mateo-Sagasta et al. 2017) observes that "In most high-income countries and many emerging economies, agricultural pollution has already overtaken contamination from settlements and industries as the major factor in the degradation of inland and coastal waters (e.g. eutrophication)….In low-income countries and emerging economies, the large loads of untreated municipal and industrial wastewater are major concerns. Nevertheless, agricultural pollution, aggravated by increased sediment runoff and groundwater salinization, is also becoming an issue (Mateo-Sagasta et al. 2017, p. 3)."

This book is about public policies for managing water pollution from agriculture. We are especially interested in the challenge of reforming policies to improve the efficiency and effectiveness of water quality protection. The topic is of fundamental importance to the quality of global water resources, the ecosystem services they provide, and the health and well-being of people and societies as they are affected by water quality. The topic is also of fundamental importance to the structure and performance of agricultural systems that are essential to the health and well-being of people and societies. Measures to significantly reduce pollution from agriculture will necessarily change the structure and increase the cost of agricultural production. The challenge is to implement policies that achieve water quality goals without unduly harming the capacity of agriculture to supply affordable food, fiber, and feedstock for bioenergy and to cope with global environmental change. Meeting this challenge requires policies that restructure agricultural systems to provide a balance between production efficiency, agricultural prosperity, and environmental protection. It is this challenge that motivates this book.

This book is intended for students, policy analysts, researchers, educators, and other professionals who are interested in economic, political, and scientific challenges for designing effective and efficient policies for addressing water pollution problems in which agriculture has a large role. The approach is that of economics, but the book is designed to be accessible to readers who are not trained in economics or environmental economics. Necessary concepts and theories are developed so that the book also serves as an introduction to environmental economics. While based on economics in its approach, the book provides a multidisciplinary understanding of the agricultural problem that is essential to economic and policy analysis. The book could serve as a text for a course on water quality policy for agriculture that emphasizes the economics and policy of the problem, a supplementary book for courses on environmental policy, or a resource for policy analysts or researchers.

1.2 Choices, Trade-offs, Economics

A common definition of economics is the study of the allocation of scarce resources among competing uses. When resources are scarce and can be put to competing uses, decision makers, whether private or public, face trade-offs. Choices must balance competing ends. Underlying the societal causes of water pollution are economic choices made by farmers, consumers, and others in the agricultural supply chain. Choices in this food and agricultural system are strongly influenced by agricultural and other policies. The societal consequences of water pollution are mediated by economic choices of water supply companies, retail water consumers, commercial and recreational fishermen, and others who make use of the ecosystem services provided by water resources. Economics provides concepts and tools for measuring the private and social costs of choices, characterizing trade-offs, and explaining choices. It also provides concepts and tools for evaluating policy interventions.

Water pollution from agricultural production is most directly the result of producers' economic choices about what to produce (e.g., grains, fruits, vegetables, livestock), the use of potential pollutants in production (e.g., pesticides, fertilizers, pharmaceuticals), farming practices that affect the formation and fate of pollutants (e.g., tillage types, pest management practices, manure management practices), and the location of production (e.g., adjacent to or set back from streams). Producers must decide how to allocate scarce productive land between alternative crops, and between crops and pasture. The choices matter to water quality outcomes as profitable production systems for crop choices vary in their pollution potential. They must also choose allocations between productive (crops and pasture) and unproductive uses that protect water resources such as riparian buffers. In irrigated agriculture, they must similarly decide how to allocate scarce water. Producers must also decide how to allocate their scarce time and financial resources between alternative inputs, farming practices, and conservation and pollution control activities.

Shaping producers' choices are the physical (soils, topography, climate), economic (input and output prices, taxes, subsidies), and institutional (laws, regulations, cultures) environments in which they operate. The economic and institutional environments are constructed by choices by others within complex systems. Moving backward in the supply chain from farms are markets for energy resources, labor, fertilizers, pesticides, materials, equipment, and land. Moving forward in the supply chain from the farm to final consumers is almost always an integrated system of markets that moves products from farms to processor to retailers. Decision makers throughout the system decide on the allocation of scarce resources, with consumers at the top allocating income to competing uses including food budgets and allocating food budgets to specific products.

The societal consequences of water pollution also embed economic choices. Consider nutrient pollution, generally regarded as the leading contemporary water quality challenge for agriculture and a problem of global significance (Mateo-Sagasta et al. 2017; OECD 2017). Nutrient pollution is caused by various forms of nitrogen and phosphorus. These compounds are essential nutrients for living organisms and healthy aquatic systems. Water pollution problems occur when the concentrations of these nutrients are elevated by human activity to levels that significantly increase the rate of primary production in aquatic ecosystems. Excessive growth of algae and other photosynthetic organisms disrupts aquatic ecosystems resulting in loss of submerged aquatic vegetation that serves as natural pollution filter, food for waterfowl, and habitat and nursery ground for various aquatic species. Decaying algae consume dissolved oxygen, leading to hypoxic conditions that harm or kill aquatic organisms. Some species of algae are toxic to aquatic life and humans.

Consequences of nutrient pollution with economic implications include reduced stocks and increased harvesting costs for fish and shellfish sought by commercial and recreational fishermen, degradation of beaches and open waters for recreational uses, and degradation of drinking water supplies. The economic costs of these harms will depend on opportunities for those who are affected to avert adverse consequences and the costs of those actions. In other words, pollution creates economic choice problems for those affected. The costs they ultimately incur depend on their decisions. In the case at hand, averting behaviors available to fishermen include changing the time and resources devoted to harvesting preferred species, and reallocating time and resources to alternative target species or to alternative fisheries. In the case of degraded drinking water supplies, averting behaviors for water suppliers include the development of new sources and treatment of existing sources. Depending on suppliers' choices, retail customers may choose to respond to changes in quality with home water treatment or bottled drinking water. Water recreationists contending with algae blooms have options for alternative recreational sites and activities.

Environmental policy makers also face trade-offs. Setting water quality goals for polluted resources requires implicitly, if not explicitly, trading off benefits from better water quality against the costs of water pollution control. Pollution control planning for water bodies affected by multiple sources implicitly, if not explicitly, entails allocating pollution load reductions between different sectors, agriculture, industry, and municipalities, and between members of those sectors. Consider again nutrient pollution. Major sources of nutrients include municipal wastewater treatment plants, urban stormwater runoff, and intensive agricultural production. Reducing municipal discharges requires increased capital and operating costs for wastewater treatment. Reducing urban stormwater requires capital and operating costs for systems to capture and treat stormwater. These costs are typically borne by taxpayers or system rate payers. Reducing nutrients from agriculture generally involves reductions in fertilizer use, decreasing crop yields, and capital and operating costs for practices to reduce nutrient runoff. Depending on agricultural price increases, the burden of increased production costs may be borne by farmers or by consumers paying higher food costs. Additional costs include those of administering and enforcing pollution control policies.

Economic research shows that the willingness to pay for additional pollution control decreases as water quality is increased. Another way to say this is that the incremental or marginal benefits from additional pollution diminish as water quality improves. Conversely, research shows that the cost of improving water quality increases with the level of pollution control. Policy makers setting water quality goals must therefore contend with a benefit-cost trade-off that becomes increasingly sharp as water quality is improved. Because the costs of pollution control differ between sectors and sources within sectors, the exact nature of this trade-off will depend on how load reductions are allocated across sources. For nutrients, agriculture is often the comparatively low-cost source of additional control, which would imply that agriculture should be allocated greater responsibility for additional pollution control to minimize overall social costs of pollution control.

Societal choices of water quality targets and sectoral pollution allocations include societal considerations beyond economic costs and benefits. Society has multiple concerns and multiple objectives for agricultural productivity, agricultural landscapes, and rural economic well-being, for example, that should be considered in policy choices. But sound policy is informed by economic trade-offs.

1.3 Policy Design and Economics

Given a water quality goal, whether chosen by policy makers with or without consideration of economic benefits and costs, the policy challenge is how to bring about changes in agricultural production and in pollution loads from other sectors that achieve the goal. In economics this problem is one of policy instrument choice and design. Pollution policy instruments are the interventions (e.g., educational programs, regulations, taxes, subsidies) that are used to align private choices with public policy goals.

Agricultural pollution is often framed as a technology problem rather than a systemic economic problem. Correspondingly, solutions are commonly framed as technological fixes involving the implementation of pollution prevention (e.g., reduced fertilizer and pesticide use) or pollution treatment (e.g., riparian buffers to intercept pollutants) technologies on agricultural land. These technologies are commonly referred to as Best Management Practices (BMPs). BMP implementation is essential for managing pollution from agriculture, but its implementation as "fix" is an engineering approach that neither embraces the scope of farmers' choices for reducing water pollution loads, nor explicitly addresses the underlying systemic economic and institutional causes of the problem (see Box 1.1).

Economics sees the solution to water quality problems not in technical fixes, but through the choice and design of policy instruments that modify the economic and institutional environment to align economic choices with societal objectives.

Box 1.1 The Anthropocene

A current environmental buzzword is the Anthropocene, referring to a proposed new geologic epoch to follow the Holocene beginning sometime between the beginning of the industrial revolution and the mid-twentieth century. The epoch is proposed to characterize the time in which human societies have transformed the Earth's surface, atmosphere, oceans, and biogeochemical systems. Anthropocene means the "recent age of man" in Greek. The Anthropocene is not yet recognized as a geologic epoch and there is substantial disagreement about whether it should be. But the concept conveys something very important for environmental management. Predicting and managing environmental change requires predicting and managing coupled human and natural systems in which societal drivers are fundamentally influential.

With global population nearly 8 billion, the impact of our food production on natural ecosystems is massive. World's land surface is about 149 million square kilometers (29% of Earth's surface). Out of this, approximately 29% is not habitable (glaciers, barren mountains, deserts). Out of the habitable 104 million square kilometers, about half is designated to agriculture. Most of the 51 million square kilometers on agriculture is designated to animal husbandry, either directly as pastureland or as crops grown for animal feed. About 11 million square kilometers is on crops for human consumption (Ritchie 2019).

The global average flow of rivers and recharge of aquifers due to precipitation is around 42,000 cubic kilometers. Withdrawal for agricultural use is about 6% of this. The global average is thus not that high but in certain regions, such as Western, Central, and South Asia, agriculture uses on average 51% of the renewable water resources (FAO 2011).

A striking way to look at the ecological scope of modern societies is to compare the quantity and composition of past and present biomass of Earth's wild mammals, us, and our livestock:

(*continued*)

Box 1.1 (continued)

The two bars in Fig. 1.1 represent the estimated biomasses before modern civilization and today. The vertical axis denotes the biomass in gigatons of carbon. Today, the largest share of biomass is in our livestock. It is more than twice that of wild mammals before the Anthropocene. Our existence has quadrupled the biomass living on the face of the Earth. At the same time, we have decreased the species diversity by allocating vast land areas to food production and by using pesticides and fertilizers. Eutrophication is not merely a nuisance for humane recreation: it is also one of the key threats to biodiversity (Sud 2020).

The urge of improving the governance of agricultural environmental impacts is not a temporary fad. It is a matter of existence for our entire way of life.

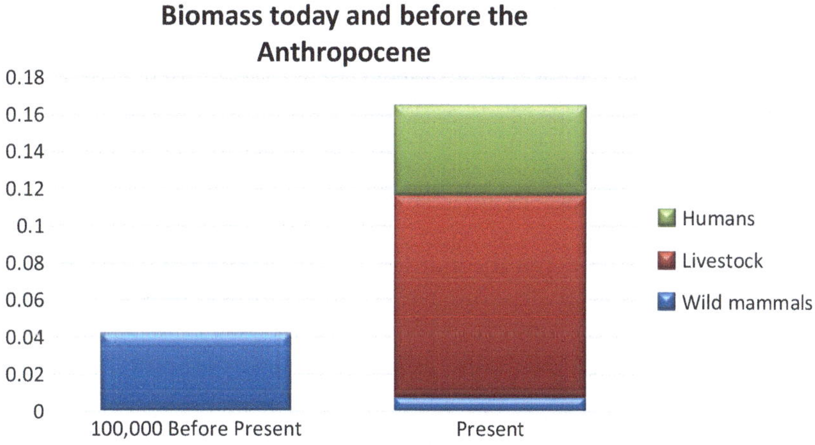

Fig. 1.1 Wild mammal, human, and livestock biomasses before civilization and in the present. (Source: Adapted from Bar-On et al. (2018))

Economic research on instrument choice and design blends two economic approaches to policy analysis. One is positive economics, which is concerned with describing and explaining economic phenomena and predicting how economic systems respond to policy interventions or other drivers. The other is normative, which is concerned with setting social goals and evaluating policy to inform and support the selection of policies with preferred social outcomes.

In the case at hand, positive economics explains and quantifies the economic choices of farmers that result in water pollution and how they are shaped by external economic and institutional conditions. Positive economics offers insights about what kinds of instruments can produce desired responses and provides tools and research findings to predict responses.

An obvious first criterion for a pollution control policy instrument is that it results in a reduction in pollution. If not, it is not effective and should not be considered. An economic analysis of this question would begin by analyzing how individual farmers would respond to application of the instrument. For example, one approach proposed for reducing nutrient pollution is taxing fertilizer use. Economic theory supported by empirical research evidence indicates that increasing the cost of fertilizer will reduce the use of the input, making the tax a candidate policy instrument.[2] However, some policies that seem likely to be effective turn out not to be so when the economics of their application are fully considered.

Economic research in several domains demonstrates that predictable economic responses to resource conservation and environmental policies can undermine their effectiveness or even make them counterproductive. An example of such a response is the *rebound effect*. Rebound effects occur with mandates or subsidies for resource-conserving technologies (Paul et al. 2019). The first-order effect of the use of the technology is to reduce the use of the resource. But the induced economic response—the second-order effect—offsets the first-order effect and if strong enough may lead to an overall increase. A much-observed example is automobile fuel efficiency standards (Gillingham et al. 2016). Increased fuel efficiency reduces the cost of driving, which leads to increased driving and fuel use, partially offsetting the fuel efficiency effect. Rebound effects have also been observed to result from increases in land productivity and irrigation water use (Paul et al. 2019).

In the fertilizer example, suppose a new corn variety is developed that requires less nitrogen to obtain a given yield. If a farmer who has adopted the variety applies fertilizer to achieve the same yield as the previous

[2] Later we will describe that though potentially effective, fertilizer taxes are generally not considered the best economic choice for nutrient pollution control.

variety, fertilizer use will be decreased. But the improved productivity increases the returns from fertilizer use, which will lead to an offsetting increase in fertilization. This increased intensity of input use on crop land is described as an *intensive margin* response. In addition, the new variety by increasing the profitability of corn production may create incentives to increase the amount of land in cultivation, further increasing fertilizer use. This expansion of land use is described as an *extensive margin* response. And still further, if enough farmers adopt the technology, the costs of suppling corn to the market will decrease leading to reduced crop prices and an increase in the quantity demanded, leading to yet another increase in fertilizer use. The overall effect could be an overall increase in nitrogen fertilizer use, a case of *Jevons Paradox* (Abler and Shortle 1995).[3] Economics provides other useful concepts to describe induced economic responses with potentially significant consequences for policy effectiveness such as *slippage* and *leakage* that are relevant to the agricultural water quality problem.

A key role then of positive economic analysis is to provide insights and predictions about how farm decisions affecting water quality outcomes will respond to alternative types or mixtures of policy instruments considering not only responses at the farm level, but also market-level responses that feed back to affect farm choices. Another role for positive economics is to understand and predict the effects of various public policies affecting agriculture on water pollution and the costs of water pollution control from agriculture. Government intervention in agriculture is pervasive. Interventions include agricultural price and income supports, agricultural trade policies, bioenergy policies, and land and water use policies. Although some may be complementary to environmental protection, policies that encourage surplus production, increase production intensity, and encourage expansion of agriculture into environmentally sensitive lands have the opposite effect. Policy reforms to address perverse policy outcomes must be considered in the overall policy mix (OECD 2012).

Normative economic analysis introduces social criteria to environmental policy analysis. While effectiveness is a first criterion for environmental policy evaluation, analysis of social welfare outcomes and policy preferences

[3] The first description of rebound effects is due to British economist W.S. Jevons who reasoned that increased efficiencies in coal use would result in increased coal use (Jevons 1865). The net increase in resource use resulting from the application of productivity-increasing technologies is now referred to as *Jevons Paradox*.

requires consideration of others. Standard considerations in normative environmental policy analysis include the economic costs of pollution control, incentive effects for green technology innovation, administration and enforcement costs, and distributional outcomes. While normative economics can help inform the selection of environmental goals, economic research on policy design generally takes goals as exogenous political choices. Policy design research focuses on the selection and design of instruments for achieving established goals taking into account private and public sectors costs, and innovation incentive effects, and distributional outcomes.

Controlling pollution from agriculture is what Rittel and Webber (1973) describe as a wicked problem (NRC 2012; Shortle and Horan 2017). Features of wicked environmental problems include many complex, and often imperfectly understood, ecological and anthropogenic interactions contributing to the problem; complex spatio-temporal interactions, operating at different scales that necessitate unique strategies over space and time; and economic, political, and institutional complexity affecting potential solutions (NRC 2012; Shortle and Horan 2017). A consequence of complexity and uncertainty is that attempted solutions will often reveal misunderstood or unrecognized aspects of the problem, and have unintended consequences (DeGrace and Stahl 1990; NRC 2012; Rittel and Webber 1973; Shortle and Horan 2017). These are all features of the agricultural pollution control problem.

1.4 Outline

The material in this book describes why pollution control in agriculture is a wicked problem, approaches that have been developed to address the problem, the need for policy changes to improve the water quality outcomes, and the economic and budgetary efficiency of water pollution controls. We describe changes in existing policy approaches that can improve their performance, but also describe new approaches that are of significant interest.

Chapter 2 presents an introduction to key concepts and theories used in economic analysis of environmental policy, introduces current policies and institutions for water pollution control in agriculture, and develops the need for policy reform to improve the effectiveness and efficiency of water pollution control. Background economic topics include the theory of environmental externalities, the need for public policy to address environmental externalities, policy instruments for pollution control, and criteria for policy evaluation. The institutional and policy background includes a review of the evolution of water quality policy and current approaches to pollution control for agricultural and nonagricultural sources. Chapter 2 also introduces the importance of agricultural policies intended to promote rural prosperity, renewable energy, or other goals to water quality impacts of agricultural production.

The remainder of the book establishes unique challenges that water pollution control poses for policy makers, and develops concepts, analytical frameworks, policy options, and research results that can guide initiatives to improve policy. Chapter 3 describes the kinds of water quality problems that result from agricultural production, the significance of these problems in various parts of the world, and the significance of agriculture's contribution. Effective water pollution control is a multidisciplinary problem. Chapter 3 also introduces essential concepts from water science and engineering about the physical processes that link agricultural production and land use to water quality and technological options for pollution control. Key concepts, relationships, and tools needed to understand the physical processes involved in the movement of pollutants from farms to water resources are introduced.

Chapter 4 examines how farm-level decision making affects environmental outcomes. The chapter shows how economic decisions on outputs, inputs, and farming practices influence pollution loads, and how environmental and other agricultural policy interventions can influence those decisions and water quality outcomes.

Chapter 5 builds on the material presented in Chaps. 3 and 4 to examine water pollution control instruments for agriculture. This chapter begins with a discussion of unique features of the agricultural problem and their implications for the choice of water quality goals and policy instruments for agriculture. An essential point is that the kinds of instruments routinely recommended for efficient pollution control are not plausible for agriculture. Alternatives are required. The chapter presents a menu of policy instruments for managing pollution from agriculture based on implied property rights to pollute, choices of policy targets, and regulatory mechanisms. The menu includes instruments that are now used in practice, but it also includes novel approaches that are of interest for improving ecological and economic outcomes. With this menu in hand, the chapter turns to the selection and design of policies. This discussion begins with essential attributes of policies for effective and efficient control of agricultural pollution. Concepts of policy instrument optimization are presented and illustrated using case studies. The last part of this chapter presents a review and evaluation of contemporary agricultural nonpoint pollution controls given the criteria and the principles for efficient policy developed in previous sections. The discussion identifies systemic flaws that limit the effectiveness and efficiency of current policy architectures. The chapter concludes with a discussion of policy reforms to improve policy performance and sets the stage for subsequent chapters on market-based approaches.

Chapters 6, 7, and 8 examine three types of policy innovations that are of substantial current interest for improving the effectiveness and efficiency of water pollution control in agriculture—water quality trading, water quality auctions, and credit stacking. The chapters discuss the theory of the instruments, applications, and the outlook of their use. We conclude in Chap. 9 with a set of recommendations for policy improvement. Our recommendations emphasize: (1) the importance of political ambition, exhibited through spatially explicit water quality goals, accountability frameworks, and realistic timetables for achieving results; (2) that policy instruments must be selected for the specific place-based problems they are meant to address considering the relevant water science and the economic, institutional, and cultural contexts; and (3) that there is great merit in the development of solutions through local innovation that can best craft place-based solutions. Within these general guidelines, we recommend suites of instruments that include a mix of "softer" elements providing nudges and technical assistance to producers, with "harder" elements needed to induce significant change in farming practices. The mix would vary with the severity of the problem and the agricultural contribution. Key attributes for "harder" elements are that they target environmental performance rather than specific farming practices, utilize economic incentives rather than regulatory restrictions to induce producer behavioral change, are implemented to accommodate heterogeneity in cost and benefits within and across watersheds, and consistent with the polluter-pays-principle.

References

Abler, D.G., and J.S. Shortle. 1995. Technology as an agricultural pollution control policy. *American Journal of Agricultural Economics* 77 (1): 20–32.

Bar-On, Y.M., R. Phillips, and R. Milo. 2018. The biomass distribution on Earth. *Proceedings of the National Academy of Sciences* 115 (25): 6506–6511.

DeGrace, P., and L.H. Stahl. 1990. *Wicked problems, righteous solutions*. Vol. 2. Upper Saddle River, NJ: Yourdon Press.

Food and Agriculture Organization (FAO). 2011. *The state of the World's land and water resources for food and agriculture (SOLAW) – Managing Systems at Risk*. Rome: Food and Agriculture Organization of the United Nations.

Food and Agriculture Organization of the United Nations (FAO). 2020. FAOSTAT Statistical Database.

Gillingham, K., D. Rapson, and G. Wagner. 2016. The rebound effect and energy efficiency policy. *Review of Environmental Economics and Policy* 10 (1): 68–88.

Jevons, W.S. 1865. *The coal question*. Macmillan and Co.

Mateo-Sagasta, J., S.M. Zadeh, H. Turral, and J. Burke. 2017. *Water pollution from agriculture: A global review. Executive summary*. Rome: FAO Colombo: International Water Management Institute (IWMI). CGIAR Research Program on Water, Land and Ecosystems (WLE).

National Research Council (NRC). 2012. *Science for environmental protection: The road ahead*. National Academies Press.

Organization for Economic Cooperation and Development (OECD). 2012. *Water quality and agriculture: Meeting the challenge*. OECD Publishing.

———. 2017. *Diffuse pollution, degraded waters: Emerging policy solutions*. OECD Publishing.

Paul, C., A.K. Techen, J.S. Robinson, and K. Helming. 2019. Rebound effects in agricultural land and soil management: Review and analytical framework. *Journal of Cleaner Production* 227 (1): 1054–1067.

Ritchie, H. 2019. Half of the world's habitable land is used for agriculture. Published online at OurWorldInData.org. Retrieved from: https://ourworldindata.org/global-land-for-agriculture [Online Resource]. Original data source: http://www.fao.org/faostat/en/#home

Rittel, H.W., and M.M. Webber. 1973. Dilemmas in a general theory of planning. *Policy Sciences* 4 (2): 155–169.

Rosenberg, D.M., P. McCully, and C.M. Pringle. 2000. Global-scale environmental effects of hydrological alterations: Introduction. *Bioscience* 50 (9): 746–751.

Shortle, J., and R.D. Horan. 2017. Nutrient pollution: A wicked challenge for economic instruments. *Water Economics and Policy 3* (02): 1650033.

Sud, M. 2020. Managing the biodiversity impacts of fertiliser and pesticide use: Overview and insights from trends and policies across selected OECD countries.

Tyagi, A., and J.S. Shortle. 2019. The agriculture—water policy nexus. (Chapter 10). In *Global challenges for food and agricultural policies*, ed. D. Blandford and K. Hassapoyannes. World Scientific.

CHAPTER 2

Economics and Policy for Water Pollution Control

2.1 Introduction

This chapter has two broad objectives. One is to introduce basic theoretical concepts and methods used in economics to explain market determinants of environmental pollution, the need for pollution control policies, inform the setting of environmental policy targets, and inform the selection of policy instruments to achieve them. The second is to establish the policy context that leads to a need for water pollution policy reform for agriculture. These topics are connected.

Economics is not needed to understand that current policies are not meeting water quality objectives, but it can help to explain why current policies fall short and assess options for improvement. Economics can also identify dimensions of policy failure that are missed by a sole focus on water quality outcomes relative to water quality goals. Unnecessarily high costs of pollution control and inadequate incentives for green technology innovation are such dimensions of policy failure.

The economic theory of environmental pollution emphasizes the role of market failure as a cause of water quality problems in market economies. But as we note in the introduction, the large contribution of agriculture to contemporary water quality problems is not the result of policy makers ignoring the market failure. There has been significant policy

J. Shortle et al., *Water Quality and Agriculture*, Palgrave Studies in Agricultural Economics and Food Policy,
https://doi.org/10.1007/978-3-030-47087-6_2

development and enormous resources devoted to the problem. The high costs and limited environmental success of existing policy architectures are the result of the types of policies that have been pursued and also unintended effects of policies intended to boost farm incomes, promote rural development, expand the use of renewable fuels, and achieve other societal goals related to agriculture. A comprehensive understanding of the agricultural problem must therefore consider both market and government policy failures in explaining outcomes and identifying strategies for moving forward. This chapter introduces the economic and institutional foundations for our journey.

We begin with the economic theory of pollution control policy. Topics include market failures, environmental externalities, benefits and costs of water quality protection, environmental target setting, pollution control policy instruments, criteria for instrument selection, and standard prescriptions for policy instruments. With basic theory, concepts, and research results in hand we turn to the agricultural problem. Our focus is on policies and institutions that have evolved for managing water quality impacts of agriculture, and the instruments of water quality protection in agriculture. This discussion is not concerned with policy theory but with policy practice focusing on the US and the EC. The chapter will introduce an agenda for learning that guides the remainder of the book.

2.2 Efficiency, Markets, Market Failure, Externalities

Humans want many things. As consumers and citizens, people want plentiful, nutritious, and varied foods grown on the land and harvested from rural landscapes that appear beautiful and prosperous; low-cost housing; clean water for drinking, cooking, cleaning, swimming, fishing, and boating; cheap electricity for powering home electronics and for home heating and cooling; affordable and flexible transportation; low-carbon energy, clean air; humane treatment of animals. This very partial list of wants is for things that are affected directly or indirectly by the organization and performance of the agricultural sector. And in this short list are numerous trade-offs. Agricultural lands can be used for biofuels production but with adverse effects on food prices and water quality. Production systems that supply low-priced livestock products are often inhumane, and significant sources of water pollution and carbon emissions. Reserving lands for agricultural use can drive up housing costs, adversely affect water quality, and diminish the wealth of farmers.

A fundamental question is how trade-offs between agricultural prosperity and water quality, or affordable food and water quality should be resolved? Peoples' connections and interests in agriculture vary. On almost any issue it will surely be the case that people with different preferences, different economic interests, in different economic circumstances, and living in different locations have very different views. High prices for agricultural products benefit some and harm others. Biofuel mandates benefit some and harm others. Agricultural land preservation programs benefit some and harm others. Subsidies for irrigation water benefit some and harm others. Vigorous water pollution controls benefit some and harm others.

2.2.1 *Pareto Efficiency*

In dealing with contentious issues, we are often advised to "do the right thing" or to do the "common sense thing," implying the morally correct or rational course is clear. But different people with the best of intentions have different concepts of what the right thing means. Some of the most contentious societal issues stem from strongly held differences of opinion about what moral behavior requires. And well-informed and rational people can come to different conclusions about best courses of action depending on their values, weighing of evidence, and mental models of cause and effect.

One method for resolving contentious issues is voting. Voting is a mechanism for people to express their preferences on issues on which there is disagreement and to select courses of action that correspond to the weight of those preferences. Consider a referendum on a proposed societal action with all who are affected eligible to vote. If one or more people vote for the action and none vote against it, then the action is reasonably viewed as preferred by those with a stake in the outcome. Such an action is known as a *Pareto Improvement* after Italian Economist Vilfred Pareto (1848–1923) who proposed the concept for analyzing economic efficiency.[1] In a society with heterogenous preferences and interests, the

[1] The "80/20" rule, roughly attributing 80% of consequences to 20% of causes, is sometimes known as the Pareto Rule or Pareto Principle. The idea that a small part of population explains a large part of a problem is considered to hold in various ways in agricultural water

number of proposed actions that would pass a test requiring no opposition is small. Actions that serve most people could be vetoed by a lone dissenter. Alternatively, suppose that those who benefit from a proposed action and would vote for it could compensate those who would be harmed and vote against it with the result that the no votes become pro votes. If after compensation some or all vote for the proposal and none vote against, then the action after compensation is again reasonably viewed as preferred by those with a stake in the outcome. In such a case, the action is a Pareto Improvement if the required compensation is paid. It is a *Potential Pareto Improvement* if the required compensation could be paid but is not. A *Pareto Optimum* is defined as a situation in which there are no Potential Pareto Improvements. It is then the case that any change can only make one or more better off at the expense of another.

Economics utilizes these concepts to examine economic alternatives. An economically efficient allocation of resources is one from which there are no Potential Pareto Improvements. An efficiency-improving action is one that provides a Potential Pareto Improvement. Economic benefit-cost analysis provides a mechanism to test whether actions with economic consequences are Potential Pareto Improvements. Actions with positive net benefits, i.e., benefits minus costs, are Pareto Improvements. Otherwise they are not. The difference in benefits and costs is a social surplus that can be used in principle to win the support or acquiescence of those who would otherwise be opposed. If the net benefits are zero or negative, there is no surplus. Payment of compensation by those who favor the action adequate to win the support or acquiescence of those who would otherwise be opposed would turn their pro votes to no votes.

Resource allocations in which net benefits cannot be increased, and must therefore be maximized, are *Pareto Efficient.* Normative economic analysis assumes that Pareto Efficiency is a desirable property of scarce resource allocation, essentially meaning making the best use of scarce resources. This proposition does not mean that economics excludes other normative values, such as equity and justice. Distributional considerations receive significant attention in economic research and policy analysis. But when considering how to allocate scarce resources, economics places analytical priority on the economic efficiency.

pollution, e.g., the number of storm events that cause runoff events, the number of acres that contribute pollution.

2.2.2 Pareto Efficiency and Markets

A simple description of an economy consists of a set of consumers, a set of producers, a set of goods available to consumers, a set of technologies for converting resources to goods, a list of occupations, and a list of primary resources. Economists define *resource allocation mechanisms* as procedures that determine which goods to produce, how they are produced, how they are distributed across consumers, and the jobs that people do. With many people, goods, occupations, and primary resources, even a small economy entails untold numbers of choices. Markets are decentralized resource allocation mechanisms in which choices are made by individuals pursuing their own ends. Individuals or private institutions representing individuals (e.g., unions, corporations) interact through exchange of property rights in goods and services. Exchange can occur in formal exchange markets (e.g., auctions) or through other mechanism (contracts, informal markets). The opposite of market economies are centrally planned economies in which private property rights are limited and decisions about what to produce, how to produce, and allocations across consumers are made by the state.

One proposition that is not contentious in economics is that people freely pursuing their economic self-interest in markets with well-defined private property rights can do a better job than central planners in allocating scarce resources efficiently. This proposition does not deny a role for policy. Much of economics is concerned with inefficiencies in resource allocation that result from missing or defective markets. Environmental externalities are one. Understanding what markets do well and what they do not is fundamental to identifying causes and remedies for inefficiency.

The proposition that free exchange among free people advances social well-being is mostly famously attributed to Adam Smith (1776) with his metaphor of the "invisible hand." In an economy with well-defined property rights and perfectly competitive markets, businesses seeking to maximize profits will choose to produce goods that are sought by consumers. Producing goods that no one wants to buy will produce no profits. Competition among sellers for buyers will push prices down to levels that just cover costs. Sellers seeking to maximize profits have incentives to minimize their costs, which will lead them to seek out efficient production technologies and to use low-priced resources. Scarce resources are used efficiently in production and allocated to sectors where goods are in demand. Resource suppliers seeking to maximize their income will lose out in competitive markets if they overcharge producers. Competition among resource suppliers will push down costs.

The problem for the best intended central planners is information. They can never know what they need to know to do a good job. Consumers' preferences are complex, heterogenous, and evolving. Effective and efficient use of people in production requires knowledge of their skills, talents, labor-leisure preferences, ambitions, and attitudes toward risk (e.g., policing). Effective and efficient production requires knowing technologies and the organization of production and distribution. Even well-intended planning will likely provide people with things they do not want and fail to provide with things they do. It will put them in jobs to which they are ill-suited or find little satisfaction. The goods produced will be produced inefficiently. Markets provide people with both the opportunity and incentives to utilize their specialized knowledge. Price signals serve to coordinate activities and allocate resources to their best use. Markets also provide opportunities for innovators and entrepreneurs to expand the opportunities for improving life. Technocrats are rarely innovators. The primary role of bureaucrats is process, not innovation.

Adam Smith's proposition was formalized by twentieth-century research, including that of Nobel Economics Prize winners Kenneth Arrow and Gerald Debreu, to demonstrate that a perfectly competitive market economy would result in a Pareto Optimal allocation of resources to the production of goods, goods among consumers, and a pattern of production that matches the goods supplied with the preferences of consumers provided that certain conditions are met. This result is known in economics as the "First Theorem of Welfare Economics."

2.2.3 An Illustration

We illustrate the concept of market equilibrium and the efficiency of market equilibrium with a single market example. We take a simplified corn market as a case. Corn suppliers are producers who plant fields in the spring and harvest in the fall. Very little corn is sold for direct human consumption. Almost all is sold into a supply chain that processes corn into animal feeds, processed foods for humans, and ethanol. Demanders are therefore mostly industrial firms.

Producers develop corn production plans based on expectations of the price they will receive for their product when harvested. For simplicity we abstract from weather and other events that make the fall price and harvest uncertain. We assume we know the price they will receive. We denote the price per bushel p_c. Also influencing production decisions are the prices of seed, fertilizer, pesticide, fuel, labor, irrigation water in arid agricultural regions, and other inputs farmers must purchase to produce.

The relationship between the price producers receive and the quantity they supply is depicted graphically by a supply curve in Fig. 2.1. The market supply curve is a schedule that describes the quantity of corn, denoted Q_s, that will be supplied at different corn prices. The supply curve is the upward sloping line in Fig. 2.1.

The quantity supplied to the market is the sum of the quantities supplied by individual producers. In deciding how much corn to produce, producers compare the price they receive for selling an additional bushel to the cost they incur from producing it. The cost of an additional unit is defined as the ***marginal cost*** (MC). If the price exceeds the marginal cost, then the increase in revenues from selling another bushel exceeds the cost of producing it. Profits are then increased by the additional production. If the price is less than the marginal cost, the reverse would be true. Profits would then be reduced by the additional production. Producers therefore maximize profit by producing the quantity at which price is equal to the marginal cost (see Box 2.1). At this quantity no additional profit is possible from additional production and profits would be reduced by producing less.

Box 2.1 Marginal Analysis

Marginal analysis is a technique used in economics to assess trade-offs and deduce optimal choices. Economic choices of all types involve benefits and costs. If benefits and costs are quantified by equations, optimal choices can be calculated by computer algorithms. But that method does not provide insight into the logic of the choice. Marginal analysis highlights the logic of choice. Marginals are the incremental changes in benefits or costs that result from incremental changes in the choice variable. Marginal analysis instructs to do more when marginal benefits exceed marginal costs, to do less when marginal costs exceed marginal benefits, and to stand pat when they are equal. The task for insightful application of the technique is to properly frame the choices, benefits and costs, and the marginals.

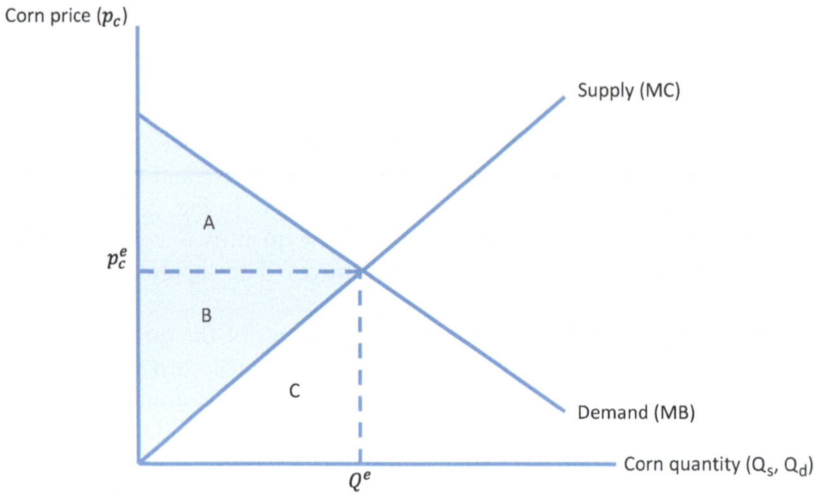

Fig. 2.1 Competitive equilibrium. (Source: Authors' creation)

The quantity supplied will vary from farm to farm with variations in farm size and with variation in soils, climate, and other variables affecting productivity and cost across locations. But all maximize profit by following the same rule. That the supply curve represents profit-maximizing behavior in which producers choose quantities so that price equals marginal cost means the price-quantity combinations along the supply curve can be read as marginal cost-quantity combinations. Since all producers are equalizing price and marginal cost, the price read off the supply curve at any quantity is the marginal cost of that quantity.

The positive slope of the supply curve means that the marginal cost of corn production is increasing. Increasing marginal cost reflects a few economic phenomena. The supply response to higher prices reflects changes in production at the field scale, referred to as the *intensive margin*, and the number of fields in production, referred to as the *extensive margin*. At the field scale, increasing output requires adding inputs to a fixed land base. Corn and other agricultural production technologies exhibit *diminishing marginal product*. Marginal product is the increase in output that results from the increase in an input holding other inputs constants. A diminishing marginal product means that the incremental increase in output that results from an incremental increase in an input decreases with the overall level of input use holding other inputs are fixed. In the case at hand, for example, greater application of fertilizer to a field increases output at a diminishing rate. This implies that the cost of an additional output from the field is increasing.

Corn supply is also increased in response to higher prices by increasing the area of land planted. Generally, profit maximization means that the most productive land is used first. Production moves to less productive land as production increases. Lower productivity land means higher production costs. We will return to these concept as they pertain to the analysis of water pollution policy for agriculture in Chap. 4.

Demanders similarly decide how much corn to buy based on the price of corn. The market demand curve shows the quantity of corn, denoted Q_d, that would be purchased in the market at various prices. The demand curve is the downward sloping line in Fig. 2.1. The quantity demanded at any price is the sum of the quantities demanded by individual buyers at that price. The downward slope means the demanders will buy more at lower prices. Because the demand for corn is derived from the demand for products produced with corn, the curve reflects the economics of the supply chain. In many processes using corn as an input, there are other inputs that can be substituted for corn if corn becomes expensive. For example, sweeteners derived from corn could be obtained alternatively from beets or cane. Higher corn prices would give food processors incentives to reduce the use of corn and increase the uses of substitutes, resulting in a decrease in quantities of corn demanded. Final consumer demands for goods produced with corn will be downward sloping in their prices for various reasons. For example, a reduced corn price will lead to a reduction in the production costs and therefore prices for corn-fed beef, which will lead to an increase in the quantity of corn-fed beef demanded by final consumers, which will lead to an increase in the quantity of corn demanded.

Like supply curves, demand curves reflect pursuit of self-interest and competition. When choosing whether to buy more corn, a livestock producer or food processor using corn as an input will compare the increase in profit, excluding the cost of corn, that results from utilizing additional corn to the cost as determined by the corn price. The increase in profit, excluding the corn cost, that results from additional corn use is the maximum amount the producer would be willing to pay for the additional unit. This profit increase is the producer's or processor's marginal willingness to pay for the additional unit. The producer or processor benefits from buying more corn if the marginal willingness to pay exceeds the corn price as more is added to overall profit than is added to the expenditure on the corn input. This would lead a profit-seeking corn demander to choose the quantity where the marginal willingness to pay is equal to the corn price. The price-quantity combinations along the demand curve will reflect this profit-seeking behavior. For any quantity, the corresponding price on the demand curve will equal the corresponding marginal willingness to pay. The demand curve can therefore be interpreted as giving the marginal willingness to pay for corn. A short-hand term for the marginal willingness to pay is the demanders' marginal benefit of corn. The downward slope means a decreasing marginal benefit as the quantity consumed increases. Behind the corn demand curve is profit-seeking competitive behavior through the supply chain from the farm to final demanders. Profit-seeking implies cost minimization and efficiency in production. Competition drives prices in the system to cost.

A competitive market equilibrium is defined as the price-quantity combination at which the quantity supplied and demanded are equal. Suppliers seek to sell no more at this price than they do. Buyers seek to buy no more at this price than they do. There is no excess demand or supply at the market price. In Fig. 2.1 the equilibrium occurs where the supply and demand curves cross at the price-combination (p_c^e, Q^e). Interpreting the demand curve as a marginal benefit (MB) curve and the supply curve as a marginal cost (MC) curve, an important feature of the market equilibrium is that it equates the marginal benefit of corn consumption with the marginal cost of producing corn as we have p_c^e = MB = MC. This is a condition for maximizing the social net benefit of corn. For corn quantities less than Q^e, MB > MC. When this is true, increases in corn consumption add more to social benefits than to social costs. Increasing corn consumption is then socially beneficial. For quantities of corn greater than Q^e, MB < MC. When this is true, more corn adds more to cost than to benefits. Reducing corn consumption is then socially beneficial. When MB = MC the net benefits from corn consumption have been maximized. More corn or less corn than Q^e reduces social net benefits. This property means that a Pareto Improvement cannot be had by a different price-quantity combination if costs and benefits are fully captured by the supply and demand curves. The market equilibrium is therefore economically efficient (or Pareto Efficient).

The supply and demand curves in Fig. 2.1 can be used to calculate social net benefits. Adding up the marginal benefits of consumption under the demand curve at each quantity between $Q_d = 0$ and $Q_d = Q^e$ gives the total benefit to demanders from corn consumption in the market equilibrium.[2] This sum is equal to the area under the demand curve $Q_d = 0$ and $Q_d = Q^e$. In Fig. 2.1 this area is the sum of the triangle areas labeled A, B, and C. Actual expenditures are $p_c^e Q^e$, which is equal to the area of the rectangle formed by adding the triangle areas labeled B and C. Subtracting actual expenditures from the total benefit gives the net benefit of corn to consumers as the area of triangle A. This net benefit to consumers is defined as the *consumers' surplus.*

Similarly, adding up the marginal cost of production at each quantity under the supply curve between $Q_s = 0$ and $Q_s = Q^e$ gives the total cost to suppliers. This sum is equal to the area below the supply curve between $Q_s = 0$ and $Q_s = Q^e$. In Fig. 2.1 this area is triangle C. The revenue received by producers is $p_c^e Q^e$, which as above is equal to the area of the rectangle formed by adding the triangle areas labeled B and C. Subtracting cost from revenues gives the net benefit to producers as the area of triangle B. This net benefit is defined as the *producers' surplus.*

The sum of consumers' and producers' surplus is the sum of the triangles A and B (the shaded area in Fig. 2.1). This sum is equal to the net benefit from production and consumption (total consumer benefit—total producer cost). A property of a competitive market equilibrium is that it maximizes the sum of consumers' and producers' surplus.

2.2.4 *Market Failure*

So, what might go wrong that results in failure of markets to maximize net benefits? Economists differentiate goods according to two characteristics that affect the capacity of markets to provide them efficiently: *divisibility* and *excludability*. A good is divisible (or *rival*) in consumption if consumption of a unit by one consumer eliminates the unit from consumption by others. Corn and other agricultural commodities are divisible goods. The supply is divided up between consumers. A good is non-divisible (or non-rival) if a unit of the good can be simultaneously consumed by more than one consumer. Many benefits from water resources are indivisible. Absent congestion, groups of people can simultaneously

[2] The sum of marginal benefits (costs) under a marginal benefit (cost) curve between any two quantities, say a and b, along the horizontal axis is equal to the change in the total benefit (cost) resulting from a change in the quantity from a to b. Mathematically the sum under marginal is the integral of the marginal functions between the two points.

enjoy the use of a lake resource for swimming, boating, fishing, wildlife watching, or the aesthetics of the water environment. Aquatic ecosystems support carbon cycling, nutrient cycling, biodiversity maintenance, wildlife habitat, and pollutant filtering, the benefits of which are indivisible.

A good is excludable if access to it can be denied. Excludability is determined by two things. One is property rights. If property rights do not exist (formally or informally) or do not grant the right to exclude, then the good is non-excludable. A second is the cost of exclusion. If excludable property rights exist, a good will still be non-excludable if exercising the right is prohibitively expensive. Classic examples of non-excludable goods are open access natural resources such as migratory birds, groundwaters, grazing lands, and ocean fisheries.

The First Theorem of Welfare Economics assumes that goods in an economy are excludable and divisible.[3] Excludability is essential for a market to function for without it a price cannot be charged. Indivisible goods can be provided by markets if the goods are excludable as a price can be charged for access and use. An efficiency loss arises if the price required to cover the cost of providing the goods results in the exclusion of some potential consumers. This efficiency loss results because there is no opportunity cost of another consumer when a good is non-rival. Price exclusion results in forgone consumption benefits with no benefit to offset the loss. Public policy sometimes leaves the provision of non-rival but excludable goods to markets (e.g., movie theaters, private gyms, and golf facilities) and sometimes supplies the good (e.g., public parks, public gyms and golf facilities, beaches, and highways).

Public goods are both indivisible and non-excludable. The non-excludability of public goods mitigates against optimal provision by markets because people can receive whatever is available at no cost. Self-interested actors will be inclined to *free ride*. Optimal provision generally requires public provision. But non-excludability also poses a problem for public provision. For example, an agency charged with managing an aquatic system might like to have those served contribute to its budget amounts consistent with the benefits they receive. But non-excludability provides people with incentives to free or easy ride, resulting in them contributing nothing or less than the benefits they receive. Thus, taxes are required to allocate resources to the provision of public goods.

[3] There are some exceptions for non-divisible goods that are excludable. Such goods are called club goods (Cornes and Sandler 1996).

Externalities occur when the actions of one decision maker (a person or organization) have a direct, unintentional, and uncompensated effect on the condition (e.g., preference satisfaction, costs, profits) of others (Baumol and Oates 1988). Externalities are positive when the effect is beneficial to the recipient(s) and negative when the effect is detrimental. Environmental pollution is negative externality. People and businesses do not pollute with the intent of causing harm. It is a consequence of actions they pursue for other reasons. Similarly, actions that result in positive externalities may be done without the intent to benefit others. Uncompensated in this definition means that the source of a negative externality does not compensate those who are harmed and that the source of a positive externality is not compensated by those who benefit. The absence of compensation implies that the exchange does not involve bargaining, contracting, buying, selling, or other market transactions. Externalities are sometimes described as the result of *missing markets*. Their exchange occurs outside of the market.[4]

Positive externalities are often the result of private provision of a public good. An example is private research and development (R&D). Firms undertake R&D to increase their own profitability, but discovery and innovation have spillovers that provide benefits to others. Policy-relevant negative externalities result from the private production of "public bads." Pollution that reduces public goods provided by water resources is an example.

Externalities result in a divergence between the private and social benefits and costs of economic decisions. It is this divergence that results in efficiency losses. Without rewards for generating benefits received by others, producers of positive externalities, like private R&D, have no direct economic incentive to consider those benefits in their decisions. Similarly, without penalties for causing harms to others, producers of negative externalities, like water pollution, have no direct economic incentive to consider those in their decisions. Self-interested agents will make decisions based on their own cost and benefits. This will result in a loss of social net

[4]That the exchange is outside the market does not mean it is involuntary. A recipient of a negative externality might be able to escape it by moving or by taking other actions. To do so would entail cost. The loss to net benefits to escape or mitigate and externality is a cost of externality. Similarly, the beneficiary of a positive externality might be able to choose actions that block the benefit, though as beneficiaries they have no incentive to do so.

benefits compared to the socially efficient choice. Missing markets eliminate incentives for socially efficient behaviors.

In the case of agriculture, normal farming activities result in pollutants moving into water resources. The cost of water quality degradation is external to farm decision making. The divergence between producers' private costs and benefits and the societal costs and benefits of their activities results in socially inefficient choices and corresponding efficiency losses. The social cost of the externality is illustrated in Fig. 2.2. This figure depicts the corn market equilibrium as in Fig. 2.1. The market supply curve is the private marginal cost of production. We have added a social marginal cost (SMC) curve that adds (*internalizes*) the marginal external damage costs (MDC) from water pollution to the private marginal cost (MC). That the difference between the private (MC) and social marginal costs (SMC) increases with the quantity indicates that the marginal damage cost of pollution increases with the level of pollution.

Figure 2.2 shows that MB > SMC for quantities less than Q^* but MB < SMC for quantities more than Q^*. This implies that social net benefits are maximized at $Q^* < Q^e$. The divergence of private and social marginal costs results in a market equilibrium in which too much corn is produced. The market price if marginal damage costs were internalized (i.e., pollution costs are borne by polluters) would be $p_c^* > p_c^e$. Reducing the quantity from Q^e to Q^* and increasing the price from p_c^e to p_c^* will reduce the sum of consumers' and producers' surplus by an amount equal to the sum of the triangles labeled D and E. This reduction would be the social cost of reducing pollution. Consumers' surplus is reduced by the increase in price. Producers' surplus is reduced by increased production costs, but this is offset to some degree by the higher price producers receive. If the price increase is large enough, producers' surplus could increase rather than decrease. Whether this is the case or not, the social cost of pollution control is positive as the sum of consumer's and producers' surplus is reduced. The social benefit of the reduction is the reduction in external damage costs. The reduction in damage costs is the sum of MDC at each quantity between Q^* to Q^e. This is the area between SMC and MC between Q^* to Q^e and is equal to the sum of the triangles D, E, and F. Subtracting the social cost of the pollution reduction D + E from the social benefit D + E + F leaves a social net benefit F of internalizing the externality. This is the shaded area in Fig. 2.2). If the externality is not internalized, this measures the social cost of the externality.

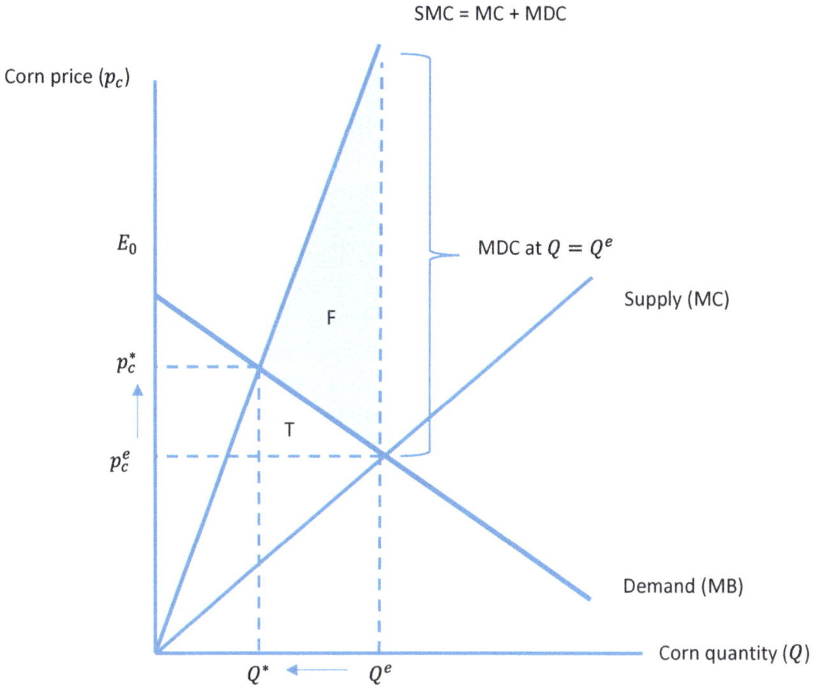

Fig. 2.2 Market equilibrium with externality. (Source: Authors' creation)

2.3 Pareto Efficient Pollution and the Role of Government

The simple market model presented in the previous section illustrates how a negative environmental externality can result in economic inefficiency. An implicit assumption in the analysis is that the level of the externality is proportional to corn production. This proportionality is a useful simplifying assumption for the illustration but limiting for a deeper analysis of the economics of pollution control. In this section we develop concepts that are used extensively in economic analysis of pollution control and use them to consider policy targets for pollution. We also further develop the case for government intervention for environmental externalities.

2.3.1 Pollution Benefits and Abatement Costs

In a model in which pollution is proportional to output, pollution control is achieved solely by managing output. But polluters generally have a range of options for pollution. Broadly, these include changes in production process to reduce the formation of pollutants and installation of treatment technologies to treat polluting emissions prior to discharge. The cost of pollution control for a profit-seeking producer is the reduction in profit that results from pollution control. This cost is commonly referred to as the pollution abatement cost.

We begin with the pollution control cost for a profit-seeking firm. Figure 2.3 shows the relationship between a polluter's emissions, denoted e, and the polluter's profit, denoted π. With an unregulated externality the baseline condition of the firm is maximizing profit without explicit measures to control pollution. Pollution is an unintentional outcome of this activity. Emissions and profit in the unregulated case are given by the point (e^0, π^0). Now suppose that the firm either decides on its own or is forced to cut its emission level to $e^1 < e^0$. The firm reoptimizes its production, perhaps by switching to alternative technologies, changing the combination of inputs, and reducing output to minimize the profit loss from the reduction. These actions must decrease the firm's profit if it was previously maximizing them. The highest possible profits that can be obtained when emissions are restricted to e^1 is π^1. Further reductions would result in new emissions-profit combinations, with profits always lower the smaller the level of emissions.

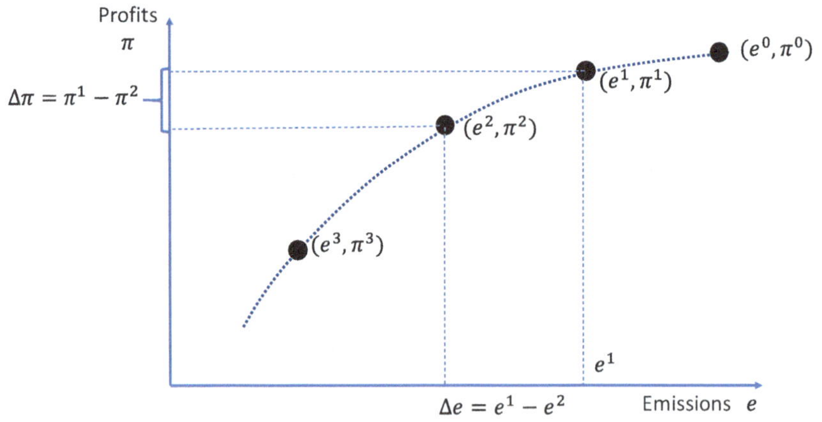

Fig. 2.3 Pollution abatement costs. (Source: Authors' creation)

In Fig. 2.3, profit increases as allowable emissions are increased from zero to e^0. Abatement and abatement costs (i.e., profit losses relative to π^0) increases as emissions are reduced from e^0. A graph of abatement cost would rise with the level of abatement. The graph would be the mirror image of the benefits of emissions.

The incremental increases in benefits that occur with incremental increases in emissions define the marginal benefits of emissions (MBE). The "mirror image" is the marginal abatement cost (MAC), which is the incremental increase in abatement cost with an incremental increase in abatement. Figure 2.4 depicts graphs of these marginal relationships. The

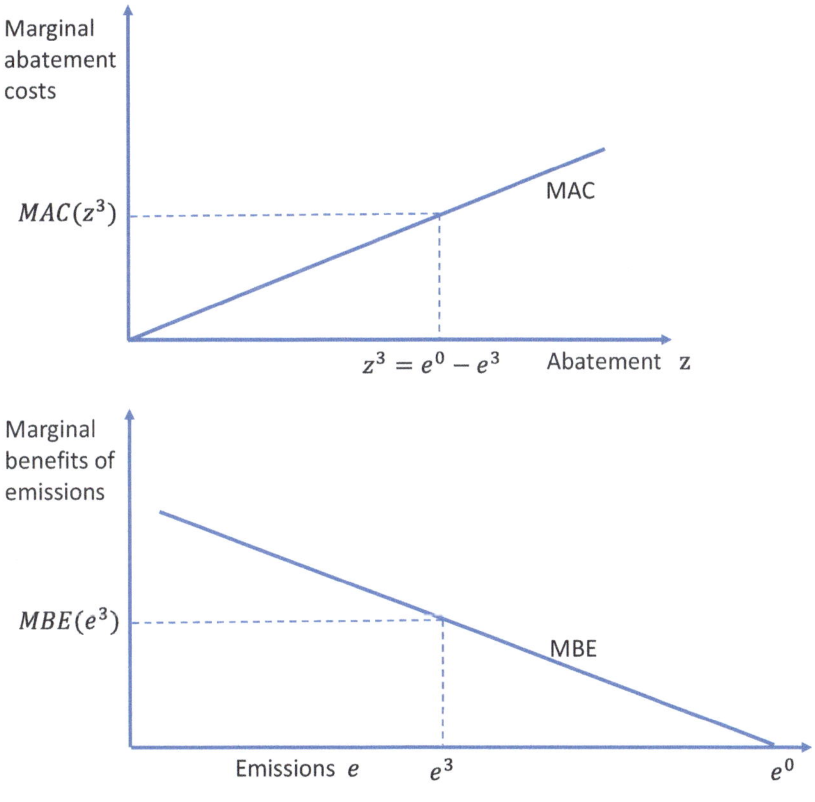

Fig. 2.4 Marginal emissions benefits and marginal abatement costs. (Source: Authors' creation)

Box 2.2 Government Pollutes Too

Environmental externalities are often described as the result of profit-seeking in markets. Yet, government entities are often significant pollution sources. Air and water pollution are pervasive in planned economies. Publicly owned wastewater treatment facilities were a primary target of the water pollution control initiatives that began in the 1970s and remain an important regulatory target. Nutrient pollution from urban storm runoff is a contemporary water quality challenge.

Abatement costs for publicly owned municipal wastewater plants are not forgone profits. The direct costs of wastewater treatment and runoff management are the operating, maintenance, and capital costs of infrastructure used to collect and treat wastewater and storm runoff. These costs in turn are borne by those who must pay, typically citizens served by the systems through water and sewer rates and municipal taxes.

downward slope of the marginal benefit curve reflects the shape of the emissions benefit curve in Fig. 2.3 showing that total benefit increases at a decreasing rate, implying that marginal benefit is decreasing. This implies the total abatement cost increases at an increasing rate and the marginal abatement cost is increasing. Both graphs thus reflect the same relationship between the costs of reducing the emissions, but from different perspectives. Notice specifically that with zero abatement (baseline emission level e^0) $MAC(0) = MBE(e^0) = 0$. The concepts are often used interchangeably in the environmental economics vocabulary.

Figure 2.4 provides another perspective on why firms pollute without regulation. The graph above tells the story from the perspective of abatement: Choosing any abatement level other than 0 would be costly for the firm. Hence, the least-cost (profit-maximizing) solution is not to do any abatement. The graph below links the emission level to profit, seeing emissions as costless inputs for the firm. Consider for instance the emission level e^3. Increasing emissions would provide benefits for the firm without any costs. Like any input, it is profitable for the firm to choose the emissions until the marginal benefit equal marginal costs. And because emissions do not cost the firm anything, it chooses the baseline level $MBE = 0 \rightarrow e^0$.

2.3.2 Pollution Costs and the Benefits of Abatement

Water pollution causes a variety of economic costs related to uses of the affected waters (Young and Loomis 2014) (see Boxes 2.3, 2.4, and 2.5). For example, disruption of aquatic ecosystems diminishes the productivity of waters used for commercial and sports fishing. Degradation of waters

Box 2.3 The Value of Water

Water provides many services that are of enormous economic value to people and economies (Young and Loomis 2014). Economic research distinguishes between *use values* and *nonuse* values. Use values are economic benefits that derive from the physical use of water in production or consumption. Use benefits may derive from *off-stream* or *in-stream* uses. Off-stream use benefits derive from water that has been removed from natural or developed raw water supplies (groundwater, streams, reservoirs). Examples of off-stream uses are drinking and food preparation, cleaning and sanitation, off-stream water recreation (e.g., swimming pools, water parks), off-stream aesthetics (e.g., landscaping, water fountains, water fixtures), irrigation and aqua-culture, food processing and other industrial processes, and thermoelectric power cooling. Irrigation and thermoelectric cooling are by far the largest off-stream uses measured by withdrawals from ground and surface water supplies. *In-stream* use values derive from water in place, whether natural or developed water supplies. Examples are fishing, boating, and swimming, waterfowl hunting, wildlife viewing, power generation, and water-based aesthetics. Some in-stream use values derive not from the direct use of water but from the contributions that water resources make to natural systems that in turn provide services to people and economies (wetlands help regulate storm runoff, filter polluted waters, provide habitat for aquatic and terrestrial wildlife, submerged aquatic vegetation provides habitat for fish). Nonuse values are benefits that people derive without making physical use of water directly or indirectly. Economic research demonstrates that people derive satisfaction (i.e., benefits) from knowing that waters they will never use are in good condition, supporting healthy ecosystems, and are of good use to others living now and in the future.

Box 2.4 Water Quality and the Benefits of Water Pollution Control

Water quality can refer to the quality of natural water in place, or the quality of "retail" water that has been withdrawn, possibly treated, and delivered to customers. In any case, the concept is a relative one. Quality depends on the intended use of the water. For example, water that may be suitable for irrigation may be unfit for human consumption. Water that is suitable for human consumption may be ill-suited for some industrial uses. Fast-moving water that is thrilling for rafters is unsafe for other in-stream uses. Chemical and other characteristics of in place natural water relevant to use can vary greatly from place to place. Degradation of natural water quality reduces the benefits derived from direct and indirect in-stream water uses (e.g., swimming, boating, fishing). Degradation of natural water also degrades the raw water used for off-stream uses. Consequences can include a mix of increased treatment costs (see Box 3.1) and reduced benefits from water quality.

The social costs of water quality degradation are the reductions in the use, direct and indirect, and nonuse values provided by water resources. The fundamental concepts used to value water quality benefits or water pollution costs are willingness to pay/accept. These concepts were introduced in Sect. 2.2.3 regarding the demand and supply of corn and concepts can be applied to water quality. The economic benefit of a water quality improvement to a user is the amount the user would be willing to pay for the improvement. The economic cost of water quality degradation is the amount the user would be willing to accept in exchange for the decline.

used for drinking or industrial uses necessitates costly treatment. Degradation of waters used for contact recreation reduces activity days when waters can be used safely and reduces the quality of recreational experiences. Some benefits of water are described as "nonuse" benefits. These derive not from the direct use of waters but from the value people place on the condition of waters that they make no use of. For example, people who will never visit the Chesapeake Bay still value the condition of the Bay (Moore et al. 2015).

As the costs of pollution abatement are the forgone benefits that result from restricting emissions, the costs of pollution are the benefits forgone

Box 2.5 Measuring the Benefits of Water Quality Pollution Control

The benefits of water quality improvements, or equivalently, the social costs of water quality degradation, have been estimated for different types of water quality problems, for different locations, and at various spatial scales. Recreational benefits have been emphasized because of the significance of recreational uses to the value of water and the significant effects of water quality degradation on recreational uses (Keiser et al. 2019). But significant benefits also arise from the protection of drinking water supplies, health impacts from contaminated drinking water and exposure to pathogens and toxins at beaches, and adverse effects on the productivity of commercial fisheries and industrial users.

When water quality degradation affects the costs of production of goods sold in markets, such as harvests from commercial fisheries, estimating pollution costs begins with estimates of the effects on the costs of production, the effects of changes in production costs on market supply, and the effects of the change in market supply on market prices and quantities. The cost to consumers is the loss of consumer surplus due to increased price. The cost to suppliers is the reduction in producer surplus. Producers surplus is reduced by cost increases, but the loss is offset to some degree by the price increase.

Buying and selling in markets provide data useful for valuing pollution costs related to goods sold in markets. Quantifying cost related to nonmarket goods is a greater challenge as there are no market transactions which could be directly linked to the damages. For most water quality problems, nonmarket values dominate social costs. Research shows that swimming, boating, recreational fishing, or just enjoying a good scenery give great pleasure and generate large economic value. However, they are not readily observable in markets. There are two basic approaches for nonmarket measurement. One is the *revealed preference*. The second is *stated preference*. Revealed preference methods use observed behavior for which data are available reflecting demands for water from choices of recreation sites, recreations activities, housing locations, and other variables to infer values. Stated preference methods use surveys and questionnaires to ask prospectively about willingness to pay

(continued)

Box 2.5 (continued)
(e.g., Bateman et al. 2002; Hanley et al. 2019). Revealed preference methods are generally preferred.

In a recent assessment of benefits estimates, Keiser et al. (2019) present estimates for the annual benefits of the US Clean Water Act, the principal water pollution control legislation in the US. The estimates range from a modest 3.9 billion 2014USD to a significantly more substantial 98.1 billion 2014USD. The studies differ in when they were conducted, and the data and methods used. The range indicates large uncertainty about benefits. The authors note various reasons for this related to data and methods. They also conclude that existing studies tend to underestimate benefits because some impacts (e.g., health impacts), pollutant types (e.g., toxics), and affected resources (e.g., coastal areas, and surface–groundwater interactions) are being missed or given inadequate attention. Another study suggesting underestimation due to missed benefits is Wainger et al. (2017).

due to water quality degradation. Consequently, where benefits of water decrease as they are degraded, costs of pollution increase. Analogous to the discussion of abatement costs, increasing levels of emissions and water quality will lead water users to re-optimize for new conditions. Re-optimization can take many forms depending on the use. For example, reduced water quality for drinking or industrial processes can be mitigated by expenditures for water treatment or alternative water supplies. Economic research generally suggests that water pollution damage costs increase at an increasing rate. This means a graph of the damage costs would show rising costs with the level of pollution with the slope of the graph becoming steeper with the level of pollution. Correspondingly, a graph of the marginal damage cost of pollution (MDE) would show the cost rising with the level of pollution. The greater the level of water pollution, the greater is the cost of the next increment. A graph of water quality benefits would show benefits increasing but at a decreasing rate with greater water quality. A graph of the marginal benefit would show decreasing marginal benefits. The better the water quality, the less valuable is the next increment.

2.3.3 Optimal Level of Emissions

Our discussion of benefits and cost highlights the existence of trade-offs between these outcomes. An obvious question is, what would be the amount of pollution to strike a balance between benefits and costs? We consider this issue for an externality with a single source and a single recipient. We consider a case where effluents from a tannery degrade a stream used to provide freshwater to a commercial fish farm. Figure 2.5 examines the balance question when framed as one of optimizing the economic benefit of tannery emissions less the economic cost to the fish farm. Benefits in this case are tannery profits. Costs in this case are forgone fish farm profits. These benefit and costs of emissions are shown in the top panel. Emissions benefit (B) (tannery profits) increases at a diminishing rate as emissions increase. The environmental damage cost (forgone fish farm profit in this case) increases at an increasing rate with emissions. Net benefits are the difference between benefits and costs. The difference is greatest, meaning that net benefits are maximized, at the emissions level e^*.

The lower panel in Fig. 2.5 shows the marginal benefit and damage cost of emissions. When MBE > MDC (e.g., $e = e_0$), an increase in emission increases the profit to the tannery more than it decreases the profit of the fish farm. In the top panel, increasing e from e_0 to e^* results in a net benefit of $\Delta B - \Delta D > 0$. This gain in the lower panel is the shaded triangle, which is the area under the marginal benefit curve less the area under the marginal cost curve between e_0 and e^*. Conversely, the reverse is true when MBE < MDC. The lost profit to the fish farm from an increase in emissions exceeds gain in profit by the tannery. Net benefits are maximized when MBE = MDC. This equivalence is a condition for net benefit maximization.

2.3.4 Market Failure Revisited: Pigou Versus Coase

Market failure provides a rationale for public policy interventions in market economies to reduce the social costs of inefficient resource allocation. In the case of externalities, the market-failure government-fix paradigm was introduced by British Economist Arthur Pigou in his classic text The Economics of Welfare (1920). Pigou proposed correcting negative externalities by placing a tax on the externality equal to the marginal social cost. In the case at hand this would mean a tax on emissions. This is illustrated in Fig. 2.5. The tax rate per unit of emissions is τ. When choosing how much to pollute when subject to the tax, the firm compares the gain in

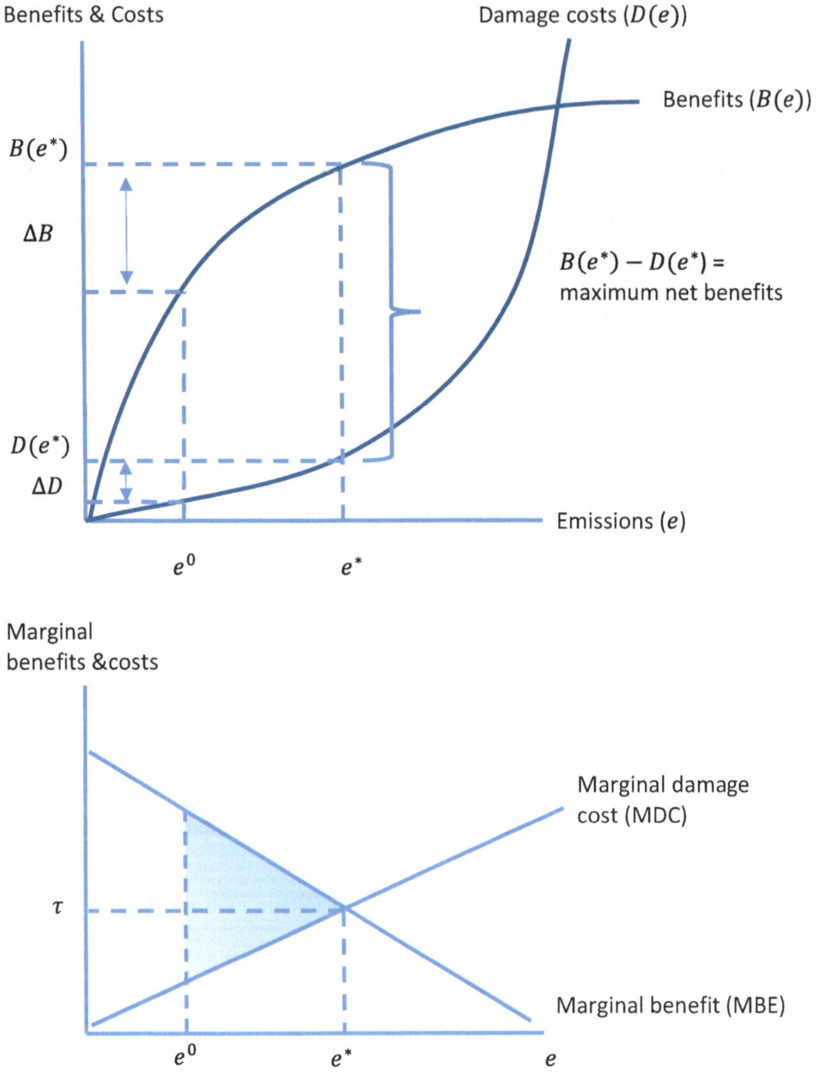

Fig. 2.5 Optimal emissions and the Pigouvian tax rate. (Source: Authors' creation)

profits, excluding the tax expenditure, from additional emissions to the increase in tax cost. At the margin the firm compares MBE to τ. If MBE > τ, the gain from additional emissions exceeds the additional tax cost and it pays to pollute more. The reverse is true if MBE < τ. The profit-maximizing firm will optimize the emissions-level setting MBE = τ. If τ = MBE(e^*) then the tax produces the efficient solution.

The Pigouvian solution was challenged by Nobel Prize winner Ronald Coase in his highly influential article, The Theory of Social Cost (1960). Coase argued that parties to an externality could work out private solutions without government intervention through bargaining and contracting. In the Law and Economics literature this is known as the Coase Theorem. In the Theory of Social Cost, Coase considers various cases involving two parties. He demonstrates that if property rights are well-defined and transaction costs (e.g., the costs of negotiating, contracting, contract enforcement) are negligible then the parties will negotiate to the efficient solution. Moreover, the allocation of property rights does not matter to the solution.

In our hypothetical case, the tannery and the fish farm need not wait for government intervention if property rights are well defined as they can negotiate a mutually beneficial solution. If the tannery has a right to pollute without regard for downstream costs, emissions without contracting would be at the level where MBE=0. If the fish farm has the right to a pollution-free stream, the tannery cannot pollute without the permission of the fish farm. Without an agreement, emissions will be zero. Because the efficient pollution level is a Pareto Optimum, moving toward the optimum provides Pareto Improvements that will leave one or both parties better off. If the tannery has the right to pollute, the amount the fish farm is willing to pay the farm to reduce pollution up to e^* is more than the amount the tannery would require for the reduction. Conversely, if the fish farm has a right to a clean stream, the amount the tannery is willing to pay to allow pollution up to e^* exceeds the amount the fish farm would require to accept the pollution. If Pareto Improvements are present, the parties have incentives to negotiate regardless of the allocation or property rights to the Pareto Efficient pollution level. No government intervention is needed.

The Coasian argument is compelling and provides a check on the presumption that government intervention is always needed. When externalities are present, a good choice may be correcting defective property rights that create barriers to private solutions. However, while there will be cases where private solutions are feasible and preferred to government intervention, this is not generally true for environmental problems. The best cases for negotiated solutions involve a small number of parties, simple and well-understood cause-and-effect relationships, and low monitoring and enforcement costs for contract compliance (Baumol and Oates 1988; Cornes and Sandler 1996).

Externalities with many parties involved create a couple of problems. One is that the larger the number the more difficult and costly it is to negotiate mutually beneficial contracts. Another has to do with the effect of large numbers on incentives to contribute to solutions. Given that water quality is a public good, free riding will be a problem if property rights lie with polluters making recipients of the externality responsible for purchasing pollution reductions. As the number of affected parties increases the incentives to free ride increase. Strategic behaviors that increase the difficulty of negotiations and increase transaction costs become more likely. Consider for example negotiating about phosphorus runoff into a small pond from a barnyard versus negotiating over phosphorus discharges in a large basin with tens of thousands of producers and millions of affected parties in the Baltic Sea.

Uncertainty about cause-and-effect relationships increases transaction costs as information is required to establish causation and to determine the outcomes of remedial actions. Uncertainty about cause and effect can also lead to strategic behaviors that increase the difficulty and cost of negotiations. Another form of uncertainty that matters is the ability to monitor the activities of the source. Monitoring and enforcement costs increase with the difficulty of observing causal actions.

It is important to recognize that regulatory agencies face many of the challenges that apply to Coasian negotiations. The demand for water quality protection remains a public goods problem in which individual preferences for water quality are imperfectly known. Uncertainty about cause and effect complicates social decision making as do high monitoring and enforcement costs. Government must contend with uncertainty, strategic behaviors, and transaction costs. Given that this is the case, can government action achieve more efficient outcomes than the alternatives? Economic research suggests the answer is yes if pollution control policies are appropriately designed and implemented. The economic case for government rests on the ability of government to utilize its police powers and its powers to tax and spend to design and implement policies that (1) reduce efficiency losses from environmental externalities and (2) entail transaction costs smaller than the resulting net social benefits of pollution control. We should note that the market-failure government-fix paradigm

Box 2.6 Government Failure

Economics research on causes of economic inefficiency is not limited to market failures. Government failure refers to efficiency losses that result from government action (Anthoff and Hahn 2010; Winston 2007). Government failure may result from poorly designed or implemented policy choices to fix market failures, or from interventions where no significant market failures exist, causing market distortions. Water pollution is a market failure in which government is needed to reduce efficiency losses. But current policy instruments fall short of being economically efficient and in some cases may well qualify as government failures as defined by imposing greater costs than benefits (e.g., Olmstead 2010; Keiser et al. 2019). Government intervention in agriculture is pervasive and is often motivated by policy concerns that are not the result of market failures (Anderson et al. 2013; Smith et al. 2017).

assumes government seeks efficient solutions, has the capacity to discover and implement them, and does so (see Box 2.6).

2.4 Efficiency in Water Pollution Control

Water pollution externalities resulting from agriculture are generally complex, with many parties being affected, highly uncertain cause-and-effect relationships, and high monitoring costs necessitating government action if efficiency-improving outcomes are to be obtained. We turn now more explicitly to the regulation of water quality. The basic question is how should it be done? The generic models of externalities presented in the previous sections are of limited use in addressing this question because they lack physical aspects of water pollution processes that are important to answering the question. These aspects include the links between polluters' choices (e.g., farming practices) and polluting emissions and between polluting emissions and water quality outcomes, over time and space. In this section, we explore physical environments encountered in water pollution control, water quality goals for these environments, and requirements for efficient pollution control.

2.4.1 *Water Quality Goals and Social Cost Minimization*

First, a small but essential reconsideration of water quality goals. The economic notion of optimal emissions is based on a set of value judgments that imply that economically efficient outcomes are socially preferred. One limitation of efficiency as a goal is that policy makers lack information on benefits and costs needed to determine efficient allocations. A second is that policy makers generally consider other normative and political criteria in choosing developing policy goals. Emissions targets are typically determined by negotiations between economic and environmental stakeholders, policy makers, governmental agencies, and natural and social scientists. When this is the case, an alternative objective for policy design is to achieve environmental goals that emerge from policy processes at least social cost.

Water quality goals often specify uses that must be supported by water resources or biological, physical, and chemical conditions that must be met. Achieving goals in either case requires limiting pollution loads at locations for which goals are set to levels that are compatible with the goal. This aspect of water quality management is not economic. Instead it requires water science to determine relationships between ambient environmental pollution and water quality conditions.

Given a limit on the allowable pollution consistent with a water quality goal, cost minimization requires two things. One is that individual polluters minimize their individual abatement costs. The second is that pollution abatement is allocated across the set of polluters to minimize the total costs of reducing ambient pollution at the location(s) for which goals are set. This second condition recognizes that the costs of reducing pollution at a downstream location will vary from one contributing polluter to another depending on the abatement costs and the relationship between their emissions. An allocation of abatement that minimizes social costs is called *cost-effective*.

Pareto Efficient pollution control has two properties. One is that it minimizes the social costs of pollution control. The second is that it balances the benefits and costs of pollution control to maximize net benefits. Least-cost allocations will not be Pareto Efficient if the water quality target is not one that maximizes net benefits; however, Pareto Efficient allocations will be cost minimizing. Accordingly, understanding how to design policies for cost-effective control is necessary whether the goal is to achieve water quality objectives at least cost or achieving Pareto Efficient outcomes.

The first step in learning how to design policies for cost-effective control is to understand the conditions required for cost minimization. These conditions provide guidelines that efficient policies must satisfy and establish tests for determining the cost-effectiveness of policies. We explore these conditions for a set of physical environments.

2.4.2 *Cost-Effective Management of Uniformly Mixed Pollutants*

To make the analysis concrete, consider a water body, for instance a lake, that receives emissions of a single type of pollutant (e.g., phosphorus) from a set of polluters. We make five assumptions: (1) polluters' emissions are completely under their control; (2) emissions can be metered accurately at little or no cost; (3) the environmental harms caused by the pollutant depend on the ambient concentration of the pollutant in the lake; (4) the impact of a unit of emissions on the ambient concentration in the lake is the same for every source regardless of the source's location; (5) the pollutant is rapidly assimilated and not persistent so that ambient concentration at any point in time is determined by the discharges at that time. We highlight assumption (4) which defines this as a case of a *uniformly mixed pollutant*. When this is true there is no reason to prefer emission reductions from one source over another from an environmental perspective as they have the same impact on water quality. And when this is true, the sole determinant of least allocations between sources is differences in polluters' abatement costs.

Suppose there are two sources, a tannery and sanitary sewer for a small town. Their emissions to the lake are denoted respectively e_T and e_S. The total pollution load determining ambient pollution in the lake is the sum of their emissions, $L = e_T + e_S$. The water quality goal set for the lake requires that total load be no more that L^*. Suppose the initial emissions levels of the two sources are e_T^0 and e_S^0. The reduction in pollution from the tannery is $a_T = e_T^0 - e_T$ and from the town sewer is $a_S = e_S^0 - e_S$ To satisfy the total load limit the combined emissions must satisfy $e_T + e_S \leq L^*$, which implies a required total abatement of $a_T + a_S \geq e_T^0 + e_S^0 - L^*$. A least-cost (or cost-effective) allocation is one in which both sources minimize their individual abatement costs and the allocation of abatement between the sources minimizes total costs.

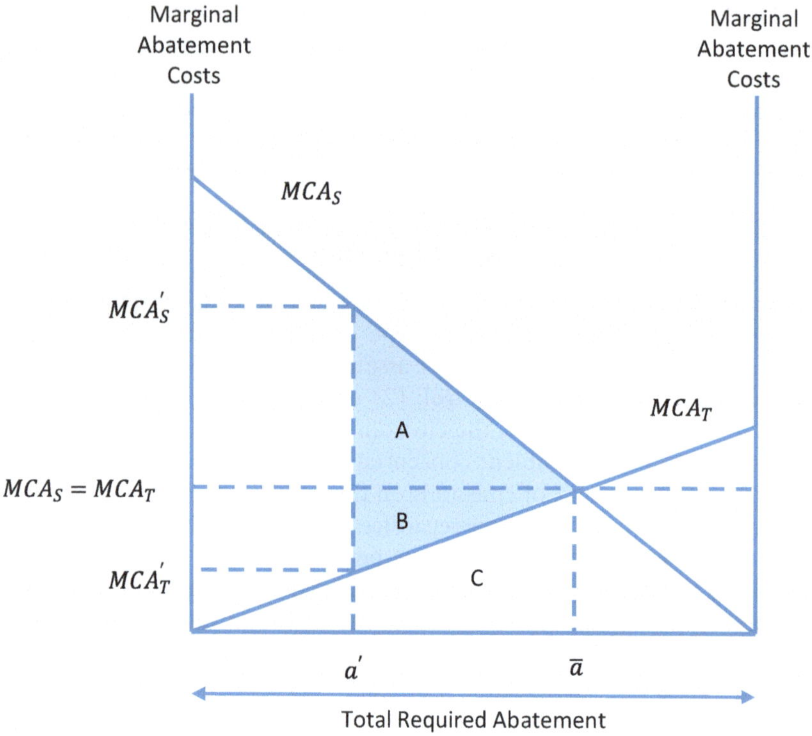

Fig. 2.6 Cost-effective control for uniformly mixed pollutants. (Source: Authors' creation)

Figure 2.6 illustrates the least-cost allocation. The horizontal axis of the graph is the total required abatement $e_T^0 + e_S^0 - L^*$. On the vertical axes are marginal abatement costs. The upward sloping line labeled MAC_T is the tannery's marginal abatement cost. The tannery's abatement increases to the right along the horizontal axis from $a_T = 0$ at the left origin to $a_T = e_T^0 + e_T^0 - L^*$ at the right origin. The downward sloping line labeled MAC_S is the town's marginal abatement cost. The town's abatement decreases to the right along the horizontal axis from $a_S = e_S^0 + e_S^0 - L^*$ at the left origin to $a_S = 0$ at the right orgin. At any point on the horizontal axis the total required abatement is divided between the two. At the mid-point, the total is split equally between the tannery and town.

The MACs for the two polluters cross at the point labeled $\bar{a}$ on the horizontal axis. Consider the abatement allocation at the point a' to the left of $\bar{a}$. At this allocation $MAC_S > MAC_T$. This means that a reallocation of abatement from the town to tannery will reduce the town's abatement cost by more than it will increase the tannery's cost. The result of a reallocation would be an overall cost savings. For instance, moving from a' to $\bar{a}$ would reduce the town's abatement cost by the sum of the areas labeled A, B, and C. The increase in the tannery's abatement cost would be the area labeled C. The net cost savings would be the shaded triangle equal to the sum of the areas labeled A and B. Conversely, to the right of $\bar{a}, MAC_T > MAC_S$. This means that a reallocation of abatement from tannery to town will reduce the tannery's abatement costs by more than it will increase the town's cost, again resulting in an overall cost savings. Similar savings exist whenever the marginal abatement costs are unequal. Accordingly, the cost-minimizing allocation is $a_T = \bar{a}$ and $a_S = \left(e_T^0 + e_S^0 - L^*\right) - \bar{a}$. Note that the least-cost allocation is to the right of the mid-point with more abatement by the tannery than the town. This is due to difference in the costs, with the tannery being able to abate cheaply compared to the town. If their MACs were identical, abatement would be split evenly.

Table 2.1 illustrates the least-cost solution for a hypothetical example where $MAC_T = a_T$, $e_T^0 = 10$ kg, $MAC_S = 2a_S$, and $e_S^0 = 15$ kg. The total baseline load is 25 kg. The target load is 15 kg requiring an aggregate reduction of 10 kg. The greater abatement by the tannery in the least-cost allocation reflects the fact that its MAC is half that of town for any given abatement level. The assumed MACs imply that tannery abatement will always be twice that of town for any feasible target.

This analysis demonstrates a condition for cost minimization for uniformly mixed pollutants known as an *equi-marginal principle*: a least-cost allocation equalizes MACs. This principle applies regardless of the number of sources.

2.4.3 Cost-Effective Management of Nonuniformly Mixed Pollutants

As we will discuss in detail in Chap. 3, uniform mixing is often not the case for water pollution problems. The location of emissions in space relative to the locations at which water quality standards are met can affect the impact of emissions on water quality. In the case of our lake, suppose that water flows into the lake from a stream and the tannery and town are located along the stream. Phosphorus moving down the stream encounter traps

Table 2.1 Least-cost abatement: uniform mixing

	Baseline emissions (kg)	*Least-cost abatement (kg)*	*MAC ($)*	*Total cost ($)*
Tannery	10	6.7	6.7	22.2
Town	15	3.3	6.7	11.1
Total	25	10	NA	33.3

Source: Authors' creation

and sinks that reduce the amount delivered downstream. A fraction of the amount released is delivered to the lake. The fraction will vary through the water system. The further a polluter is upstream, the smaller the delivered fraction.

The proportion of discharges from the sources, called delivery ratios (or transport coefficients), that reach the lake are denoted β_T and β_S. The ratios are positive but less than or equal to one. The quantity $\beta_T e_T$ is now the amount of pollution that reaches the lake from the tannery and the quantity $\beta_S e_S$ is the amount that reaches the lake from the town. The load delivered to the lake is now $L = \beta_T e_T + \beta_S e_S$. Assuming the load limit for the water quality goals is the same, it must then be the case that the delivered load satisfies the rule $\beta_T e_T + \beta_S e_S \leq L^*$. The rule can be written in terms of the required abatement of delivered pollution. The delivered pollution abatement from the tannery is $\beta_T a_T$ and from source town $\beta_S a_S$. The required reduction in delivered pollution is $\beta_T a_T + \beta_S a_S \geq \beta_T e_T^0 + \beta_S e_S^0 - L^*$. A least-cost solution in this case minimizes the cost of abatement to just achieve the limit on the delivered pollution. The prior section would suggest that the least-cost allocation should satisfy the equi-marginal rule. Indeed, it does, but in terms of the costs of delivered pollution.

Suppose the MAC for a one-unit reduction from the tannery is 1.00 USD and the delivery ratio for the tannery is 0.5. A one-unit reduction would then reduce delivered pollution by 0.5 units at a cost of 1.00 USD. The marginal cost of a one-unit delivered reduction is therefore 2.00 USD. From this example it is evident that the marginal cost of reducing delivered pollution is the MAC at point at which the pollutant is released divided by the delivery ratio. Following our previous reasoning, the cost-minimizing solution for delivered pollution abatement would require that the marginal costs of delivered pollution abatement be equal. The optimality condition for the two sources is therefore $\frac{\text{MAC}_\text{T}}{\beta_T} = \frac{\text{MAC}_\text{S}}{\beta_S}$.

This condition generalizes to many sources. Note that the marginal cost of delivered pollution varies inversely with the delivery ratio. This implies, other things being equal, a preference for pollution reduction from sources with high delivery ratios.

The example presented in Table 2.1 is repeated in Table 2.2 but with nonuniform mixing. In the example, $\beta_T = 0.5$ and $\beta_S = 0.9$. Baseline emissions are the same but because less pollution is delivered than emitted, the delivered pollution load is reduced to 18.5 kg. Taking the target load to be the same as above, $L^* = 15$ kg, the required reduction in delivered pollution is now only 3.5 kg. Contrary to the case with uniform mixing, in the cost-minimizing allocation the tannery's abatement is only 1.11 instead of two times that of the town. This reflects the effect of the tannery's smaller delivery ratio on its MAC for delivered pollution. A unit of pollution reduction at the tannery costs half as much as for the town but the town's reductions are more effective in reducing the delivered load due to the higher delivery ratio.

2.4.4 *Cost-Effective Management with Multiple Receptors*

Another feature of water pollution processes is multiple locations, referred to as receptors, at which water quality is degraded by a polluter's emissions. For example, nutrients and sediments released into streams move downstream affecting water quality in multiple downstream locations. When this is the case, emissions must be managed to meet water quality standards at multiple locations through a watershed. Similarly, water quality in lake or estuarine systems may vary across locations. The differences may reflect differences in ambient pollution in the waterbody (i.e., implying nonuniform mixing) and/or differences in water quality responses across the waterbody to pollutants. Again, emissions must be managed to meet water quality goals at multiple locations.

Multiple receptors along with nonuniform mixing result in a substantial increase in the complexity of the conditions for least-cost pollution control. To illustrate, instead of the single lake we have assumed, consider a set of connected lakes. Water flows from lake to lake from the top to the bottom of the system. Water quality standards are defined for each. Given that water moves down the system, water quality at the top is solely a function of emissions at the top. Its water quality goal can be managed completely by emissions at the top. But water quality in lakes further down the system will depend on emissions from sources discharging directly to the

Table 2.2 Least-cost abatement: nonuniformly mixed pollutants

Source	*Baseline emissions (kg)*	*Delivery ratio*	*Delivered baseline emissions (kg)*	*Least-cost delivered abatement (kg)*	*Abatement at source (kg)*	*MAC ($)*	*MAC/β ($)*	*Total abatement cost ($)*
Tannery	10	0.5	5	1.3	2.7	2.7	5.3	3.6
Town	15	0.9	13.5	2.2	2.4	4.8	5.3	5.8
Total	25	NA	18.5	3.5	5.1	NA	NA	9.4

Source: Authors' creation

lake plus emissions that arrive from upstream locations. The system is interdependent. Upstream water quality standards will affect the amount of pollution that moves downstream. Downstream standards will affect the amount of pollution that can move downstream.

We previously defined Pareto Efficient emissions for single receptors (see Fig. 2.1). In that case the optimal emissions level maximizes the net benefits of emissions less the damage cost of emissions, or equivalently, the benefits of pollution abatement less the costs of pollution abatement. With multiple receptors, damage cost functions exist for each receptor. A Pareto Efficient allocation allocates emissions across sources to maximize the sum of the resulting pollution benefits from the sources less the sum of the water quality damage costs across the receptors. The Pareto Efficient emission rate for any source equalizes the marginal benefit of its emissions and the sum of the marginal damage costs that result across the multiple downstream receptors from its emissions. Implicit in this optimization would be Pareto Efficient pollution load limits for each receptor. Equivalently, the Pareto Efficient allocation minimizes the sum of polluter abatement costs less the benefits of pollution abatement across the multiple downstream receptors. The marginal condition would equalize each polluter's marginal abatement cost with the sum of the marginal benefits from pollution abatement from the downstream receptors.

A least-cost allocation for multiple receptors will minimize the total cost of abatement while meeting the desired water quality goals set for each receptor. If the water quality goals are set to be Pareto Efficient, the result is Pareto Efficient. Otherwise the solution is cost minimizing for the goals. The conditions for cost minimization are mathematically complex but and will vary from polluter to polluter depending on their location in the system (e.g., see Hung and Shaw 2005).

2.5 Implementing Solutions

With this understanding of physical environments and corresponding conditions for cost minimization we are now able to dig into the question of how to achieve efficient water quality outcomes. We first describe types of water pollution control instruments. We then examine the efficiency properties of these instruments.

2.5.1 *Water Quality Policy Instruments*

Policy instruments are the tools that policy makers use to induce desired outcomes. Basic elements of a policy instrument are a *compliance base* (or target) and a *regulatory mechanism*. The compliance base is what the regulator monitors typically at the polluting facility (e.g., wastewater treatment plant, farm field) for management and enforcement activities. To be effective compliance bases must be connected to desired outcomes. Common compliance bases for water pollution are wastewater treatment technologies and emissions. The first manages how emissions are controlled and the second manages the actual emissions outcome. A regulatory mechanism is the form of the intervention applied to the compliance base. It is what either forces or induces the polluter to undertake compliance activities. Common mechanisms are standards (directives or mandates), charges, and fines (negative economic incentives), and subsidies (positive economic incentives).

The most common instruments for water pollution control, generally referred to as *command-and-control*, are mandated actions, practices, or prohibitions (OECD 2012). *Technology standards* prescribe the use of inputs and pollution prevention and treatment practices. *Emissions standards* impose limits on polluting emissions. To be cost-effective technology standards must organize each polluter's operations to minimize their individual abatement costs for the level of abatement that corresponds to the cost-effective allocation for the aggregate abatement target. This is a very tall order for a regulatory agency. It would require the agency to know as much about how to operate polluters' facilities profitably as do polluters. The implausibility of government agencies having the capacity to deliver polluters with the best blueprints for their operations leads economics to view technology standards as a poor policy choice for cost-effective pollution control, at least when polluters have or can acquire or develop the technological know-how.

Polluters subject to emissions standards can utilize specialized knowledge of their operations to organize production and pollution control to minimize their individual costs. Cost-effective emission standards would impose on each polluter the abatement it should provide in a least-cost allocation. Essentially, emissions standards are a polluter-specific list analogous to the abatement columns for the tannery and town in Tables 2.1 and 2.2. While emissions standards relieve the regulator of developing operational blueprints, deriving the list still requires the regulator to know polluters' abatement costs and to compute the least-cost solution for the aggregate abatement target. This is again a very tall order for a regulatory agency when there are more than a few economically complex and heterogeneous polluters. Emissions standards therefore tend not to use facility-specific information but instead to apply generally applicable limits. Accordingly, emissions standards are similarly viewed as a poor policy choice for cost-effective pollution control.

Economics research generally recommends the use of *economic incentives* for pollution control. When emissions can be measured accurately at reasonable cost, the preferred incentives put prices on emissions. This follows a "commandment" from the economics of externalities in the Pigouvian tradition to "get prices right." One reason for pricing is simply the incentive pricing provides polluters to reduce emissions. Just as pricing irrigation water is a smart approach to water conservation, pricing pollution is a smart approach to reducing pollution. Pricing pollution gives polluters continuous incentives to reduce their emissions and to control pollution in the least-cost way for their operation.

A second reason is that prices can coordinate emission reductions across polluters to minimize aggregate abatement costs. As demonstrated in our discussion of Pigouvian taxes, when emissions are priced, polluters will optimize by choosing their abatement to equalize their marginal abatement cost with the price. The efficiency conditions for the prototypical water pollution environments presented above entailed relationships between polluters' MACs. Finding that polluters will equalize their MAC's to prices indicates that efficient allocations can be induced by "getting the prices right."

The economic case for pricing pollution goes beyond the capacity for efficient pollution control. Pricing provides firms with ongoing incentives to innovate to reduce the pollution intensity of production and the costs of pollution control. The continuous incentives to reduce emissions also serve as an incentive to innovate. The incentives for innovation in command-and-control regimes are generally weaker. For example, polluters subject to technology standards do not have strong incentives to reduce emissions compared to other instruments because their compliance is determined by conformity to the standards rather than emissions reduction performance. Further, to realize the benefits of new technologies, polluters would have to await whatever political and bureaucratic procedures that are required for new technologies to be approved for use. Polluters subject to emission standards have incentives to reduce the cost of compliance, but incentives for reducing pollution below the standard will be weak if there are no economic benefits.

We note that economic research indicates that pricing pollution is not sufficient to generate optimal research and development (R&D) for environmental protection. Market failures related to the production of research result in suboptimal investments in R&D generally. Additional policies are therefore needed to support research, development, and demonstration projects (Goulder and Parry 2008).

2.5.2 Pricing Pollution: Subsidies, Charges, Markets

Standard options for pricing pollution are emissions charges, emissions reduction subsidies, and tradeable emissions permits (or allowances). The prices in charges and subsidies are administratively determined by regulatory authorities. Prices in tradeable permit systems are determined by markets. Each of these mechanisms can be designed to satisfy the conditions for cost minimization that we developed above. However, challenges in implementing the instruments to satisfy these conditions in practice can lead to differences in the expected economic and environmental outcomes. Accordingly, while potentially equally cost-effective in theory, their performance will vary in practice.

Emission reduction subsidies pay polluters to reduce pollution below a baseline. The mechanism may have appeal to reduce political resistance to implementation of pollution controls or address economic or distributional concerns related to economic harm to the regulated sector. Economic research generally does not recommend the method. When the baseline is based on a polluter's history, a perverse incentive to pollute exists prior to the implementation of the policy (Baumol and Oates 1988). Another reason has to do with the size and structure of the subsidized sector or sectors. The efficiency conditions we have developed apply across the set of polluters who are active. It does not indicate which polluters should or should not be in operation. Pollution control cost minimization may also require that polluters who are unprofitable when the external social costs of their production are factored in cease production. Emissions charges and tradeable permits can produce this outcome (Baumol and Oates 1988). Subsidies can keep polluters who are socially unprofitable in business or even encourage businesses to enter subsidized sectors resulting in an overall increase in pollution (Baumol and Oates 1988).

Emissions charges and tradeable permits are therefore the mechanisms typically recommended by economists for pricing pollution when emissions can be measured accurately at reasonable cost. Emissions charges are in the

Pigouvian tradition. The case for charges/taxes on water pollution was made compellingly in the 1960s and 1970s most notably by Alan Kneese and colleagues (Kneese 1964; Kneese and Bower 1968; Kneese and Schultze 1975). Permit trading is generally considered to be within the Coasian tradition in that it entails exchanges of property rights. The case for markets for pollution control was first made by Thomas Crocker (1967) and John Dales (1968). Subsequent research on these mechanisms identifies situations in which one approach may be preferred to the other related to difference in transaction costs, monitoring and enforcement issues, and the competitiveness of the permit markets or the product markets in which polluters operate, and regulatory or other policy distortions affecting polluters (Goulder and Parry 2008).

2.5.3 Ex Ante *and* Ex Post *Policy Assessment and Policy Criteria*

Economic assessments of environmental policies are categorized as *ex ante* or *ex post*. *Ex ante* evaluations are prospective and predictive. *Ex post* evaluations assess actual outcomes. *Ex ante* evaluation begins with the underlying economic theory. Empirical simulations using numerical models to predict outcomes are often used to test what might be realized in actual application of theoretical constructs. *Ex post* assessments examine how implemented policies work out in practice. Difference in predicted and actual outcomes are not uncommon. *Ex ante* estimates are forecasts and subject to uncertainty (Bailey et al. 2002). Common sources of uncertainty in assessing environmental and economic outcomes of environmental policies include the details of actual policy implementation, the technological options available to polluters, induced innovations in production and pollution control technologies that affect future pollution control costs, and evolving market conditions or public policies other than those under consideration that influence polluters' choices and costs (Kopits et al. 2014).

In this book we present a mix of *ex ante* and *ex post* assessments. *Ex post* assessments will be largely focused on current policy architectures for water pollution control generally and in agriculture specifically. *Ex ante* assessments will largely be focused on new approaches.

We have highlighted two criteria used in economic assessments—cost-effectiveness in achieving targets set by policy makers, and Pareto Efficiency, which considers both benefits and costs. Another economic consideration is *dynamic efficiency*. The chief considerations here are the long-run effects of policies on the structure of polluting industries and technology. Policies that encourage changes in the location of production, industrial organization, and technology that reduce pollution control costs or that reduce environmental harms are economically and ecologically desirable. We have previously mentioned concerns for distributional consequences. When considering implementation and administration, considerations include the information and other associated costs of policy development, implementation, administration, and enforcement. In this domain, another consideration is the technical capacity and resources for environmental agencies to collect and analyze data, and to monitor and enforce. Finally, we recognize that policies are constrained to various degree from place to place and over time by legal and political considerations.

2.6 Water Pollution Policy in Practice

The foregoing brief introduction to the economic theory of pollution control provides concepts and results for both policy design and policy evaluation. We will build on this theory in later chapters to consider the design of water quality policies explicitly for agriculture. This will require a deeper understanding of the relationship between agricultural production and water quality and some unique challenges these relationships pose for designing policies that are effective and cost-efficient.

We devote the remainder of this chapter to developing the institutional context. Many regions of the world in which agriculture is a major water quality problem have well-developed policies for controlling water pollution from agriculture and allocate significant resources to the task. Understanding what works and what does not is essential to considering the path forward. As we noted at the beginning of this chapter, economics is not needed to understand that current policies are not meeting water quality objectives, but it can help to explain why current policies fall short and assess options for improvement. It can also identify dimensions of policy failure that are missed by a sole focus on water quality outcomes relative to water quality goals. Unnecessarily high costs of pollution control and inadequate incentives for green technology innovation are examples.

2.6.1 A Brief History of Water Quality Protection Policy

The agricultural nonpoint pollution management problem in high-income countries takes place within the context of historical water pollution policy development. The historical paths and endpoints are not identical for all but there are generally shared elements.

The agglomeration of people into cities enabled by the development of agrarian societies has since ancient times created the dual challenges of supplying good quality water and disposing of wastewater. With some notable exceptions in time and place, the norm until comparatively recent times was for people to dispose of wastewater and solid wastes where they lived. Epidemics of waterborne and other diseases associated with unsanitary living conditions were common in urban places. Scientific understanding that living in such conditions was not only unpleasant but seriously harmful to human health did not emerge until the mid-1800s. A landmark event in the development of epidemiological science was London physician John Snow's research in 1854 tying cholera outbreaks in a London neighborhood to the consumption of infected water from a specific public well. This kind of understanding led cities in Europe and North America to invest in sewers and other public infrastructure to remove solid wastes and wastewater from the places where people lived. Where available, wastes were discharged to streams, lakes, and coastal waters. Until the last half of the twentieth century, wastewater was rarely treated prior to discharge to remove pathogens or other contaminants, resulting in degradation of water supplies for downstream users. Periodic outbreaks of cholera, typhoid, and other diseases were a significant consequence. The first response of public authorities was to develop public infrastructure to deliver potable water from new uncontaminated sources when available.

Continued population growth, urbanization, and industrialization led to increased volumes of water pollution and new types of water pollutants. Water supply development and treatment to provide good quality water to people and industry became increasingly costly, and did nothing to protect aquatic life downstream, recreational and commercial in-stream uses of water, and amenity values of water that were severely damaged by pollution from municipal and industrial wastewater discharges.

Public initiatives to invest in municipal water treatment works to clean wastewater prior to discharge began in the early 1900s. The benefits of sanitation to public health were enormous. Yet, economic and political disruptions during and following World War I delayed progress in the industrialized world. With these disruptions in the rear view mirror and the water quality degradation that followed the industrialization after World War II, it became widely apparent that water pollution control to protect natural water quality was essential for high levels of human well-being and environmental health in industrialized societies. In the United States this realization was punctuated by the Cuyahoga River in Cleveland, Ohio catching fire 1969. It was not the first time! The river had become a toxic flammable sewer for municipal and industrial waste in the heavily populated and industrialized region of northeast Ohio.

Reducing water pollution became a top environmental policy priority in the 1970s in the world's economically advanced democratic nations, leading to regulation of polluting discharges and enormous public and private spending. For example, Keiser and Shapiro (2019) find government and industry spending for water pollution control in the US since 1972 exceeded 1 trillion USD, or over 100 USD per person-year. These initiatives have had significant accomplishments. Highly degraded waters have been improved in many places to levels that restore fisheries, permit water to be used for swimming and boating, and for a variety of uses. But significant water quality problems remain and trends in some pollutants have been in the wrong direction, largely because of unregulated pollution from agriculture and other nonpoint sources. Prominent examples of the policy problem are found in the United States and the European Union.

2.6.2 Water Pollution Policy and Conditions in the US

The current water pollution control policy architecture in the US was created by 1972 Clean Water Act (CWA) and subsequent amendments. The ultimate objective of the CWA was "to restore and maintain the chemical, physical, and biological integrity of the Nation's waters." Toward this end it established national goals of "fishable and swimmable waters" by July 1, 1983 and the elimination of all discharges of pollutants into navigable waters by 1985 (Freeman 2000). The primary instrument for water pollution control established by the legislation is effluent standards implemented by permits required of all point sources of water pollution. Point sources discharge wastewater at discrete points (e.g., pipes, ditches, outfalls) directly into receiving waters. They are exemplified by municipal and industrial wastewater treatment plants. The legislation nationalized the control of point sources by prohibiting the discharge of any pollutant into waters of the US without a National Pollutant Discharge Elimination System (NPDES) permit.[5] NPDES permits specify set limits (effluent standards) on the levels of pollutants in discharges (e.g., a certain level of bacteria, nitrogen, or phosphorus). The CWA provided substantial funding for municipal wastewater treatment plants. Keiser and Shapiro (2019) cite a US Environmental Protection Agency (USEPA 1975) technical report indicating that federal grants to municipalities for wastewater treatment plants were the largest publics works program in the US in the mid-1970s. The implementation of the CWA is generally undertaken cooperatively with the states. Permitting, inspection, and enforcement activities are delegated to and implemented by state, local, and sometimes tribal governments subject to USEPA approval, regulatory guidance, and oversight.

Under the CWA, authority for nonpoint pollution control is delegated to the states. Pollutants from nonpoint sources move to receiving waters through diffuse and complex pathways over land, through soils, and in some cases through the atmosphere (see Box 2.7). Nonpoint sources are exemplified by agriculture. The dominant approach to agricultural nonpoint pollution control adopted by the states relies on voluntary adoption of Best Management Practices (BMPs) encouraged and facilitated by education programs, and technical and financial assistance for BMP implementation (Ribaudo 2012a, b, 2013; Ribaudo and Horan 1999; Ribaudo et al. 1999). Limited results from this approach has led some states to include some generally modest regulatory elements (Kling 2013; Ribaudo et al. 1999; Shortle et al. 2012). The most common mechanism is technology standards (Ribaudo and Caswell 1999). Laws directed at crop production generally allow voluntary adoption at first, with regulatory backup. Enforcement is generally triggered by citizen complaint.

While legal authority for agricultural nonpoint pollution resides with the states, the federal government has implemented significant programs through the US Department of Agriculture (USDA). The CWA names

[5] NPDES mechanism is largely a program for regulating industrial and municipal dischargers, but large Concentrated Animal Feeding Operations (CAFO) are regulated as point sources and are required to have NPSDES permits even if they do not discharge directly into water resources.

Box 2.7 Regulating Agricultural Sources as Point Sources
Water collection systems can convert diffuse runoff that would ordinarily be considered nonpoint source pollution into the functional equivalent of point sources of pollution. One example is urban runoff. Runoff from streets, parking lots, and other urban surfaces is often collected in storm drains and then discharged by pipes. Another is agricultural drain water. In some regions drainage systems are used to remove excess water from fields. Drains can convert nonpoint source pollution into point source pollution.

The definition of point sources and nonpoint for regulatory purposes in the United States is a legal matter, however. Urban stormwater runoff in the US was originally regulated as nonpoint pollution but certain categories of stormwater runoff were subsequently reclassified as point source pollution requiring discharge permits under the Clean Water Act. Large confined animal operations are also legally defined as point sources requiring permits.

The Clean Water Act expressly exempts "agricultural stormwater discharges and return flows from irrigated agriculture" from the definition of point source. Whether this exemption applies when an agricultural drainage system is in place was the subject of a legal challenge in the state of Iowa in 2017. Iowa law enables the creation of drainage districts to maintain and manage drainage networks. The districts have the power to levy assessments on members to pay costs. In 2015 the Des Moines Water Works (DMWW), the water utility serving the city of Des Moines, the largest city in and the capital of the state, filed a federal law suit against thirteen drainage districts discharging nitrates into the Racoon river, which is used by DMWW for source water. Nitrate concentrations in the Racoon exceed drinking water standards. DMWW must therefore undertake costly treatment to remove nitrate. In 2015, the utility incurred 1.5 million USD in operational costs and 2016, 80 million USD in capital cost for nitrate removal (Des Moines Register, March 20, 2017).

The federal lawsuit claimed that the drains should be classified as point sources of water pollution for the purposes of regulation, which under Clean Water Act would require that the Districts obtain pollution discharge permits and comply with effluent standards.

(*continued*)

Box 2.7 (continued)

DMWW also sought a ruling from the Iowa Supreme Court on the liability of the Districts for the nitrate removal costs incurred by DMWW.

The Iowa Supreme Court ruled that the districts were not liable under Iowa law for the damages incurred by DMWW. The federal court dismissed the lawsuit. Among the reasons were that the districts had no power to regulate the nitrates entering the drainage systems. The court did not consider the Clean Water Act claims.

Success of the DMWW legal claims would have been a landmark event, shifting property rights from a regime in which producers have an implied right to pollute to one in which they are regulated and liable for water quality damages from routine farming activities. The lawsuit thus proposes an approach to the existing paradigm of voluntary compliance and pay-for-practices that limit the effectiveness and cost-efficiency of agricultural water pollution management. Imposing liability for pollution management on the producer-owned districts does not eliminate the nonpoint source challenge but relocates it to the collective. This aligns with collective penalty mechanisms that have been proposed for nonpoint pollution control (see Box 5.3). The now inactive California Grassland Area Trading Program (see Box 6.5) in which seven irrigation districts traded selenium pollution offers a case study demonstrating that such farmer-owned organizations can be utilized as decision-making units for effective pollution management using market mechanisms.

the USDA as the primary federal source of financial and technical assistance to reduce agricultural nonpoint pollution, and the USDA administers several programs that provide producers with financial assistance to adopt BMPs (Ribaudo 2012b, 2013; Ribaudo et al. 1999). The largest "working-land" program is the Environmental Quality Incentives Program (EQIP), which was established in 1996 to provide financial assistance in the form of cost shares to producers to install and maintain conservation practices on eligible agricultural and forest land. Water quality is one of several EQIP objectives. Other USDA programs that can have positive water quality impacts include the Conservation Reserve Program (CRP) and the Conservation Stewardship Program (CSP). Assessments of USDA conservation programs indicate that they have had positive effects on water quality but have not been adequate to achieve water quality goals that are unmet due to agricultural loads (Osmond et al. 2012). Further, these programs are not cost-effective in water quality protection. Effectiveness and efficiency are diminished by inadequate targeting of resources to environmentally sensitive regions with high-priority water quality problems, inadequate targeting of resources within such regions to agricultural source areas that are disproportionately large contributors to water pollution, provision of financial assistance to producers based on practices adopted rather than the reductions in pollution loads achieved, and a reliance to voluntary producer participation (Lichtenberg 2014; Shortle et al. 2012; Ribaudo and Shortle 2019).

Evaluations of the CWA are mixed. Considerations include impacts of the legislation on water pollution, impacts on water quality conditions, and economic costs and benefits. The CWA is credited with improving water quality in many places, restoring highly degraded waters to conditions where they can be used for various in-stream uses (Bingham et al. 2000). It is also credited with protecting waters from degradation associated with population growth, urbanization, and industrialization (Bingham et al. 2000). A recent landmark study found that most types of water pollution declined over the period 1962–2001, though the rate of decrease slowed over time (Keiser and Shapiro 2019). The research found no evidence of or trend break in water pollution around 1972 when the CWA was enacted. This may not be surprising as it takes time for the effects of water pollution control investments to show up in water quality conditions.

Yet assessments indicate that over half of US river and stream miles violate state water quality standards (USEPA 2016). A recent USEPA assessment finds that 46% of US rivers and streams are in poor biological condition, 25% are in fair condition, and only 28% are in good condition (USEPA 2016). Of high significance to the condition of the US surface waters is nutrient pollution, which the USEPA describes as “one of America’s most widespread, costly and challenging environmental problems” (USEPA 2009). Nutrients are the cause of many of the high-profile water quality problems in the US—the Gulf of Mexico Dead Zone, eutrophication of the Chesapeake Bay, and algae blooms in Lake Erie. But they are also pervasive causes of smaller-scale problems. US EPA’s statistically based water resource assessments find nitrogen and/or phosphorus to be the leading stressors of rivers and streams, lakes, and coastal waters (USEPA 2017).

The 2017 USEPA National Water Quality Inventory lists agricultural nonpoint source (NPS) pollution as the leading cause of water quality impairments on rivers and streams, the third-largest cause for lakes, the second largest for wetlands, and a major contributor to contamination of estuaries and groundwater (USEPA 2017). The nation's nutrient pollution problems are to a very large degree the result of the large-scale direct (fertilizer) and indirect (animal feed) use of nitrogen and phosphorus in agricultural production (Shortle and Horan 2013). The importance of agriculture to remaining water quality problems reflects in part the strong effect that land use has on water quality and the fact that agricultural production is an extensive land use that can have adverse water quality consequences. Yet, technological means for substantially reducing pollution from agriculture are well known. Missing are the policy mechanisms necessary to bring about needed changes in farming practices in critical locations (Ribaudo and Shortle 2019).

An issue apart from the effectiveness of the current water quality protection policy architecture is the cost relative to the benefits (Shortle 2017a, b). Olmstead (2010) concludes the incremental benefits of the CWA exceeded the incremental costs through the late 1980s, but the reverse has been true since then. A review of 20 benefit-cost assessments of the water pollution policies in the US found benefits to be much smaller than their costs (Kaiser et al. 2019). Only 2 of the 20 studies estimate benefits that clearly exceed costs. The median benefit-cost ratio for the studies is 0.37. The poor economic performance found in these studies can result from various causes. One is that benefits are underestimated due to limitations of data, methods, and missing impacts. This is very likely the case, though it is not known whether better estimates would change the outcome (Kaiser et al. 2019) (see also Box 2.5). Another is that costs are overestimated. A third is that the current policy architecture is not cost-effective, with the result that society is devoting more resources to water quality protection than required to obtain the benefits received. In this case, policy changes that improve the cost-effectiveness of pollution control are sure to produce net benefits. That the current policy architecture is not cost-effective is well established (Shortle 2017a, b). One reason is an overreliance on comparatively expensive point source pollution controls to achieve water quality goals. The overreliance on point source regulations in the US both limit water quality gains and increase the costs of water quality protection (Shortle 2017a, b; Ribaudo and Shortle 2019).

2.6.3 Water Pollution Policy and Conditions in the European Union

The state of European Union (EU) water quality is comparable with the US. Point source water pollution controls have produced improvements, but goals remain unmet in large part because of diffuse agricultural pollution. According to the most recent report on the state of the European environment, surface water and marine ecosystems are largely not on track to reach established environmental targets (European Environment Agency 2020). The Fitness check of the Water Framework Directive by the European Commission (2019) found only 40% of EU's surface waters were in good ecological status (38% had good chemical status). Nutrient pollution leading to eutrophication is a widespread problem and largely the result of nitrogen and phosphorus from animal manure and fertilizers (OECD 2017). Nitrates in drinking water supplies is an issue in some regions (OECD 2017).

Each member state has its own institutions and history in water protection policies. Southern European countries face problems stemming from competing uses of water and water scarcity, problems which will be aggravated by climate change. Water quality problems are the greater issue in Central and Northern European countries, particularly eutrophication of surface waters (Andersen et al. 2017; Martins et al. 2013).

Environmental legislation setting technical standards and end-of-pipe measures for point sources via mandatory permits began in the 1970s. There are differences in the evolution of across point source regulations across countries but common for all countries is that point source pollution has declined.

The Water Framework Directive (WFD 2000/60/EC) unified the patchwork of national water protection policies. Adopted in 2000, it mandates the member states to identify River Basin Management Plans, updated every six years to help reach a good ecological status of surface and groundwaters. The definition of good ecological status is set for each River Basin following certain guidelines. The multiple objectives are linked to target values of various ecological and chemical water quality indicators. The member states are required to develop plans to reach the required ecological status in a cost-effective manner.

The WFD thus sets the overarching guidelines for national water protection policies within the EU. The minimum standards for environmental permitting of point sources were set in the Urban Wastewater Treatment directive (1991) and the Directive for Integrated Pollution and Prevention Control (1996). A key mandatory EU-wide regulatory element is the Nitrates Directive adopted in 1991. It set upper limits for nitrogen fertilization that are primarily meant to protect groundwater resources in areas designated as vulnerable by the member states.

The Common Agricultural Policy (CAP) sets the framework for the main instruments used to support and steer agricultural production. It consists of two "Pillars." The main role of Pillar I is to provide farm income support, but it has become "greener" over time with the addition of conditions that require producers receiving income support to utilize environmental protection practices without additional financial assistance. These conditions are mandatory cross-compliance rules. The current CAP program requires member states to designate 30% of the direct payments to farmers to a compulsory set of farming practices (crop diversification, maintenance of permanent grassland, and establishing ecological focus areas) to address climate and other environmental goals. Crop diversification aims to improve soil quality by mandating the cultivation of two or three crops on arable land. Biodiversity and carbon sequestration are promoted by maintaining the permanent grasslands. The ecological focus areas should cover at least 5% of arable land and follow management practices set by the member states

Specific agri-environmental practices are included in Pillar II. The selection of the practices is the responsibility of member countries but subject to EU Commission approval. Adoption of the practices by individual famers is voluntary. Adoption is typically incentivized by a uniform subsidy that covers the costs of adoption. The producer may choose from a set of alternatives. The compensations are determined separately for each practice. In addition, extra compensation of at most 20% of the costs may be paid to cover transaction costs (Environment 2017). In Finland, for instance, the adoption rate has been between 88% and 96% of farmland area during the EU membership of the country, starting in 1995 (Huttunen and Peltomaa 2016). By design, the uniform subsidy results in payments greater than adoption costs and therefore serves as a form of income support. The cost of the support may be a loss in cost-effectiveness in water quality protection, a topic taken up in Chap. 7.

Because of its coupling to income support, cross-compliance requirements are powerful in establishing wide-ranging environmental regulation in the EU. There are some problematic features in the existing policies which make it hard to succeed in managing water pollution from nonpoint sources. Water quality targets and the agricultural policy tools that are intended to serve them are almost completely separated. The targets are spatially specific, expressed in river basin management plans. The primary tool to control agricultural pollution, however, is embedded in the CAP where: (a) environmental aspirations are but one of the set of goals of CAP; (b) the instruments are not connected to WFD River basin management plans; and (c) the instruments provide uniform payments for the agri-environmental practices.

In the EU as in the US, the effectiveness and efficiency of water pollution control are diminished by inadequate targeting of resources to environmentally sensitive regions with high-priority water quality problems, inadequate targeting of resources within such regions, provision of financial assistance based on practices adopted rather than achievements, and a reliance to voluntary participation. Another problem in both the US and the EU is that the utilization of individual policy instruments is asked to serve a package of policy goals that are not always complementary and, in some cases, may be conflicting. These packages include agricultural income support, rural development, and multiple environmental objectives. What is done to serve one goal can diminish the effectiveness and efficiency for others. The policy failures result from entangled policy goals, underutilization of environmental and economic understanding, limited or nonexistent environmental targeting, and neglect of economic incentives and overreliance on voluntary participation, and thus the need for policy reform that motivates this book.

2.6.4 Transboundary Pollution

Transboundary pollution poses a particularly vexing problem for the design and implementation of environmental protection policies. With transboundary problems the affected environment is shared by multiple political jurisdictions. Water pollution is often transboundary as the contributing and affected watersheds are often under the control of different political jurisdictions. The jurisdictions may be different counties, states, or countries.

Differences in benefits and costs of pollution controls in different jurisdictions give rise to differences in incentives for pollution control resulting in a coordination problem. Jurisdictions generally have an incentive to free ride, enjoying the benefits of collective pollution reductions but avoiding costs. This incentive will be especially strong in jurisdictions that are sources of pollution but receive little or no benefit from water quality improvements (Olmstead 2010; Sigman 2005).

The Baltic Sea is a leading example of the challenges of transboundary water pollution in a multinational context in which agriculture is a significant source. The Baltic Sea receives nutrients from 14 countries that vary significantly in their levels of economic development (Iho et al. 2015). Nine of the 14 countries (the littoral group) together with the EU comprise the Baltic Marine Environment Protection Commission, referred to as the Helsinki Commission or HELCOM. It is the governing body of the Convention on the Protection of the Marine Environment of the Baltic Sea Area (Helsinki Convention) which was originally signed in 1974 by the seven littoral countries of the time.[6] The members have agreed, among other things, to country-specific nutrient abatement targets as defined in the Baltic Sea Action Plan. Like most international environmental agreements, the targets are nonbinding as HELCOM does not have enforcement mechanisms. The most important nonparticipant is Belarus, a significant nutrient source with no Baltic coast. Eight of the nine littoral countries are European Union members, the exception being Russia.

The measures taken so far have resulted in reductions in nutrient inputs but not in significant environmental improvements (Savchuk 2018). Ahlvik and Pavlova (2013) find strong free-riding incentives for some Baltic countries precluding Pareto Efficient control in the absence of a supranational enforcing authority. They suggest that efficient abatement and full participation could be achieved by using the power of the European Union to enforce such an agreement among its members and, simultaneously, negotiating with Russia within the Helsinki Commission.

Transboundary pollution can also occur within countries. For example, the Chesapeake Bay and Gulf of Mexico hypoxia problems in the US do not involve the movement of pollutants across international borders, but they do involve pollutants moving across state borders. Analogous to Belarus in the Baltic Sea case, Pennsylvania contributes to pollution, but does not have any coastline on the Bay and hence does not benefit directly from the water quality improvements. The context is a natural one for free riding. However, unlike the Baltic Sea, there is an authority, the US Environmental Protection Agency, above the states that can do much to bring states to the table to find cooperative solutions.

[6] In 1974 there were two German states and the Soviet Union but no Russia, Estonia, Latvia, or Lithuania.

2.7 Policy Interactions and Conflicts

Water pollution control in agriculture is one of many domains in which agriculture is subject to public policy interventions. Long-standing agricultural policy goals in industrialized countries include supporting farm incomes, fostering rural economic development, food affordability, and stabilizing food supply and prices. Specific policy concerns have evolved over time. Soil and water conservation along with farm income support emerged as national policy concerns in the US following the 1930s Dust Bowl. Water quality protection and other environmental quality issues joined the US agricultural policy mix in the 1990s, and renewable energy production became a major policy goal in the twenty-first century. The 1992 CAP reform brought a shift from price supports to area-based payments in the EU. It also introduced requirements for green set-asides to reduce overproduction and promote environmental goals. EU Agenda 2000 policy reforms increased the emphasis on environmental aspects of farming, with agri-environment schemes made compulsory for every member state. Significant cross-compliance conditions were introduced by the EU in 2004 (OECD 2011). Since then, the rural landscape structure, cultural heritage, biodiversity, and to some extent water quality have been the major concerns in Europe (Lankoski and Ollikainen 2013). OECD (2011) assessed that the overall impacts have been beneficial for biodiversity but the effects on water quality and soil health are unclear.

Like water quality goals, agricultural policy goals can be pursued through a suite of instruments that vary in their effectiveness and efficiency with respect to particular goals, and that also vary in the degree to which they complement or conflict with others. A challenge for water quality protection and resource conservation in agriculture is interventions that encourage expansion or maintain agricultural production in environmentally sensitive locations, pollution-intensive commodity production, and overproduction (Hellerstein and Malcolm 2011; Housh et al. 2015; Khanna 2017; Lichtenberg 1989; Searchinger et al. 2008; Secchi et al. 2011). US Department of Agriculture crop insurance programs, which account for about a third of all US farm subsidies, encourage producers to plant crops on highly erodible soils, while the Conservation Reserve Program (CRP) pays producers to take such land out of production (Smith et al. 2017). Another example from the US is not an explicit agricultural policy but an energy policy with a large agricultural impact. With the intent of reducing greenhouse gas emission and reducing reliance on imported fuels, the Renewable Fuel Standard (RFS) established by the US Congress in 2007 mandated the production of renewable fuels to replace or reduce the quantity of petroleum-based fuels. The act led to an increase in the demand for corn to produce ethanol resulting in the conversion of grasslands in areas providing feedstock for ethanol plants (Wright et al. 2017). Conversion of land to a pollution-intensive crop—corn is nitrogen-fertilizer- and tillage-intensive—has led to increased nutrient loads to the Gulf of Mexico worsening the dead zone in the Gulf of Mexico (Donner et al. 2002; Hendricks et al. 2014).

The RFS is an energy policy that operates on the demand for agricultural commodities with adverse environmental consequences. Nonagricultural policies can also operate with adverse effects through interventions in agricultural input markets. Explicit or implicit subsidies for agricultural irrigation water increase water quality externalities associated with irrigated agricultural production (Lichtenberg 1989; Weinberg et al. 1993). Local land use policies can also have an effect. For example, land use controls in Maryland have been found to increase nutrient loads to the Chesapeake Bay by reducing the conversion of farmland to urban development subject to stronger nutrient pollution regulations (Wrenn et al. 2019).

There are many other agricultural interventions in the EU and OECD countries. These include output quotas and trade protections for specific commodities that prop up output prices, area payments for specific crops, and payments related to specific products, such as milk. Not all are environmentally adverse. The effect of policy choice is illustrated by the EU Common Agricultural Policy (CAP) area-based support payments, the most important of which is the single-farm payment. This single-farm payment is decoupled from production and has some environmental cross-compliance conditions. To receive the payment, producers must agree to comply with these cross-compliance conditions. The general aim is that the market guides the producers' choices and single-farm payment helps to provide producers a secure part of income.

2.8 Summary

This chapter presents a variety of topics that are essential foundations for the remainder of the book. The first part is devoted to the meaning and causes of environmental problems, criteria for policy solutions, and instruments for pursuing environmental protection. The approach to these topics is that of economics. Specific topics included environmental externalities, benefits and costs of water quality protection, environmental target setting, pollution control policy instruments, criteria for instrument selection, and standard prescriptions for policy instruments. The second part of the chapter introduces the agricultural problem. Our focus is on policies and institutions that have evolved for managing water quality impact of agriculture, the instruments of water quality protection in agriculture, and the status of agriculture's impacts on water quality. The focus in this discussion is the US and EU. Agriculture has long been recognized as significant cause of water quality problems in these regions and has not suffered from policy neglect. Yet, it is a major cause of persisting water quality problem because of policy architectures that have failed to achieve water quality objectives. The same is true in other economically developed regions with advanced pollution control policies (OECD 2012, 2017). The conclusion is that policy reform is essential for effective and efficient policy. Economics provides essential theories and tools to improve both the effectiveness and efficiency of water quality protection in agriculture and to better integrate agriculture into the broader water quality protection initiatives of national and subnational water quality authorities. It also provides tools to understand and modify policies to address distributional and other concerns in water pollution policy design. We have introduced the basic theories and tools, but unique features of the agricultural problem require adaptations and refinements, which we develop in subsequent chapters.

References

Ahlvik, L., and Y. Pavlova. 2013. A strategic analysis of eutrophication abatement in the Baltic Sea. *Environmental and Resource Economics* 56 (3): 353–378.

Andersen, J.H., J. Carstensen, D.J. Conley, K. Dromph, V. Fleming-Lehtinen, B.G. Gustafsson, A.B. Josefson, A. Norkko, A. Villnäs, and C. Murray. 2017. Long-term temporal and spatial trends in eutrophication status of the Baltic Sea. *Biological Reviews* 92 (1): 135–149.

Anderson, K., G. Rausser, and J. Swinnen. 2013. Political economy of public policies: Insights from distortions to agricultural and food markets. *Journal of Economic Literature* 51 (2): 423–477.

Anthoff, D., and R. Hahn. 2010. Government failure and market failure: On the inefficiency of environmental and energy policy. *Oxford Review of Economic Policy* 26 (2): 197–224.

Bailey, P.D., G. Haq, and A. Gouldson. 2002. Mind the gap! Comparing ex ante and ex post assessments of the costs of complying with environmental regulation. *European Environment* 12 (5): 245–256.

Bateman, I.J., R.T. Carson, B. Day, M. Hanemann, N. Hanley, T. Hett, M. Jones-Lee, G. Loomes, S. Mourato, D.W. Pearce, and R. Sugden. 2002.

Economic valuation with stated preference techniques: A manual. Cheltenham/ Northampton: Edward Elgar.

Baumol, W.J., and W.E. Oates. 1988. *The theory of environmental policy*. Cambridge university Press.

Bingham, T.H., T.R. Bondelid, B.M. Depro, R.C. Figueroa, A.B. Hauber, S.J. Unger, G.L. Van Houtven, and A. Stoddard. 2000. *A benefits assessment of water pollution control programs since 1972: Part 1, the benefits of point source controls for conventional pollutants in rivers and streams*. Prepared for US Environmental Protection Agency, Office of Water. Water Quality Assessment.

Coase, R. 1960. The problem of social cost'..... *Journal of Law and Economics* 3: 1–44.

Cornes, R., and T. Sandler. 1996. *The theory of externalities, public goods, and club goods*. New York: Cambridge University Press.

Dales, J.H. 1968. *Pollution, property and prices*. Toronto: University of Toronto Press.

Donner, S.D., M.T. Coe, J.D. Lenters, T.E. Twine, and J.A. Foley. 2002. Modeling the impact of hydrological changes on nitrate transport in the Mississippi River basin from 1955 to 1994. *Global Biogeochemical Cycles* 16 (3): 16–11.

Environment, A. 2017. The Thünen Institute (2017) Evaluation study of the payment for agricultural practices beneficial for the climate and the environment. European Commission, DG Agriculture and Rural Development, Brussels.

European Commission. 2019. Fitness check of the water framework directive, groundwater directive, environmental quality standards directive and floods directive. *Commission Staff Working Document* 10 (12): 2019.

European Environment Agency. 2020. *The European environment-state and outlook 2020; [SOER 2020]*. Publications Office of the European Union.

Freeman, A. Myrick, III. 2000. Water pollution policy. In *Public policies for environmental protection*, ed. Paul R. Portney and Robert N. Stavins, 2nd ed., 169–213. Washington, DC: Resources for the Future.

Goulder, L.H., and I.W. Parry. 2008. Instrument choice in environmental policy. *Review of Environmental Economics and Policy* 2 (2): 152–174.

Hanley, N., J. Shogren, and B. White. 2019. *Introduction to environmental economics*. Oxford University Press.

Hellerstein, D., and S. Malcolm. 2011. The influence of rising commodity prices on the conservation reserve program. Economic Research Service, Paper No. ERR, 110.

Hendricks, N.P., S. Sinnathamby, K. Douglas-Mankin, A. Smith, D.A. Sumner, and D.H. Earnhart. 2014. The environmental effects of crop price increases: Nitrogen losses in the US Corn Belt. *Journal of Environmental Economics and Management* 68 (3): 507–526.

Housh, M., M.A. Yaeger, X. Cai, G.F. McIsaac, M. Khanna, M. Sivapalan, Y. Ouyang, I. Al-Qadi, and A.K. Jain. 2015. Managing multiple mandates: A system of systems model to analyze strategies for producing cellulosic ethanol and reducing riverine nitrate loads in the Upper Mississippi River Basin. *Environmental Science & Technology* 49 (19): 11932–11940.

Hung, M.F., and D. Shaw. 2005. A trading-ratio system for trading water pollution discharge permits. *Journal of Environmental Economics and Management* 49 (1): 83–102.

Huttunen, S., and J. Peltomaa. 2016. Agri-environmental policies and 'good farming' in cultivation practices at Finnish farms. *Journal of Rural Studies* 44: 217–226.

Iho, A., M. Ribaudo, and K. Hyytiäinen. 2015. Water protection in the Baltic Sea and the Chesapeake Bay: Institutions, policies, and efficiency. *Marine Pollution Bulletin* 93 (1–2): 81–93.

Keiser, D.A., and J.S. Shapiro. 2019. Consequences of the clean water act and the demand for water quality. *The Quarterly Journal of Economics* 134 (1): 349–396.

Keiser, D.A., C.L. Kling, and J.S. Shapiro. 2019. The low but uncertain measured benefits of US water quality policy. *Proceedings of the National Academy of Sciences* 116 (12): 5262–5269.

Khanna, M. 2017. Nexus between food, energy and ecosystem services in the Mississippi River Basin: Policy implications and challenges. *Choices* 32 (4): 1–9.

Kling, C.L. 2013. State level efforts to regulate agricultural sources of water quality impairment. *Choices* 28 (316-2016-7675).

Kneese, A.V. 1964. *The economics of regional water quality management*. Baltimore.

Kneese, A.V., and B.T. Bower. 1968. Managing water quality: Economics, technology. *Institutions* 61.

Kneese, A., and C.L. Schultze. 1975. *Pollution, prices, and public policy*. Washington, DC: The Brookings Institute.

Kopits, E., A. McGartland, C. Morgan, C. Pasurka, R. Shadbegian, N.B. Simon, D. Simpson, and A. Wolverton. 2014. Retrospective cost analyses of EPA regulations: A case study approach. *Journal of Benefit-Cost Analysis* 5 (2): 173–193.

Lankoski, J.E., and M. Ollikainen. 2013. Innovations in nonpoint source pollution policy—European perspectives. *Choices* 28 (316-2016-7685).

Lichtenberg, E. 1989. Land quality, irrigation development, and cropping patterns in the northern high plains. *American Journal of Agricultural Economics* 71 (1): 187–194.

———. 2014. Conservation, the farm bill, and US agri-environmental policy. *Choices* 29 (316-2016-7746).

Martins, G., A.G. Brito, R. Nogueira, M. Ureña, D. Fernández, F.J. Luque, and C. Alcácer. 2013. Water resources management in southern Europe: Clues for a research and innovation based regional hypercluster. *Journal of Environmental Management* 119: 76–84.

Moore, C., D. Guignet, K. Maguire, C. Dockins, and N. Simon. 2015. *A stated preference study of the Chesapeake Bay and Watershed Lakes* (No. 2168-2018-8163).

OECD. 2011. *Evaluation of agricultural policy reforms in the European Union.* Paris: OECD Publishing. https://doi.org/10.1787/9789264112124-en.

———. 2017. *Diffuse pollution, degraded waters—Emerging policy solutions.* Paris: OECD.

Olmstead, S.M. 2010. The economics of water quality. *Review of Environmental Economics and Policy* 4 (1): 44–62.

Organisation for Economic Co-operation and Development (OECD). 2012. *Water quality and agriculture: Meeting the policy challenge.* Paris: OECD.

Osmond, D., D. Meals, D. Hoag, M. Arabi, A. Luloff, G. Jennings, M. McFarland, J. Spooner, A. Sharpley, and D. Line. 2012. Improving conservation practices programming to protect water quality in agricultural watersheds: Lessons learned from the National Institute of Food and Agriculture–Conservation Effects Assessment Project. *Journal of Soil and Water Conservation* 67 (5): 122A–127A.

Pigou, A.C. 1920. *The economics of welfare.* London: Macmillan.

Ribaudo, M.O. 2012a. Non-point pollution regulation approaches in the US. In *The Management of water quality and irrigation technologies*, 99–118. Routledge.

Ribaudo, M. 2012b. Working-lands conservation programs. *Agricultural Resources and Environmental Indicators*, 2012 Edition, p. 47.

———. 2013. Policy instruments for protecting environmental quality. www.ers.usda.gov/topics/natural-resources-environment/environmental-quality/policy-instruments-for-protecting-environmental-quality.aspx#.Uh4f_j_leq0. Accessed 28 Aug 2013.

Ribaudo, M., and M.F. Caswell. 1999. Environmental regulation in agriculture and the adoption of environmental technology. In *Flexible incentives for the adoption of environmental technologies in agriculture*, 7–25. Dordrecht: Springer.

Ribaudo, M.O., and R.D. Horan. 1999. The role of education in nonpoint source pollution control policy. *Review of Agricultural Economics* 21 (2): 331–343.

Ribaudo, M., and J. Shortle. 2019. Reflections on 40 years of applied economics research on agriculture and water quality. *Agricultural and Resource Economics Review* 48 (3): 519–530.

Ribaudo, M., R.D. Horan, and M.E. Smith. 1999. *Economics of water quality protection from nonpoint sources: Theory and practice* (No. 1473-2016-120718).

Savchuk, O.P. 2018. Large-scale nutrient dynamics in the Baltic Sea, 1970–2016. *Frontiers in Marine Science* 5: 95.

Searchinger, T., R. Heimlich, R.A. Houghton, F. Dong, A. Elobeid, J. Fabiosa, S. Tokgoz, D. Hayes, and T.H. Yu. 2008. Use of US croplands for biofuels increases greenhouse gases through emissions from land-use change. *Science* 319 (5867): 1238–1240.

Secchi, S., P.W. Gassman, M. Jha, L. Kurkalova, and C.L. Kling. 2011. Potential water quality changes due to corn expansion in the Upper Mississippi River Basin. *Ecological Applications* 21 (4): 1068–1084.

Shortle, J. 2017a. Policy reforms needed for better water quality and lower pollution control costs. *Choices* 32 (4): 1–7.
———. 2017b. Policy Nook: "Economic incentives for water quality protection". *Water Economics and Policy 3* (02): 1771004.
Shortle, J., and R.D. Horan. 2013. Policy instruments for water quality protection. *Annual Review of Resource Economics.* 5 (1): 111–138.
Shortle, J.S., M. Ribaudo, R.D. Horan, and D. Blandford. 2012. Reforming agricultural nonpoint pollution policy in an increasingly budget-constrained environment. *Environmental Science & Technology* 46 (3): 1316–1325.
Sigman, H. 2005. Transboundary spillovers and decentralization of environmental policies. *Journal of Environmental Economics and Management* 50 (1): 82–101.
Smith, A. 1776. *An inquiry into the nature and causes of the wealth of nations: Volume One.* London: printed for W. Strahan; and T. Cadell, 1776.
Smith, V.H., J.W. Glauber, and B.K. Goodwin. 2017. *Agricultural policy in disarray: Reforming the farm bill.* American Enterprise Institute.
USEPA. 1975. Clean water construction grants program news. Technical Report.
———. 2016. *National rivers and streams assessment 2008–2009: A collaborative survey.* Washington, DC: U.S. Environmental Protection Agency, Office of Water and Office of Research and Development, EPA/841/R-16/007, March.
———. 2017. National water quality inventory. Available online: https://www.epa.gov/waterdata/2017-national-water-quality-inventory-report-congress. Accessed 156 June 2020.
———. 2009. An urgent call to action: Report of the state-EPA nutrient innovations task group. Washington, DC. Available online: https://www.epa.gov/sites/production/files/documents/nitgreport.pdf.
Wainger, L.A., D. Secor, C. Gurbisz, M. Kemp, P.M. Glibert, E.D. Houde, J. Richkus, and M. Barber. 2017. Resilience indicators support valuation of estuarine ecosystem restoration under climate change. *Ecosystem Health and Sustainability* 3: e01268.
Weinberg, M., C.L. Kling, and J.E. Wilen. 1993. Water markets and water quality. *American Journal of Agricultural Economics* 75 (2): 278–291.
Winston, C. 2007. *Government failure versus market failure: Microeconomics policy research and government performance.* Brookings Institution Press.
Wrenn, D.H., H.A. Klaiber, and D.A. Newburn. 2019. Price based policies for managing residential development: Impacts on water quality. *Resource and Energy Economics* 58: 101115.
Wright, C.K., B. Larson, T.J. Lark, and H.K. Gibbs. 2017. Recent grassland losses are concentrated around US ethanol refineries. *Environmental Research Letters 12* (4): 044001.
Young, R.A., and J.B. Loomis. 2014. *Determining the economic value of water: Concepts and methods.* Routledge.

CHAPTER 3

Agricultural Land Use, Production, and Water Quality

3.1 Introduction

Eutrophication of surface waters, sedimentation of rivers and coastal areas, groundwater contamination from nitrates and pesticides, and emerging contaminants are affecting the environment and human well-being. These are problems driven by waterborne pollution, mainly from agriculture. This chapter takes a deeper look at water quality problems resulting from agricultural production and the physical processes involved. It describes the pollutants that cause them, their ecological significance, and the significance of agriculture's contribution. The chapter also introduces essential concepts from water science and engineering about the physical processes that link agricultural production and land use to water quality and technological options for pollution control. Key concepts, relationships, and tools needed to understand the physical processes involved in the movement of pollutants from farms to water resources are developed.

3.2 Water Pollution Problems: Nutrients

Nutrient pollution refers to water quality problems that result from elevated levels of plant nutrients, mainly nitrogen and phosphorus, in water resources. Sources of nutrient pollution include agriculture, industrial and

J. Shortle et al., *Water Quality and Agriculture*, Palgrave Studies in Agricultural Economics and Food Policy,
https://doi.org/10.1007/978-3-030-47087-6_3

municipal wastewater, and fossil fuel combustion (Selman and Greenhalgh 2010). Agriculture has become increasingly important over time and is generally the most significant source. Since the 1970s, developed nations have done much to reduce nutrient loads from industrial and municipal point sources but have not been successful in reducing pollution from nonpoint sources of nutrients, particularly agricultural sources (OECD 2013, 2017). This limited success reflects the political and institutional challenges of environmental regulation of agricultural sectors and the inherent complexity of the economic and ecological systems involved (Shortle and Horan 2017). Reducing nutrient loads from municipal and industrial point sources with known technologies has proven comparatively simple. This has not been the case for agriculture.

There are two main manifestations of nutrient pollution. One is degradation of drinking water supplies due to nitrates. We address this topic in the next section. The second is eutrophication of rivers, lakes, coastal waters, and open sea areas, which is generally considered the most significant water quality problem caused by agriculture globally.

3.2.1 Nitrates in Drinking Water

From 1900 to 2000, the global use of chemical nitrogen fertilizers increased from one million tons to 83 million tons and the growth is expected to continue. The use of N manure has grown threefold (Bouwman et al. 2013). Crops can utilize nitrogen in various compounds containing nitrates (NO_3) and ammonia (NH_3). Nitrates are highly soluble and are easily leached to groundwater when more nitrogen is applied than utilized by crops. Nitrates are common groundwater pollutants globally (Mateo-Sagasta et al. 2018).

Elevated nitrate concentrations in drinking water pose health problems particularly in areas relying on local groundwater. Nitrates are especially harmful for infants whose gut flora converts nitrates into nitrites. Elevated nitrite levels in blood are linked to "blue-baby" syndrome, which involves oxygen deprivation and can lead to death.[1] With centralized water utilities the elevated nitrate concentrations can be detected and purified, turning the health problem into a cost problem.

[1] https://www.who.int/water_sanitation_health/diseases-risks/diseases/methaemoglob/en/

Concern for nitrates in drinking water has led environmental agencies to establish drinking water standards. These vary substantially. The EU Groundwater Directive (2006/118/EC) and WHO set a safety limit of 50 mg per liter. In India, the national limit is 45 mg per liter. In the US, the US Environmental Protection Agency (USEPA) set the limit to 10 mg per liter. Even the relatively lax EU limit is exceeded in many aquifers of the member states (Mateo-Sagasta and Burke 2010). Concentrations of up to 140 mg per liter have been found in the groundwater in the Danube Basin, and up to 230 mg per liter in East Aegean Basin in Bulgaria (Aloe et al. 2014).

California's Central Valley is one of the world's most productive agricultural regions. Central Valley crops are irrigated with surface and groundwater. Elevated nitrate levels in groundwater are pervasive, particularly in the San Joaquin river basin. Nitrates are one of the most severe groundwater contamination problems. In 2012, a University of California study analyzed nitrate concentrations in groundwater in Tulare Lake Basin and parts of Salinas Valley where about 2.6 million people use groundwater as drinking water supply (Harter and Lund 2012). More than half of the nitrogen applied eventually finds its way to groundwater. One in ten sampled wells exceeded the limit set by the California Department of Public Health (45 mg per liter). If the sample is representative, this would indicate that nearly 300,000 Californians relying on groundwater were exposed to unhealthy levels of nitrates in their drinking water.

Box 3.1 Economic Damages from Nutrients: Drinking Water

Adverse impacts of nutrients on the quality of drinking water supplies are often mitigated by water treatment. The costs can be substantial.

Eutrophication can adversely affect the quality of water for human use to the extent that treatment is required. For example, Lake Waco in Texas supplies water to 166,000 residents of the city of Waco. Eutrophication of the lake adversely affects the smell and taste of the water. Between 2002 and 2012, the water utility invested more than $70 million in treatment technology. In addition, it may have lost as much as $10 million in lost revenues due to drinking water quality issues (Dunlap et al. 2015).

Toxins generated by some types of algae blooms can be harmful when ingested in drinking water. For example, the Collins Park

(*continued*)

Box 3.1 (continued)
Water Treatment Plant provides drinking water to 500,000 residents in the city of Toledo from Lake Erie. In 2014 the plant was closed due to dangerous levels of toxins and the city issued a "do-not-drink" order for the water supply system. The toxins were generated by the 2014 mass blooms of blue-green algae in Lake Erie which serves as water supply for the city (Roberson 2014). Since then, the water utility has invested in detection and algal toxin treatment technologies (for instance, $6 million 2014–2015; $41 million in 2016) and is undergoing large ozone treatment investments for use in future mass bloom situations. Altogether, it estimates that $500 million will be invested by the year 2023 (Toledo 2020). Of course, the costs should not be allocated to algae problems alone as they include other modernizations that can be done at the same time.

Nitrates in drinking water are a pervasive problem in regions where water supplies are sourced from aquifers or surface waters in intensively farmed places. Complying with nitrate limits is costly for individual households relying on groundwater sources. In California, the annual additional costs of providing safe drinking to 2.6 million inhabitants in Tulare Lake Basin and Salinas Valley were estimated to be between $20 and $36 million (Harter and Lund 2012). The Des Moines water utility must remove nitrates from its water supply due to upstream loading from agricultural areas. The operational costs in 2015 were $1.5 million with investments worth $80 million following next year (see Box 2.4 in Chap. 2). Keeler et al. (2016) provide a rare analysis on the social cost of nitrogen, with an application for Minnesota. They estimate that the total costs caused by nitrates to rural households and water utilities serving urban residents is about $37 million. The state of Minnesota has 5.6 million inhabitants.

3.2.2 *Eutrophication*

Producers apply nitrogen and phosphorus to crops to promote crop growth. These plant nutrients are also essential to healthy aquatic ecosystems. Elevated levels of these nutrients, however, cause eutrophication.

Eutrophication is defined as elevated primary production and the ensuing changes in aquatic ecosystems. Primary production is the scientific term for the processes in which algae capture and convert solar energy into chemical energy. The energy is transmitted through the food chain. First to zooplankton that eat algae, then to planktivorous fish that eat zooplankton, then to predatory fish, and on up the chain. Algae is therefore essential for any water ecosystem to thrive. However, excessive algae growth—eutrophication—has multiple negative effects. It changes the composition of aquatic species, decreases water clarity, and increases the occurrence of harmful mass blooms of blue-green algae. It may also cause large areas of bottom sediments to become deficient of oxygen, which results in multiple problems for aquatic life. Eutrophication is one of the most pervasive water pollution problems worldwide (OECD 2017; European Commission 2012; Diaz and Rosenberg 2008; Sutton et al. 2013).

Although specific eutrophication problems are local or regional in scope (e.g., Lake Erie, Chesapeake Bay, the Baltic Sea), they have global drivers. First, the amounts of bioavailable nitrogen and phosphorus in the global biogeochemical cycle have increased dramatically. Before the twentieth century, plant-available nitrogen compounds were introduced to the ecosystems from atmospheric nitrogen (N_2) only by nitrogen-fixing bacteria: blue-green algae in the surface waters and plants hosting the bacteria in their roots on land. Agriculture relied on the latter to provide crops with nitrogen. Today, industrial nitrogen production and fixation introduce about 150 Tg (one Tg is equivalent to million metric tons) of bioavailable nitrogen every year into global nitrogen cycle (Steffen et al. 2015). This is too much: several assessments view that the amount crosses the safe planetary boundaries (de Vries et al. 2013; Steffen et al. 2015; Rockström et al. 2009).

The same kind of disruption holds for phosphorus. The annual natural background phosphorus flow to oceans is estimated to be around 1.1 Tg (Steffen et al. 2015). The current flow is between 8 and 9 Tg. The rate at which mining introduces phosphorus into the global economy and subsequently into ecosystems is much higher than the flow of phosphorus to oceans. Around 23.5 Tg was mined in 2008. The quantity of land applications either in the form of inorganic phosphorus or as manure was almost identical, 22.6 Tg (Carpenter and Bennett 2011).[2] That is, future pressure

[2] Livestock feed provides essential nutrients including phosphorus. Animals are not perfectly efficient in the utilization of nutrients with the results that a fraction of the intake is removed in urine or manure.

on oceans from phosphorus is increasing. Upstream of this flow is increasing phosphorus flows to rivers, lakes, and coastal waters.

On the global scale, agriculture is driving at a dangerously high speed in terms of nutrient use. Nutrients fuel ecosystems and speeding can have severe consequences. Anoxic areas in the world's oceans are increasing at a troubling rate. The world's oceans have lost about 77 billion tons of oxygen since the 1960s. This accounts for a 2% loss in oxygen content (Schmidtko et al. 2017). In addition to increasing nutrient loads, eutrophication is also accelerated by warming water due to climate change (Breitburg et al. 2018). The development may bring about unpredictable changes on a global scale.

Future global-scale calamities aside, eutrophication is already causing severe problems. Attention is often drawn to headline-making examples. These include Lake Taihu in China; Lake Biwa in Japan; Lake Victoria in Africa; Moreton Bay in Australia; the Baltic Sea in Europe; and the Chesapeake Bay, Lake Erie, and the Gulf of Mexico in the US. The high-profile cases are significant but can also obscure the pervasiveness of the problem. In Europe, for example, 30% of water bodies show signs of eutrophication (European Commission 2012). In the US, more than two out of five river and stream miles have nutrient levels that are too high for healthy aquatic ecosystems (USEPA 2016).

The Baltic Sea, Chesapeake Bay, and Lake Erie cases are useful to illustrate the nature of eutrophication problems, the relevance of agriculture to these problems, and the policy problem. The Baltic Sea is a very large (surface area 377,000 km^2) brackish water body. It is surrounded by nine countries and freshwater flows to the Baltic come from 14 countries. The Baltic suffers from severe eutrophication punctuated by massive algae blooms in the spring and summer. Nitrogen and phosphorus contribute to what is called the "vicious cycle" of eutrophication. High nutrient concentrations intensify algae growth. This increases the consumption of dissolved oxygen by decaying algae, leading to hypoxia. Hypoxia in turn leads to phosphorus releases from bottom sediments that fuel algae blooms.

Given the vicious cycle, both nitrogen and phosphorus concentrations must be reduced to diminish the eutrophication problem. Toward this end, the Baltic Sea Action Plan, a joint policy recommendation adopted by all coastal countries and the EU in 2007, sets abatement targets for phosphorus and nitrogen. The loading of both nutrients was at its highest level in the 1980s. During that decade, the annual total phosphorus loading was on average around 70,000 tons. Today it is about 30,000 tons

(Savchuk 2018). Investments in point source abatement have decreased phosphorus loading to the levels of the 1950s. However, phosphorus concentrations have not diminished correspondingly.

An important consideration in phosphorus management in the Baltic is the extent to which phosphorus comes from annual inputs from agriculture and wastewater plants, how much is internally released from sediments, and how much is already in the water column. Today, the "inventory" or stock of phosphorus contained in Baltic Sea water column is around 700,000 tons, more than twenty times higher than the annual external load (Savchuk 2018).

The continuing release of phosphorus from bottom sediments is even more troubling than the amount of phosphorus presently in the water body. Phosphorus has three ways out of the Baltic: a very slow exchange with the Atlantic via the Danish Straits, fisheries which remove an amount of phosphorus equal to about 10% of the annual external loading, and permanent burial in sediments. In recent decades, the sediments have turned from phosphorus sinks into phosphorus sources. The reason for this is the large anoxic area of the Baltic (see, e.g., Ahlvik and Iho 2018). In anoxic conditions, phosphorus bound to iron in the sediments is released and the dissolved phosphorus re-enters the water body. If there is oxygen in the water column right above the sediment, the iron-phosphorus bound is formed again. This is the simplified phosphorus loop of a healthy sediment. However, in anoxic conditions the iron can find a new partner to bond with. This leads to formation of ferrosulfides. Abandoned by iron, phosphorus is free to enter the productive upper layers of the water body and accelerate algae growth. The decaying algae consumes oxygen and worsens the oxygen deficiency which fosters the release of phosphorus from iron—the vicious cycle of phosphorus is ready (Canfield et al. 2005).

The substantial impact of internal releases can be seen from the rapid change of ratios of external phosphorus input and phosphorus already in the water column. In the 1980s, the stock varied between 500,000 and 600,000 tons while the external load was as high as 70,000 tons. Today, the stock of total phosphorus in the Baltic Sea is more than twenty times the annual loading (Savchuk 2018). At the same time, the oxygen deficiency has been getting worse (HELCOM 2018). The cycle is difficult to break. The amount of phosphorus already in the water is so much larger than inflows that even zero emissions would not provide a fast cure for the

Baltic. Indeed, the frequency and severity of algal blooms have increased with the evolution of the phosphorus stock (Kahru et al. 2020).

Whereas the Baltic is brackish, Lake Erie is a freshwater lake. It is a large lake (25,667 km^2) but small by comparison to the Baltic. Typical of freshwater bodies, algae growth is limited by the availability of phosphorus. Prior to the 1970s, excess nutrient loads from municipal and industrial sources caused severe eutrophication, greatly reducing the lake's value for fishing, water supply, and aesthetics. Regulations and investments in the US and Canada, which share the lake, beginning in the 1970s reduced phosphorus loads, which exceeded 25,000 tons annually in the late 1960s, by more than half by the mid-1980s. However, nutrient problems returned in the early 2000s and have become more severe since, punctuated by widespread harmful algal blooms. Massive blooms of blue-green algae occurred in 2011 and 2015. The intensity of mass blooms seems to follow the April–July concentration of dissolved reactive phosphorus very closely. While a variety of factors contributed to the relapse, increased dissolved phosphorus (a highly potent form) from nonpoint sources, particularly agricultural land in the Lake Erie watershed, plays a leading role (Ho and Michalak 2017; Lake Erie LaMP 2019; Maccoux et al. 2016; Ohio-EPA 2010). Only 12% of the phosphorus loading comes from upstream lakes on the Great Lakes system, majority of it from the Lake Erie watersheds (Ohio, EPA 2010).

Where the Baltic problem highlights the significance of the existing inventory and the internal release of phosphorus, the Lake Erie problem highlights the importance of dissolved phosphorus and provides a warning about potential unintended consequences of BMP implementation. Point source pollution controls and adoption of reduced tillage practices in agriculture have reduced total phosphorus loads. But the composition of phosphorus inputs has changed with dissolved fractions, which are immediately and fully algal available, increasing dramatically in the biggest tributaries to the Western Basin of Lake Erie (Jarvie et al. 2017). The increase in dissolved phosphorus is attributed to a substantial increase in soil-conserving cultivation practices such as no-till.

The Chesapeake Bay is the largest estuary in the US and the third largest in the world. Its surface area (11,600 km^2) is less than half that of Lake Erie, but the surrounding watershed spans more than 165,759 km^2, encompassing parts of six states—Delaware, Maryland, New York, Pennsylvania, Virginia, and West Virginia—and the entire District of Columbia. The land-to-water ratio of the Chesapeake Bay is 14:1, the largest of any coastal water body in the world. This highlights the importance of land use in the watershed and land-based pollution sources like agriculture to water flowing into the Bay.

As an estuary the Chesapeake Bay has portions that are fresh, salt, and brackish. Eutrophication problems result from both nitrogen and phosphorus. Sediment is also significant cause of degradation of the Chesapeake's aquatic ecosystems. The Chesapeake Bay Program (2017) estimates that agricultural nonpoint pollution contributes 42% of nitrogen and 55% of phosphorus entering the Bay. Agriculture is also the leading source of sediments entering the Bay, contributing 60% of the total loads.

The Chesapeake Bay has been a focus of research on nutrient pollution and a focus of significant nutrient pollution policy initiatives and spending for nutrient pollution control in the US since the 1980s. The six states in the watershed along with the District of Columbia and the USEPA established the Chesapeake Bay Program in 1983 and entered a series of subsequent agreement to reduce nutrient and sediment loads to meet water quality goals set for the Bay. While progress has been made for point sources and agricultural nonpoint sources, the reductions have been insufficient to achieve water quality goals. This failure led the USEPA to issue a Total Maximum Daily Load (TMDL) for the Bay in 2011. TMDLs are limits on pollution loads required by the Clean Water Act (CWA) for waters that do not meet water quality standards. The TMDL requires the six states and the District of Columbia to reduce nitrogen, phosphorus, and sediment from managed sources by 26%, 24%, and 15% by 2025 compared to 2009 levels. The reductions required for agricultural nonpoint source pollution for nitrogen, phosphorus, and sediment are 36%, 30%, and 29%, respectively. From 1999 to 2018, the ten-year moving averages of point source loading of nitrogen and phosphorus have decreased by about 60%.[3] At the same time, the ten-year moving average of total nitrogen loading has decreased by 17% and that of phosphorus has increased by 14% (CBP 2020). The increase is driven by very high loads in 2011 and 2018. The story is familiar: point sources have substantially reduced their emissions while nonpoint loads remain on a high and highly variable.

[3] Load shares are estimated for different source types from the nine monitoring stations downstream. These loads contribute to 40% of all nutrient loads. The developments of relative shares for the entire watershed are likely to be very similar.

The Chesapeake Bay, Lake Erie, and the Baltic Sea are significant examples of water bodies in which the total nutrient load has declined over the decades with effective regulation of point source emissions. Agricultural sources have not been similarly regulated nor have they decreased in the same way (Ohio EPA 2010; OECD 2017; HELCOM 2018). Agriculture's share of nitrogen and phosphorus loading in OECD countries varies between 20% and 80% (OECD 2013). This, together with the facts that the total loading in these countries has declined and the water quality is not improving means that the pressure to find effective ways to curtail nutrient pollution from agriculture will keep on increasing.

Box 3.2 Economic Damages of Eutrophication

Expressing the environmental damages in monetary terms helps society and policy makers understand what is at stake from water pollution and can help policy makers target and design policies. Some of the costs of water pollution are straightforward such as the capital and operations costs of water treatment by water utilities (see Box 3.1). Eutrophication contributes to these costs but mostly affects nonmarket values from water quality.

Eutrophication decreases the aesthetic values by decreasing the water clarity, it affects the foodwebs, typically by increasing the amount of cyprinid fish, which have low value as commercial or recreational catch. Mass blooms of blue-green algae may be toxic and cause health effects to humans and can be lethal to their pets and production animals. They emit putrid odors at the late stage of the bloom and are unsightly—unless seen from space in which case they are colorful and rich in curious patterns of green and yellow. Eutrophication promotes the growth of macrophytes, which may pose concrete obstacles to boating, swimming, and fishing. In addition to these, it inhibits using freshwater lakes and rivers as sources of drinking water or makes it more costly. It is obvious that all these have negative impacts on the value of the water ecosystems. But how much is the negative effect worth?

Hyytiäinen et al. (2013) estimated the benefits of Baltic Sea protection and compared them to their costs. The nonmarket benefits were estimated using contingent valuation methods (Ahtiainen et al. 2013). Reaching the targets set by the Baltic Sea Action Plan was

(*continued*)

Box 3.2 (continued)
estimated to generate annual nonmarket benefits between 3.6 and 4 billion EUR.

The total costs of achieving the targets cost-effectively were estimated to be between 1.4 and 2.8 billion EUR. The net benefits of protection are positive and the benefit-cost ratio indicates more than 2.6 EUR in benefits for each 1.00 EUR in cost. The study also showed that benefits and costs would be unevenly distributed. The highest beneficiaries are found in countries like Sweden and Finland while the highest costs fall upon the areas draining to Baltic proper such as Latvia, Lithuania, and Poland. This uneven distribution has been also been observed in other studies that note the political challenges it creates (Ollikainen and Honkatukia 2001; Gren 2001). While we focus largely on economic efficiency, the uneven distribution of costs and benefits is a reminder that fairness in the distribution of burdens and benefits is important. An appeal of emissions trading is the potential for cost-effective pollution control plus flexibility in design choices that control the distribution of costs.

Some other cases: Dodds et al. (2009) estimate the damages from freshwater eutrophication in the US. The largest sources of damages are the losses of recreational benefits and lowered property values. Overall, they estimate that annual combined costs of eutrophication were approximately 2.2 billion USD. This may be a substantial underestimate. For example, Moore et al. (2015) estimate a suite of use and nonuse benefits to Chesapeake Bay watershed residents of reductions in nutrient and sediment loads to meet water quality goals for the Bay. The estimated benefits are between 1.20 and 6.49 billion USD annually. These benefits include improvements in the Bay's water quality and improvements in the quality of freshwater lakes located in the Bay's watershed that would benefit from measures to improve the Bay. The Moore et al. estimates exclude some significant sources of benefits (Wainger et al. 2017). Pretty et al. (2003) analyze the costs of freshwater eutrophication in England and Wales. They quantify economic losses due to, for instance, waterfront property values, losses for tourist industry and amenity values. The annual damages were estimated to be in the range of 105–160 million USD.

3.3 Sediments, Salinity, Pesticide, Emerging Contaminants

Nutrient pollution is the leading contemporary water quality problem for agriculture. However, there are other significant problems related to or transmitted by water. Sedimentation, salinization, and damage caused by pesticides and a wide class of emerging contaminants will be discussed in the following sections.

3.3.1 Pesticides

Pesticides are chemical or biological substances that are used to protect crops from insects, weeds, predators, and diseases. They include insecticides, herbicides, fungicides, bactericides, rodenticides, and plant growth regulators. While promoting crop yields, pesticides also harm nontarget species, decrease biodiversity, and impose threats to human health both directly and through effluents. Biodiversity is the core of the global life support system. Declining populations and weakening biodiversity are warning signs of permanent losses of species (Ceballos and Ehrlich 2002). Agriculture is the most important driver of biodiversity loss (Dudley and Alexander 2017).

Pesticides have been used in agriculture since ancient times. Sumerians used compounds of sulfur as insecticides 4500 years ago and arsenic was used to protect seeds from insects, mice, and birds in China over thousand years ago (Pavela 2014). Two thousand years ago, Roman scholar Varro recommended using *amurca,* a substance obtained as a by-product from olive oil production, for weed and pest control:

> ".. [*amurca*] *is an enemy of grass and a poison to ants and to moles.*"

The development of pesticides was a key factor in agriculture's rapid productivity growth in the decades after World War II. The flip side of the coin has been environmental impacts from the use and misuse of pesticides.

Today, the value of global pesticide industry is estimated to be around $45 billion (Pavela 2014). The use of pesticides is the highest in fruit and vegetable industry. The three biggest users (China, US, and Argentina) account for about 70% of global applications (Pavela 2014). The highest increases in pesticide use are in middle- and low-income countries. Many have experienced 20-, 50- or even 100-fold increases in pesticide use during the last 20 years (Mateo-Sagasta et al. 2018). On the other hand, pesticide sales in OECD countries have been slightly declining, in 2000–2010 at the annual rate of -1.1% (OECD 2013).

Pesticides harm the environment in many complex ways. The damage may be caused by the active ingredients of the pesticides, their additives or by their transformation products. Generally, the toxins enter the ecosystem via two interconnected routes. First, the organism in contact with a substance obtains a higher concentration in its tissue than initially in the surrounding environment. This is referred to as bioaccumulation. Then, certain persistent, often fat-soluble substances, such as the currently banned insecticide DDT, may enrich themselves as they move up the food chain. This phenomenon of biomagnification nearly caused the extirpation of the white-tailed eagle from the Baltic Sea in the 1970s (Helander et al. 2008).

The overall effect on biodiversity is one of the most worrying effects of pesticides. Neonicotinoids are insecticides which are assumed to have contributed to the overall decline in insects in Europe (Goulson 2014). However, the impact is not restricted to insects. Bird populations have been most severely hit in regions most polluted with neonicotinoids (Goulson 2014). In 2013, the European Food Safety Authority stated that neonicotinoids pose an unacceptable risk to pollinator populations. The EU commission then restricted the use of certain neonicotinoids in 2013 (Regulation No 485/2013) and in 2018 banned the outside use of three active substances entirely: clothianidin, imidacloprid and thiamethoxam (Regulations 2018/783, 2018/784, 2018/785).

Some pesticides are toxic to certain nontarget species. One of the most used herbicides globally, glyphosate is harmful for frogs, impairing their growth and increasing mortality. Another widely used herbicide, atrazine causes male frogs to obtain female characteristics (Mateo-Sagasta et al. 2018).

Pesticides may couple tightly with soil particles or they may be easily dissolved. Erosion detaches soil particles and may expose them to environments where they dissolve and find their ways into the ecosystem. Dissolved substances may leach to groundwater polluting drinking water, or they may be lost to surface waters with runoff. They may also be vaporized from soil and brought to water by rainfall. Whatever the pathway and the receiving media, pesticides carry on with their toxic job until they are degraded by microbial processes, ultraviolet radiation, and/or time.

Therefore, pesticides are often harmful when they enter the water ecosystems. By harming aquatic plants and animals, they influence ecosystem health and eventually humans. Contaminated groundwaters cause human health problems directly. The US Geological Survey, a US federal agency with water management responsibilities, compared the occurrence of pesticides in groundwater across the US between 1992–2001 and 2002–2011. Of the twenty most common pesticides, the occurrence of herbicides had remained about the same while the occurrence of insecticides had declined (Stone et al. 2014). The three most common ones—Atrazine, Deethylatrazine, and Metolachior—were detected in more than 75% of the samples in both periods.

Another way for direct exposure is obtained from eating contaminated food, mainly fruit and vegetables or being exposed to pesticides at work. In developing countries there are three million pesticide poisonings each year resulting in 300,000 deaths. These deaths are partly intentional, comparable to drug overdose suicides in developed countries (Gunnell and Eddleston 2003).

The decrease in pesticide use in Europe and in the US is related to effective regulation enabled by functioning regulatory institutions. Functioning institutions are a necessary, although not sufficient, condition for management of agricultural externalities. They provide oversight to use of pesticides and can react if the damages of using certain substances clearly outweigh the benefits. As early as 1972, the US and other countries banned the use of DDT for this reason (Mateo-Sagasta et al. 2018). The list of five countries where the pesticide use has increased the most in last 20 years is illustrative: Burkina Faso, Ghana, Ethiopia, Argentina, and Cameroon. Except for Argentina, the rankings of these countries in the UN human development index (out of 189 countries) were very poor: 182, 142, 173, 48, 150.[4]

[4] http://hdr.undp.org/en/content/2019-human-development-index-ranking

Box 3.3 The on Economic Value of Pollinators
Agriculture generates externalities, both beneficial and detrimental, and it also is a recipient of externalities, both beneficial and detrimental (Zhang et al. 2007). Pesticides are an example. Agriculture is a source of pesticide externalities but is at the same time affected by ecological harms that result from pesticides.

Pollinators provide one of the most crucial ecosystem services for agriculture. Estimates of the value of these services can be used as a yardstick for the damages if we were to lose them. Southwick and Southwick (1992) estimated these services to be worth ranging from $1.6 billion to $5.7 billion in the US alone. Levin (1983) estimated the value to be $18.9 billion. Gallai et al. (2009) take a shot at the global value: the pollinating insects provide a service worth $153 billion. It is substantially higher than the value of global pesticide industry ($45 billion; Pavela 2014). There is thus a trade-off involving externalities and public goods. Pesticides help an individual farm to promote crop yields. Collectively, the use of pesticides weakens the ecosystem services provided by the insects to all farms. Optimal management would balance between these two, and optimal policies would implement optimal pesticide use rates to individual farms.

The commercial pollinator services is a large industry. In the US, the total volume of pollinator services sales was $320 million in 2017. California is dominating the demand, with the costs of almond pollination alone reaching $253 million (USDA 2017). In addition to pollinating services, the production of honey is an important economic component for beekeepers. In California, honey sales comprise about 25% of their revenue, the national average being 50% (Champetier and Sumner 2019).

3.3.2 Sedimentation

Wind and water erosion detach and transport soil particles consisting of minerals and organic matter from agricultural land. These particles become sediments when deposited in water resources. Sedimentation is part of a natural cycle, in which erosion removes soils and biogeochemical processes create new soil through weathering of rocks, organic growth, and transport of soil material. Agricultural production can accelerate erosion rates, leading to economic losses to producers in the form of soil productivity and environmental consequences of sedimentation.

The key reason for increased erosion is the clearing of land for cultivation. Removal of natural vegetative cover and ploughing leave the soil bare exposing it to weather events. The effect of ploughing is greater if done up and down slopes on steep fields (instead of ploughing along the slope contours). Poor soil structure, resulting for instance from compaction by heavy farm machinery, also increases the risk for erosion by making the runoff water move along the soil surface instead of infiltrating in the soil profile. The effect of agricultural practices on erosion risk varies strongly with soil types, climate conditions and topography. For instance, steep slopes, heavy rainfall events, and periods of frost and thaw increase the risk of erosion.

In most water bodies, erosion is the dominant source of phosphorus loading as it carries phosphorus bound to the soil particles to water. In addition to nutrients, soil particles carry other pollutants such as heavy metals and pesticides with them.

Sediment has direct physical effects as well as chemical effects on water quality. It decreases the clarity of water and, by changing the light conditions, it influences the entire foodweb: plants, algae, fish species, etc. (Henley et al. 2000). Fine sediments can destroy fish spawning habitats, weakening reproduction. Large volumes of sediments can clog rivers and lakes, hindering water-based transportation, fouling irrigation equipment, increasing flooding risk, and more. Sedimentation also decreases recreational value of waters for boating, fishing, and swimming (Kosenius 2010).

Increased use of soil conservation practices and removal of highly erosive land from production in North America have decreased erosion substantially and diminished the sedimentation problem. Soil-conserving cultivation practices such as no-till have been adopted widely. One of the indicators of agriculture's environmental pressure used by the Canadian Ministry of Agriculture and Agri-Food is soil erosion. The indicator has shown a reduction in environmental pressures from erosion (Clearwater et al. 2016). Underwater grasses are an indicator of ecosystem health followed in the Chesapeake Bay. They are particularly vulnerable to sediment loading. In 1984, the total area of underwater grasses was 38,000 acres. From this all-time low, the area has gradually increased so that the ten-year average during 2008–2018 was about 79,700 acres. In 2019, the area was again lower: at 66,400 acres.[5]

[5] Chesapeake Bay Program Media Release 07-08-2020. https://www.chesapeakebay.net/images/press_release_pdf/Media_Release_SAV_7.8.20.pdf

3.3.3 *Salinity*

Salts refer to ions such as sodium, chloride, sulfate, or magnesium. They occur naturally in waters in dissolved forms. Plants and animals are adapted to certain salinity levels. Salinity becomes an environmental problem when salt concentrations increase making the habitats unsuitable for some species. Elevated salinity also decreases agricultural productivity and influences human health.

Clearing of land for agriculture affects its hydrology. Replacing native plants with cultivated crops decreases the evaporation and elevates the groundwater tables. In arid areas this may lift the naturally saline groundwater closer to the surface, leading to salinity problems. This has been a persistent problem in, for instance, the State of Victoria, Australia (Hart et al. 1991).

Irrigation causes salinity problems in arid and semiarid areas (Mateo-Sagasta et al. 2018). Irrigation may elevate saline groundwater tables. As evaporation exceeds average precipitation, salts tend to accumulate in the soil. Irrigating such soils leaches salts to drainage systems. On the other hand, utilizing groundwater for irrigation causes saltwater intrusion in coastal areas as the aquifers are partly recharged by the seawater (Barlow and Reichard 2010).

Globally, 24% of irrigated soils are degraded by salinization, half of these severely (Mateo-Sagasta and Burke 2010). The exact global scope of freshwater salinization is not that well understood (Mateo-Sagasta et al. 2018). However, there are numerous watersheds that have suffered from salinization for a long time. One of the much-studied cases is the Murray-Darling River in Australia (see, e.g., Wheeler et al. 2014; Grafton et al. 2012).

In the Colorado River segment flowing through Mexico, salinity problems emerged in the 1960s. This was a result of upstream withdrawals and pumping of naturally saline aquifers into the River in Arizona (Brownell and Eaton 1975). The overdraft of water from the Rivers is so large that it routinely runs dry before reaching its mouth in the Gulf of California.

Salinization of freshwaters threatens biodiversity of rivers, lakes, and wetlands. Invasive saline-tolerant species replace the saline-sensitive ones, eventually influencing the entire water ecosystem. The decline and salinization of the Aral Sea, a brackish lake in Central Asia, is a dramatic example of such development. Heavy upstream irrigation infrastructure development and the ensuing increase in water usage during the 1960s started the process which eventually extirpated a vital fishing industry altogether—along with the lake (Micklin et al. 2020).

Salinity also has indirect environmental effects via eutrophication as salts may change the biogeochemical cycles of nutrients. Sulfate-mediated eutrophication of freshwaters is a process where internal phosphorus loading from anoxic sediments increases with increasing sulfate concentration. Formation of ferrous sulfates replaces the chemical bonds between phosphorus and iron (Smolders and Roelofs 1993).

Salinized drinking water affects human health. High intakes of sodium or chloride are related to blood pressure problems. About 1.1 billion people live in areas with salinized groundwater, irrigation being the most important driver of the problem (Van Weert et al. 2009). The demand for irrigation will be increasing with climate change, further aggravating salinity problems in irrigated areas (Döll 2002).

3.3.4 Emerging Contaminants

Emerging contaminants are a large group of substances which are either new or whose potential adverse environmental effects have been learned only recently. Their effects on human or environmental health are not well understood but are concerning. Many of them interfere with the hormonal systems of humans and animals, posing grave risks for our health. Emerging contaminants also tend to lack adequate environmental monitoring systems (Geissen et al. 2015).

Hundreds of artificial or naturally occurring chemical compounds are considered emerging contaminants. They are grouped into broad classes, two of which are relevant for agriculture: pharmaceuticals and pesticides. As pesticides were discussed above, we turn our attention here to pharmaceuticals.

Antibiotics are substances that kill or prohibit the growth of bacteria. Thousands of plants have antimicrobial properties, many of which have been used for wound treatment since prehistorical times (Forrest 1982). The role of bacteria in causing diseases and the fact that some microorganisms are developing substances blocking bacterial growth was understood only in the late nineteenth century (Zaffiri et al. 2012). However, medical utilization of antibiotics was launched only after large-scale extraction methods for penicillin (a substance secreted by a mold, *Penicillium chrysogenum*) were developed in the 1940s.[6]

The development of antibiotics has been rapid and today they are used extensively in medicine. However, their wide usage has caused antimicrobial resistance in certain pathogens. The World Health Organization warns that antimicrobial resistance is "*threatening our ability to treat common infectious diseases, resulting in prolonged illness, disability, and death.*"[7]

Antibiotics have been used in animal farming since the 1950s. The antibiotic resistance is linked particularly strongly to antibiotic use in animal agriculture (Schechner et al. 2013; Mateo-Sagasta et al. 2018). In the livestock industry, antibiotics are used to treat sick animals and to prevent the diseases, but also to promote growth. Antibiotics are used extensively in pig production (Lekagul et al. 2019), in poultry production (Mund et al. 2017), and in dairy production (Ronquillo and Hernandez 2017). Antibiotics are removed from animals in urine and manure. Therefore, the primary mechanisms introducing antibiotics into the water environment are the same as those introducing nutrients: barnyard runoff and land application of manure.

Antibiotics are not the only pharmaceuticals of concern. In 111 monitored rivers by HELCOM, traces of 58 different pharmaceuticals were detected (Vieno et al. 2017). Veterinary use of pharmaceuticals is a source of high importance for pharmaceuticals (Mateo-Sagasta et al. 2018).

3.4 Agricultural Production, Past, Present, Future

Agricultural water quality problems vary in scale from small, localized problems to problems involving large international waters. As we have described with the Baltic Sea, Lake Erie, and the Chesapeake Bay, problems of a common type, like nutrient pollution, differ significantly in details relevant to policy solutions. Efficient solutions are always place based. However, broader economic and technological drivers operating at macro scales are very important to agriculture's impact on water quality at all scales. In this section we look at some macro-level drivers. On the demand side is the structure and growth of demand for agricultural

[6] The findings earned Ernest Chain, Howard Florey, and Alexander Fleming a Nobel Prize in Medicine in 1945.

[7] https://www.who.int/news-room/fact-sheets/detail/antimicrobial-resistance

commodities. Demand is the primary driver of agricultural production and a key determinant of the magnitude of agricultural externalities. Demand determines the overall level on land use and the use of polluting inputs. It also affects the composition of the commodities produced. This is important as commodities vary in the pollution intensity. For instance, almonds consume a lot of water, corn requires high nitrogen fertilization rates, and fruits and vegetables are associated with high pesticide use (Goldhamer and Fereres 2017; USDA 2019; Mateo-Sagasta et al. 2018). On the supply side are developments in the technology of agricultural production.

3.4.1 *Demand for Agricultural Commodities*

Agricultural products provide food for humans, renewable fuels, fiber for clothing and other issues and inputs for a variety of industrial products.

Food demand is driven by population and income. In the past fifty years the world's population has nearly doubled from 3.7 to 7.8 billion. The UN estimates it will continue to grow to nearly 10 billion by 2050 (United Nations 2019).

Box 3.4 New Developments in Protein Production

What if we could de-link protein production from land, fertilizers, environmental resources, and climate conditions? That would be a revolution in the traditional way of producing food. Winston Churchill (1932) was the first to raise this possibility. Professor Mark Post presented in 2013 the first hamburger made out of cultured meat. Currently, there are researchers and start-up companies saying that this revolution is not far away. Industrial production of both plant and (artificial) meat protein is offered as solution to the twin challenge of feeding growing populations while reducing the environmental harms of conventional crop and livestock production. As the production is an industrial process, production facilities could be built close to cities irrespective their location and climate conditions. Moreover, proteins could be produced by constant costs in contrast to decreased returns and increasing costs of agriculture. The most optimistic predictions suggest that new technologies will be in place in the near future. Singapore approved cultured, lab-grown "chicken" meat in 2020.

(*continued*)

Box 3.4 (continued)

Cultured meat

The notion of cultured meat (also called artificial meat, lab meat, synthetic meat, or vitro meat) refers to a cultured muscle that has been produced by taking skeletal muscle stem cells (myosatellite cells) from live animals (Post 2012; Kadim et al. 2015). These cells are induced to grow and differentiate to form muscle fibers in vitro. While this process works in a laboratory, the development of commercially viable solutions is still underway.

The culture media surrounding the growing cells and fibers must contain nutrients, oxygen, growth factors, and bioactive compounds required for normal muscle development. Growth in vitro of a single layer of myocytes and muscle fibers on a base of collagen fibers has been achieved. However, the formation of steak-like three-dimensional (3D) structures will require a 3D framework or scaffold and a means of ensuring that every cell/fiber has a continuous and adequate supply of nutrients and oxygen, as well as a means of removing waste products such as CO_2 (Kadim et al. 2015).

Cultivated meat, if successfully produced, could play an important role alongside conventional meat products in meeting predicted increases in the global demand for meat. It has many advantages relative to conventional meat: efficiency of resource use (land, energy, and water), lower greenhouse gas emissions, improved animal welfare (no killing), and in the ability to manipulate the nutrient composition of the product (Tuomisto and Teixeira de Mattos 2011; Post 2012).

A still open question is how cautious consumers will be in accepting such products due to perceptions of "unnaturalness" and "artificialness" in different cultures. The first tests have been promising (Hamdan et al. 2018; Rolland et al. 2020).

Protein as solar food

With the latest development, it is also possible to grow protein at the microbial level using air and electricity. A natural protein produced from carbon dioxide and renewable electricity is much more climate friendly than meat and crop production, and possibly even more friendly than cultured meat production. The bioprocess is similar to making wine and beer but uses instead of sugar, bubbles of carbon dioxide and hydrogen.

(*continued*)

Box 3.4 (continued)

The "neutral-flavor" protein can be used widely as an ingredient in food production. It provides a cheaper source of protein than the usual meat and plant-based protein. Thus, like cultivated meat, solar food protein may become competitor to the production of meat and soy. The main cost component of solar protein is electricity.

Like cultivated meat, dietary protein can be produced in areas where food production has previously been impossible, such as the desert, the Arctic or even space. Favorable places for protein made from thin air can also be those locations where food production has previously been impossible. Environmental benefits from solar food are similar to those of cultivated meat. The production of microbial proteins is already scalable and can be expected to enter the market soon.

Human populations have also grown wealthier, leading to increased food consumption, and changing diets. In 1970, 35% of people living in developing countries were not obtaining enough calories to "cover the energy requirement for an active and healthy life," a definition for undernourishment. In 2015 the share was 13% (Roser and Ritchie 2013). Food demand has grown substantially with growing populations and incomes. Particularly notable is the increase in overall meat consumption driven by increased population and increased per capita meat consumption. The demand for meat is income elastic, meaning people demand more as their incomes rise. The global average daily meat consumption per capita was 74 grams in 1970 and 118 grams in 2013. In that time, consumption in China had grown almost sevenfold: from 25 to 169 grams. In the US, the same numbers were 290 grams in 1970 and 315 grams in 2013 (Ritchie 2017). If the US values are taken as a benchmark for future development, we may expect the global per capita meat consumption to increase further as per capita incomes continue to grow. The demand for feed in meat production is a major driver of crop production. About one-third of global cereal production is used as livestock feed.

Also noteworthy is the production of biofuels. During the last decade, the global production of biofuels has increased at an annual rate of about

[8] This accounts for about 2% of the global oil production of about 81 million barrels a day.

7%. The daily production in 2019 was about 1,840,000 barrels of oil equivalents (BP 2020).[8] Roughly 38% of the production was in the US. Other important producers are Brazil (24%) and Indonesia (7%). Agricultural feedstocks are crucial for biofuels production. Bioethanol utilizes coarse grain and sugarcane while biodiesel uses various vegetable oils. In 2018–2019, about 38% of corn production in the US was used for bioethanol and about 32% of soybean harvest was used for biodiesel (USDA 2020). Biofuels clearly now comprise a substantial part of agricultural commodity market demand in the US.

The textile industry utilizes natural fibers from plants and animals: sheep, goats, cotton, hemp, etc. Cotton is the most important single product of these. The FAO (2014) estimates that its global revenues are more than $50 billion annually. In 2013/2014, about 2.3% of world's arable land was on cotton. The biggest producers are China and India, which account for more than half of global production with approximately equal shares; the US, Pakistan, Brazil, and Uzbekistan, which account for about 29%. In thirty years, the global average production per hectare has increased from 411 kg/ha in 1980 to 790 kg/ha in 2013. About 6% of global pesticides and 14% of insecticides are used for cotton. In India, Pakistan, and Egypt the entire production is from irrigated land, and in China almost the entire production (FAO 2014).

3.4.2 *The Technology of Agricultural Production*

Historically, increasing demand for agricultural output has been met by increasing the cultivated area. Agricultural impacts on the environment were largely a consequence of land conversion and the amount of land in production. However, since the mid twentieth century expansion at the extensive margin has stalled. Increased output has come not from more land but from increased yields. One reason is the land constraint: most suitable farmland is already taken into production. The other is rapid technological change which has made the increasing per hectare yields possible. Consider, for instance, the period from 1960 to 2010. During that time, the world population grew from three to seven billion and the agricultural production per capita also increased. Nevertheless, the agricultural land grew only by 10% (Van Ittersum and Cassman 2013). Productivity growth has been the major source of increased supplies (Ludena et al. 2007).

Other things equal, increased output from a given set of land, water, fertilizer, pesticide, energy, and labor inputs is economically and environmentally beneficial. Higher yielding plant and animal varieties, water conservation technologies, conservation tillage practices, and improved information technologies for managing nutrients and pests are an example of technologies that can lower consumer prices and/or increase farm profits and reduce environmental harms. But technological change in agriculture has involved increased intensity of input use per unit of land area and the use of inputs that are harmful to the environment. Fertilizers and pesticides stand out.

The significance of technological change is highlighted by the global nitrogen and phosphorus cycles associated with fertilizer use discussed earlier. Globally, there are roughly two pools of nitrogen: the inert pool in the atmosphere and the pool of reactive nitrogen comprising of multiple pools of different nitrogen compounds in soil water, air, and organic matter. By manufacturing fertilizers, agriculture is introducing reactive nitrogen to the global ecosystem from the inert pool of nitrogen gas.

One of the most dramatic changes in agricultural productivity is the result of a technological innovation in the beginning of the twentieth century that allowed for industrial-scale nitrogen fixation. The Haber-Bosch process utilizes atmospheric (molecular) nitrogen and hydrogen as inputs to generate ammonia.[9] This is an infinite resource as nearly 80% of the atmosphere is molecular nitrogen. The innovation led to a significant price decrease of nitrogen fertilizers and the subsequent boom in their application. The most rapid increase of nitrogen applications was experienced between 1940 and 1980 as the annual total fertilizer consumption increased from less than 0.3 Tg to about 9.8 Tg (Cao et al. 2018).

The sustainability of global nitrogen use is typically analyzed from the perspective of the rate of anthropogenic nitrogen fixation rate. De Vries (2013) estimate that annual fixation of 60–100 million tons of nitrogen would be sufficient to feed the world and low enough not to disturb ecosystems too much. Current rates of chemical fertilizer use plus crop-driven nitrogen fixation are at a bit over 120 million tons (Bouwman et al. 2013).

While nitrogen is obtained from the atmosphere, phosphorus is obtained from rock. Phosphorus was discovered by Hennig Brand 351 years ago (Sharpley et al. 2013). In the late eighteenth century, bone ash was discovered as a source of phosphorus, replaced by bird and bat

[9] The development was prompted by the demand of ammonium nitrate for ammunition, and the embargo imposed by the Allied powers of the WWI that cut down the supply of guano, the primary source of nitrogen at the time.

guano and finally with chemical utilization of phosphate rock in the mid-nineteenth century.[10] The quantities of phosphate rock production have increased from 5000 tons in 1850 to 100,000,000 tons in1974 and finally to 240,000,000 tons of today (Van Kauwenbergh et al. 2013; USGS 2020). According to United States Geological Survey (USGS) estimates, the annual production accounts for about 0.08% of the known world phosphate rock resources (USGS 2020).

Annual nitrogen fertilizer applications are generally essential for high yields from nitrogen-intensive crops like corn. This is not the case with phosphorus. The effect of annual phosphorus fertilization on crop yields is insignificant if the pool of plant-available phosphorus in soils is sufficiently large (Sharpley 2000; Dodd and Mallarino 2005). Periodic phosphorus fertilization is therefore to maintain the pool rather than to increase the yield in a growing season. Annual phosphorus applications measured in both quantity and expenditure per hectare for crop production are substantially lower than for nitrogen.

The journey of nitrogen from agricultural lands to the oceans has many phases during which nitrogen contributes to local water pollution damages. It also undergoes changes from one form to another. Most importantly, during this nitrogen cascade, part of it is lost from the immediately biologically active parts of the system. Of these the most important ones are groundwater storage (although this may contribute to other problems), denitrification in riparian zones, and burial in and denitrification at sediments of lakes and rivers (Billen et al. 2013). Nitrogen and its effects on local water pollution problems are spatial and spatially interconnected.

Livestock production is a huge driver in the global agricultural supply system. Its effect on land use is dramatic. Of the Earth's habitable surface area (about 10.4 billion hectares), permanent grassland covers some 3.5 billion hectares. Ritchie (2019) counts this area into livestock pasture and rangeland. Mottet et al. (2017), on the other hand, suggest that only 2 billion hectares of permanent grassland are devoted to livestock. In addition to this, both estimate that the total field acreage used to produce feed for livestock is 560 million hectares. This is 40% of the global arable land. Therefore, any changes in the demand for meat will have magnified effects on all externalities related to crop production.

[10] Any geologic material containing relatively high concentrations of phosphorus such as apatite.

3.5 Agricultural Production and Water Quality

We now drill down further into specific features of agricultural production that cause water quality problems and that must be managed for water quality protection. There are macro and micro dimensions to this topic. The significance and pervasiveness of agriculture's impacts on water quality have two main dimensions. At the macro level is the spatial extent of agriculture, which means that water from rainfall and snowmelt will often cross agricultural lands before entering aquifers or streams. Agriculture reshapes the Earth's surface. Natural vegetation is replaced with species selected and developed for human uses. Tillage, filling wetlands, re-routing streams, and other land modifications dramatically change the structure of landscapes. The results include changes in the volume, timing, and temperature of water flowing to streams, the fractions of water going to surface and groundwaters, and host of other variables that affect the structure and health of aquatic ecosystems, water quality, and the utility of water for various uses.

Overall land use in agriculture explains in large part the enormous magnitude of agriculture's impact on water. But given the overall importance of agriculture to society and the corresponding importance of having land in agriculture for the foreseeable future, reducing the overall amount of land in agriculture is not the path forward for addressing water quality problems, given dominant agricultural production technologies.[11] The micro dimension and the focus of this section is the methods or intensity of agricultural production. Some agricultural practices extract natural elements from agricultural lands, like salts and eroded soils, and result in their removal to waters where they can cause harm. Some practices treat lands with inputs, fertilizers, and pesticides for example, to promote crop and livestock growth or protect them from output-reducing harms, that become pollutants. Intensive production practices tend to increase both the extraction of pollutants from agricultural lands and the application of pollution inputs to agricultural lands.

In this section, we look at the methods of production used to meet demands and the physical processes that link agricultural production and land use to water quality impacts. These links include the formation of pollutants on agricultural lands and their transport from farm boundaries

[11] At the micro scale, removing environmentally sensitive lands from agricultural production is a very cost-effective method in some cases. It has been a policy priority in the US with beneficial impacts.

(edge-of-field) to water resources. We introduce concepts and analytical tools that are commonly used for water quality management from relevant disciplinary perspectives including environmental science, engineering, and economics.

3.5.1 Nutrients and Crop Production

The elements that crops need for growth are phosphorus, potassium, nitrogen, carbon, hydrogen, oxygen, and various trace elements such as sulfur, calcium, iron, magnesium, sodium, and manganese. Von Liebig (1859) established the concept of limiting factors for growth. There is always something that is constraining plant growth: temperature, nutrients, water. Increasing the supply of the limiting factor, be it water or nutrients, is a defining feature of agricultural production from ancient times.

Nature provides many of the necessary elements in sufficient quantities depending on the location and soil type. Nutrients that are limiting can be provided in fertilization. The key inputs are nitrogen, phosphorus, and potassium. Nitrogen and phosphorus are the ones with high significance for water quality. Agricultural crops can utilize only chemical compounds on nitrogen and phosphorus that are in bioavailable forms.

Continuous farming produces nutrient deficits and requires that nutrients removed by crops be replaced. Crop rotation is a timeless method to provide crops with nitrogen, still practiced today. Certain plants host nitrogen-fixing microbes in their root zones. During the growing season, such plants fix nitrogen from the atmosphere and turn it into organic forms. Nitrogen-fixing plants include clover, soybeans, alfalfa, and other legumes. These are commonly used in rotation with corn, wheat, and other cereals.

3.5.2 Nutrients and Animal Production

Traditionally, production animals play an important role as sources of manure and thereby of crop nutrients. Feed for production animals contains nutrients, most of which end up in manure. Manure also contains organic matter and therefore is beneficial for agricultural soils. Increasing livestock production poses pressures on sustainable land management via grazing pressure and deforestation via extension of pasture and rangeland and feed production. The increasing sizes of animal facilities and the separation of crop and animal production have brought by a problem of animal waste management. From the environmental perspective this is most of all a nutrient pollution problem.

Animal farming systems can be roughly divided into those focusing solely on livestock and mixed crop-livestock farming systems (Seré et al. 1996). The rate of growth for meat production in landless production systems is growing fastest in livestock industry. Thirty-seven percent of all meat was produced in landless systems in the 1990s (Seré et al. 1996). Also, farm sizes have been increasing and continue to increase. Except for niche-market and hobby farms, very small commercial farms are vanishing while larger ones are more likely to continue and grow larger (see, e.g., Niskanen et al. (2020) for the Baltic Sea region). This changes the landscape of nutrient flows within the livestock industry, increasing pressures but also opening possibilities for large-scale manure management.

Feed produced for livestock uses 40% of the global arable land (Mottet et al. 2017). Nutrient management in feed production is one issue but it is not very different from crop production for any other purposes. However, manure management is specific for livestock industry only.

Economies of scale favor large animal farms over small ones. Together with regional agglomeration this continues to concentrate animal farms to certain regions and increase the average animal numbers of individual farms (Niskanen et al. 2020). In all OECD countries, the average farm sizes have increased, and the number of farms has decreased, especially since World War II (Huffman and Evenson 2001). MacDonald (2020) shows that the development has been particularly strong in the US. This evolution started already in 1930 and has continued ever since.

A related change has been the specialization of farms, which has separated animal and crop production to different farms and eventually to different regions. This has led to net nutrient imports to animal production regions. For instance, according to van Dijk (2016), phosphorus balances in Belgium and Netherlands are on average more than 20 kg/ha.[12] That is, on average 20 kg more phosphorus is applied to agricultural fields than removed by crops. The main reasons for the imbalance are the costly hauling of manure and the uncertain and unfavorable

[12] As noted by Einarsson et al. (2020), manure export driven by recent mandates has decreased the phosphorus balances substantially in the Netherlands.

content of crop nutrients. The economic underpinnings of this topic are discussed briefly in Chap. 4.

3.5.3 *Tillage*

The invention of ploughing was a crucial step in the development of agriculture. The earliest ploughs were simple sticks or hoes used to make furrows for seeds. Domestication of oxen to pull ploughs was a major innovation. Steam power was harnessed in the nineteenth century and internal combustion powered tractors became the standard in the twentieth century.

Tilling serves multiple purposes in farming. Originally, it was to seed. Since then two other farming purposes have become important: controlling weeds and improving soil fertility. Turning the uppermost soil layer kills the weeds. Performed at different times in the growing season, the same process serves to incorporate organic residues left after harvests, such as corn stalks and straw into soils. The organic matter decays in soil, releasing plant-available nutrient in the residues to soils.

While beneficial for production, tillage leaves the land bare for periods of time increasing the risk for wind and water erosion and for nutrient loadings from fields. The North American Dust Bowl of the 1930s is attributed to environmental conditions but also to wide utilization of ploughing of the topsoil (Lee and Gill 2015). Huge expanses of land once covered by native grasses that protected soils from the effects of wind and rain were ploughed. The removal of natural vegetative cover, disturbance of soils that increases susceptibility to erosion, combined with drought that prevented crop growth to protect soils resulted in an agricultural and environmental catastrophe.

The green revolution of the 1970s brought about more affordable and effective chemical weed control methods. Technological innovations also brought profitable new seeding technologies that enabled cultivation without ploughing. Such no-till technologies and effective herbicides thus diminished the two basic reasons for ploughing.

3.5.4 *Irrigation*

Irrigation makes it possible to grow crops in arid and semiarid regions. It can also increase profitability in humid regions experiencing occasional dry periods. From 1970 to 2012, the global area equipped for irrigation has increased from 184 to 320 million hectares (FAO 2014). The future productivity growth is expected to be particularly focused on irrigated land (Mateo-Sagasta et al. 2018).

The most common irrigation method is surface irrigation. One of three basic technologies—furrow, borderstrip, and basin irrigation—is used to move the water to the fields and wet the soil either completely or partially. Furrow and borderstrip irrigation utilize furrows, ridges, and levees to steer the water at the intended times to the intended field locations. Basin irrigation floods the entire field. For rice cultivation, the fields remain submerged. For other crops, the waterlogging typically lasts for one or two days.

These common irrigation methods have their origins in antiquity. Alternative methods have been developed that make more efficient use of water. Drip irrigation delivers water to the root zone in strictly controlled quantities, literally drop by drop. This is the most water-efficient form of irrigation. As the name suggests, sprinkler irrigation distributes the water from one or several sprinklers using high pressure. Center pivot irrigation generates the green spheres visible from airplanes when flying over (arid) agricultural areas. A wheeled system of tubes and sprinklers rotates around the center. Subirrigation is used in some areas. There, the water table is artificially elevated to moisten the root zone from below (Playán and Mateos 2006).

We have introduced irrigation as a cause of water quality problems in this chapter. It is also important to recognize that irrigation, like pesticides, is connected to significant externalities beyond those associated with water quality. Water quality externalities result from irrigation water removing pollutants (salts, pesticides, fertilizers) from fields and carrying them into ground or surface waters. Significant externalities result from the provision of water (Shortle and Griffin 2001). Damming rivers to capture and store water for agricultural and other uses destroys natural and cultural assets by inundation. Dams and diversions create physical barriers to fish migration necessary to reaching spawning grounds, resulting losses of some species and ongoing threats to the survival of others. Dams and diversions also alter streamflow regimes, water temperature, and trap sediments with adverse impacts aquatic ecosystems. The threat to aquatic species has led to dam removal in some locations. The Red Bluff water diversion on the Sacramento River, California, which endangered salmon, sturgeon, and steelhead, is an example (see, e.g., Mahmoud and Garcia 2000). The dam, constructed in 1962, blocked passage of salmon, steelhead and sturgeon runs on the Sacramento River. Declining fish populations led to a sequence of solutions, starting from fish ladders at the side of the dam, to enable migration of endangered steelhead. Later, the gates of the dam were kept open during the migration periods and finally in 2013, the dam was decommissioned (USBR 2015). Massive groundwater utilization in irrigation can also lead to land subsidence. This has been a significant and highly visible problem in the southern part of California Central Valley, in the San Joaquin Valley since 1920 (Ireland 1983). In some locations, the land had subsided nine meters by the 1980s (Faunt et al. 2016).

3.6 From Field Edge to Ambient Water Quality: Principles and Methods of Watershed Management

We have described how water quality is influenced by agricultural activities. Moving from general concepts to planning and policy for water quality management in specific places requires detailed information on water quality problems and potential solutions for the place and models for predicting how water quality conditions respond to changes in management variables.

The approach we describe is a systems-based watershed approach that contrasts with standard approaches to water quality management. Traditional approaches to water pollution control separate the control of point sources from agricultural nonpoint sources. Systems-based watershed management is integrative. Effective and efficient management must be integrative and consider trade-offs resulting from differences in pollution levels, abatement costs, effectiveness, and reliability of different pollution sources (Shortle et al. 2020). Traditional approaches also focus primarily on technology implementation. Effectiveness and efficiency require a focus on environmental water quality outcomes.

3.6.1 Place-Based Water Quality Management

To introduce place-based water quality management, consider a hypothetical water quality agency charged with achieving a water quality goal for a particular water body and wishing to do so in a way that minimizes the social cost. We refer to the locations at which water quality goals are set as receptors. A receptor may be a stream reach, a pond or lake, estuary, or coastal water. If the agency knows very little about the economic and physical environment, the plan will be crude and likely to fail in some manner. It may fail to achieve the goal. It may overshoot the goal. It may be overly expensive. Effective and efficient plans generally require substantial information about the economic and physical environment.

Pollutants flow to receptor from one or more rivers and perhaps from atmospheric sources. For simplicity, we will focus on managing the flow from one river. For example, the Susquehanna for the Chesapeake Bay, the Mississippi for the Gulf of Mexico, or the Danube for the Black Sea. The river receives water from tributaries and groundwater base flows.

Planning begins with a hydrological watershed map for the receptor. A watershed is an area of land that drains all the streams and rainfall to a common outlet. A large watershed with multiple streams can be subdivided into small watersheds with fewer streams feeding the larger river. The river system can be imagined as a branching tree with smaller branches connecting to larger branches that connect to the trunk. The trunk is the main stem of the river. The receptor's watershed is the entire land area that contributes to the flow that exits from the river to the receptor. The river's watershed can be subdivided into sub-watersheds (or catchments), one for each branch. Water flows downhill from the boundaries of the large watershed through the smaller watersheds to the main stem.

For water pollution management, the hydrological watershed map must be supplemented with other maps. One need is to locate point source discharge locations and land uses that are potential nonpoint sources. For agriculture, a map of landownership is needed to identify the set of agricultural landowners, or producers, who are candidates for water quality management actions. A soil map is needed to identify areas within farms that have differential propensities to contribute to pollution or differential responses to management activities based on soil types, surrounding topography, or other factors. These may be barnyards, fields or portions of fields, or pastures or portions of pastures. Some areas may contribute little pollution, some may contribute a lot. The latter are often referred to as critical source areas. Contributing areas are often variable in time and size, increasing or decreasing in size, or appearing or disappearing in various locations depending on time of year, rainfall, temperature, topography, and vegetation among other factors.

Prioritizing locations for management is commonly referred to as *spatial targeting*. Optimal spatial targeting considers not only the propensity of areas to contribute pollution but also the cost of control actions within areas. Spatial heterogeneity of agricultural land as it affects agricultural productivity and water quality impacts is a crucial consideration in effective and efficient water quality protection.

This mapping characterizes the economic and physical landscape. A baseline characterization of economy and environment would include the natural conditions (soils, topography), economic activities in the landscape (crops, livestock, farming inputs and outputs, etc.), and water quality conditions. These conditions can be identified in principle by surveys of various types. The description will indicate how water quality is impaired and the types and sources of pollutants causing the impairment.

Achieving water quality goal requires reducing point and nonpoint pollution loads. For agriculture, this requires changing the agricultural practices in contributing areas. Doing this effectively requires predicting the effects of changes in agricultural activities on pollution loads and water quality outcomes. Environmental science research provides biophysical modelling systems for this purpose. Such models are described in the next section.

With the landscape characterization and the modelling system an agency can predict water pollution outcomes that result from various alternative agricultural landscapes described by spatial distribution of crops, livestock, and farm inputs and practices. Given many spatial planning units and many possible farming activities, there will be many possibilities. In a central planning approach, an agency would evaluate alternatives according to the agency's criteria and mandate solutions. In a market-driven agricultural economy an agency must devise policy interventions to change farm behavior. This requires an understanding of the economic decision making, their determinants, and how choices respond to different types of interventions. Policy evaluation requires integrating economic and biophysical models.

3.6.2 *Physical Processes from Source to Receptor*

The fate of pollution originating from an individual field depends on the characteristics of its transition process. Figure 3.1. illustrates the transport stages from the farm to the waterbody and connects these stages to the sources of social welfare from agriculture: positive benefits from food production and negative from pollution. These stages are captured by biophysical pollution models.

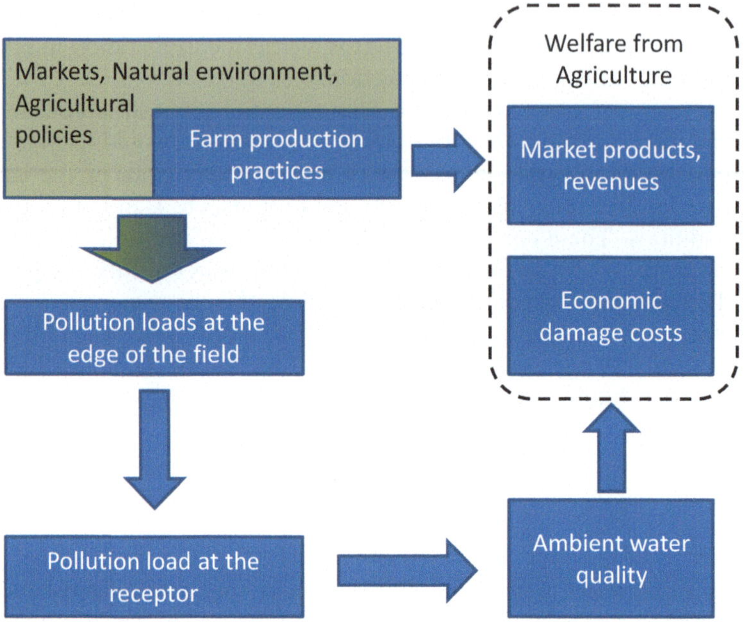

Fig. 3.1 Stages of pollution from fields to water quality, and its role in welfare outcomes of agriculture. (Source: Authors' creation)

Before arriving at the receptor point, pollutants are affected by processes in ditches, rivers, and lakes (some of which may also be receptors for water quality management). Several simultaneous riverine processes affect the flow of nutrients. Water enters the river channel from upstream, overland flow, groundwater discharge, direct precipitation, etc. Water exits the river system via groundwater recharge, bank storage, overbank flow, and anthropogenic withdrawals (Harvey 2016).

The fact that water does not flow through a river like it does through an artificial pipeline means that journey takes a longer time. The longer it takes, the more the water and the nutrients and contaminants it contains are affected by microbes and by geochemical activity in the sediments. Also, nutrient uptake of river vegetation is stronger with longer flow-through durations. Therefore, the overall status of the river affects how much of the edge-of-field pollution flows through the system and how much is retained (Ator et al. 2011).

Box 3.5 Delivery Ratios

In Chap. 2 we discussed Pareto Efficiency when pollution from different sources affects receptors (i.e., locations that are the targets of water quality protection measures) with varying intensity. We showed that the marginal abatement costs at different pollution sources *evaluated in units of pollutants arriving at the receptor point* must be equal.

The variability in the intensity can be measured with a delivery ratio, which is the ratio of pollution delivered to the receptor to the pollution emitted from a source. A ratio of one indicates that all emitted pollution arrives at the receptor point. A ratio of 0.5 indicates that half of the emitted pollution does, and half is retained in the biogeochemical processes upstream.

Nutrients in streams are utilized by aquatic life and removed by various other processes with the result that the nutrients delivered to downstream receptors are a fraction of the nutrients discharged. In general, the delivered fraction decreases with the distance transported. Mississippi watershed has a total area of 3.2 million square kilometers, reaching 31 states and parts of Canada. Alexander et al. (2008) estimate the delivery ratios for phosphorus and nitrogen when emitted to different locations in the watershed. Figure 3.2 illustrates the delivery ratios.

This has substantial policy implications. The Gulf of Mexico suffers from hypoxia caused by excessive loads of nutrients, mainly delivered by the Mississippi river. Since the first Gulf of Mexico Hypoxia Action Plan issued by the USEPA in 2001, 12 states along the main channel of the Mississippi river have defined priority watersheds and nutrient-reduction strategies. The protection efforts have focused on the states closest to the main channel and within the states, on watersheds that have a large impact on delivered loads. This prioritization is supported by the economic efficiency criteria: all else equal, the higher the delivery ratio, the higher the optimal efforts.

Lakes retain large quantities of nutrients. The basic processes are the same as in rivers with biological uptake being the key process. The organic compounds in algae are transformed into ammonium and then into nitrites and nitrates. The soluble forms may re-enter the productive layers of the

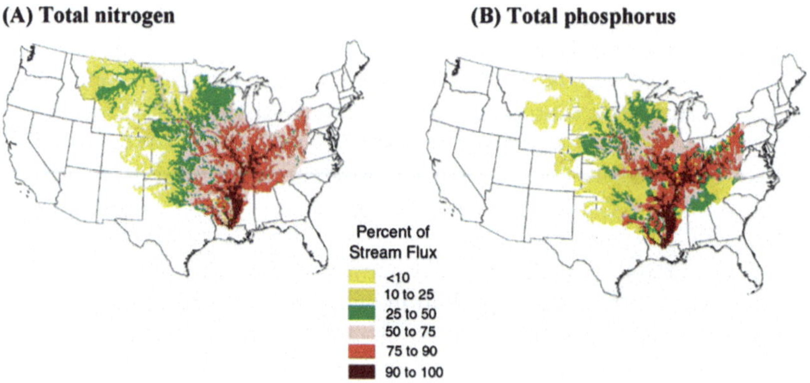

Fig. 3.2 Percentage of stream nutrient load delivered to the Gulf of Mexico from the incremental drainage of MARB reaches: (**a**) total nitrogen; (**b**) total phosphorus. (Source: R.B. Alexander et al., 2008. Differences in phosphorus and nitrogen delivery to the Gulf of Mexico from the Mississippi River Basin. *Environmental science & technology*, *42*(3), pp.822–830. pubs.acs.org/doi/abs/10.1021/e... Further permissions related to the material excerpted should be directed to ACS Publications)

water column or they may follow the process of denitrification. It is a sequential reduction process of nitrates resulting in, among other forms of oxidized nitrogen, N_2 which may evaporate into the atmosphere. The process is a by-product of anaerobic microbes utilizing (eating up!) organic matter in the sediments. It is a complex, overlapping biogeochemical process involving microbes, oxygen (and other elements microbes can use as terminal electron acceptors), and the supply of organic matter (Conley 1999; Canfield et al. 2005). In lakes, fish harvesting may account for large quantities of removed nitrogen and phosphorus.

The permanent sedimentation of phosphorus is strongly related to the oxygen conditions in the bottom sediments.[13] The organic phosphorus in decaying algae is mineralized by microbial activities bringing dissolved phosphorus into the water column, which can be again uptaken by algae.

[13] The geological cycle of phosphorus eventually leads to permanent burial to sediments. After very long time, these sediments will form the future deposits of sedimentary phosphate rock. Sedimentary phosphorus forms 95% of current global phosphate resources (Pufahl and Groat 2017).

Part of the dissolved fraction is adsorbed to iron hydroxides. If the water just above the sediments contains oxygen, only a small fraction of phosphorus is released back to water column. However, summer typically increases water temperatures and increases the consumption of oxygen as algal blooms decay. Anoxic sediments release most of the iron-bound phosphorus back to circulation. This is the damaging phenomenon observed in the present-day Baltic Sea.

Ambient water quality is determined by the quantity and type of pollutants eventually reaching the receptor. Eutrophication is mainly driven by the concentration of the nutrients most critical for algae growth. Various algae species need nutrients in fixed proportions, and one of these is always limiting for algae growth. Water ecosystems have algae populations consisting of various species with different nutrient needs. This means that for eutrophication, the state of water ecosystem where algae growth triggers various unwanted changes, there may also be a clearly defined limiting nutrient (Howarth 1988; Carpenter et al. 2001).

Generally, phosphorus is the critical nutrient in freshwater ecosystems whereas coastal ecosystems are driven by nitrogen and phosphorus (Paerl 2009; Carpenter 2008; Conley 1999). Estuarine ecosystems are driven by both nutrients and open sea areas typically by nitrogen.

How ambient water quality responds to edge-of-field pollution that reaches the receptor is also influenced by the features discussed earlier in the context of the Baltic Sea and Lake Erie: the stock pollution character in the Baltic Sea case and the changes from particulate to dissolved dominated fractions (Lake Erie case). Box 3.6 discusses the importance of taking such qualitative features in pollution into account in optimal water quality management.

Box 3.6 Pollutant Forms Matter: The Case of Phosphorus

Discussions of nitrogen and phosphorus as pollutants often focus on the aggregates without considering specific forms. Specific forms can matter greatly to water quality outcomes and have significant implications for management choices. For example, farm management practices can affect whether nitrogen is lost to the atmosphere with ammonia volatilization or lost to surface water or groundwater as nitrates.

(*continued*)

Box 3.6 (continued)

Phosphorus is an especially complex problem for water quality management. One reason is trade-offs between the forms of phosphorus that cause water quality problems. The second is the sluggishness of soil phosphorus removal which makes quick remedies difficult. This is the legacy phosphorus problem.

Most of phosphorus load from agricultural catchments is in particulate form. Particulate phosphorus is attached to soil particles and moves to water with soil erosion and sedimentation. Measures to mitigate erosion are therefore commonly recommended to mitigate particulate phosphorus loads. These include permanent soil cover, no-till practices, and other soil-conserving cultivation methods.

But phosphorus is also removed from fields in runoff as dissolved (or soluble) reactive phosphorus (DPR). For instance, between 2009 and 2013 DPR was on 33% of the total phosphorus load to Lake Erie (Maccoux et al. 2016).

Although effective in reducing the sum of all phosphorus fractions, most soil-conserving practices have the tendency to increase the loading of dissolved phosphorus (Dodd and Sharpley 2016). In consequence, there is a trade-off: reductions in particulate phosphorus loads are partly offset by increases in dissolved phosphorus loads. What makes this trade-off very problematic is that dissolved phosphorus is more harmful in receiving waters than particulate phosphorus. Dissolved fractions are completely algal available whereas only a fraction of particulate phosphorus is transformed into an algal-available form.

How much of particulate phosphorus should be counted as algal available? In Lake Erie, the range is estimated to be 25%–50%.[14] The availability depends on soil types of the source areas, the hydrological characteristics, and water quality of the receiving waters (Baker et al. 2014; Ekholm and Lehtoranta 2012). In any case, if a quarter of particulate phosphorus is transformed into dissolved form, a kilogram of dissolved phosphorus has the same effect on eutrophication

(continued)

[14] Annex, G.L.W.Q.A., 4. Objectives and Targets Task Team, 2015. Recommended Phosphorus Load Targets for Lake Erie. Final Report to the Nutrients Annex Subcommittee. Available at [https://www.epa.gov/glwqa/report-recommended-phosphorus-loading-targets-lake-erie].

Box 3.6 (continued)
as 4 kg of particulate phosphorus. This has important implications for the effectiveness of erosion control measures in mitigating phosphorus mediated eutrophication. Depending on the bioavailability of particulate phosphorus, many widely used measures might be eutrophication inducing. This is the significant unintended consequence Jarvie et al. (2017) describe as a culprit for Lake Erie re-eutrophication.

The legacy phosphorus is related to the management of DRP loading. Its key driver is the potentially plant-available phosphorus in soil (Pote et al. 1996). The problem with soil phosphorus is that it cannot be chosen freely. Instead, it is gradually driven by the annual differences in phosphorus applied and removed by crops. That is, even if we were to cease phosphorus fertilization altogether, the previously accumulated soil phosphorus reserves keep on enriching the runoff waters for long periods, even decades (Sharpley et al. 2013).

3.7 Biophysical and Economic Models

Place-based planning and policy development for pollution control in agriculture inevitably turns to biophysical models. Models serve a variety of purposes. These include assessing water quality conditions and causes of water quality impairments, predicting how water quality will change in response to changes in pollution loads or changes in land use and land cover affecting watershed hydrology, and determining limits on pollution loads needed to achieve water quality objectives. These purposes apply whether the sources of water quality problems are point or nonpoint sources.

Water quality modeling takes on special significance for agricultural and other nonpoint sources because emissions from individual sources cannot be metered routinely nor regulated with precision in time or space as can discharges from industrial or municipal point sources. This is due to the diffuse and complex pathways by which nonpoint pollutants move from their origins to receiving waters, and the fact that weather events are key drivers of nonpoint pollution flows. Effective management requires quantification of the relationships between pollution loads and agricultural practices in specific places. Models of agricultural pollution processes are developed for these purposes.

Research on nonpoint pollution management has produced a suite of models for quantifying relationships between land use activities and nonpoint water pollution loads. The suite can be differentiated in several ways. One distinction is between models for urban areas and models for rural areas (Deliman et al. 1999). Models for agriculture are in the rural category. Models are also differentiated by the spatial units for which they are intended. Some models are developed for watersheds with multiple land uses while others are developed for smaller scales, in agriculture, fields or even portions of fields. Field-scale models estimate or simulate pollutants or precursors of pollutants from individual field parcels. The estimates are commonly referred to as edge-of-field runoff or losses depending on the context.

Table 3.1 lists examples of field- and watershed-scale models used for studying and managing agricultural nonpoint pollution. It is apparent that there are many options. For both types, models vary in the pollutants they simulate or estimate, the timescales at which they operate, the degree to which they represent heterogeneity within the modelled units,

Table 3.1 Field, farm, and watershed models for agricultural nonpoint pollution

Field- and farm-scale models
Universal Soil Loss Equation (USLE) (Wischmeier and Smith 1978)
Chemicals, Runoff and Erosion from Agricultural Management Systems (CREAMS) (Knisel 1980)
Groundwater Loading Effects of Agricultural Management Systems (GLEAMS) (Knisel 1991)
The Erosion Productivity Impact Calculator (EPIC) (Williams 1985)
Climatic Index for Soil Erosion Potential (CSEP) (Kirkby and Cox 1995)
European Soil Erosion Model (EUROSEM) (Morgan et al. 1998)
Water Erosion Prediction Model (WEPP) (Flanagan and Nearing 1995)
OVERSEER® (Wheeler 2015)
Watershed models
Agricultural Nonpoint Source pollution model (AGNPS) (Young et al. 1987)
Better Assessment Science Integrating Point and Nonpoint Sources (BASINS) (Whittemore 1998)
Generalized Watershed Loading Function (GWLF) (Haith and Shoemaker 1987)
Hydrologic Simulation Program Fortran (HSPF) (Bicknell et al. 1993)
Soil and Water Assessment Tool (SWAT) (Arnold et al. 1998)
Automated Geospatial Watershed Assessment Tool (AGWA) (Miller et al. 2007)
Watershed Analysis Risk Management Frame (WARMF) (Goldstein 2001)

Source: Authors' creation

the data they require, the physical processes they represent, computational complexity, and other dimensions (Arabi et al. 2012; Loucks and van Beek 2017; Shoemaker et al. 2005; Yuan et al. 2020). The degree of model complexity increases with the number and spatial detail of variables describing physical site conditions (soils, topography, climate), the number of Best Management Practices (BMPs)[15] included, capacity to analyze combinations of interacting BMPs, details on farming practices (crops, rotations, nutrient management history), sophistication of the algorithms used to estimate outcomes, and the specificity of emission estimates in time (e.g., steady-state long-term average vs. shorter timescales) or space (edge-of-field, local watershed outlet, downstream locations).

The fundamentals of many models for nonpoint pollution management were developed during the 1970s–1990s to address specific watershed issues (Yuan et al. 2020). Developments since then have expanded the scope of application and integrated advances in data availability resolution (e.g., GIS, remote sensing, and electronic sensor technology) and data-driven research methods (Yuan et al. 2020).

3.7.1 Watershed-Scale Model Applications

Earlier in this chapter we introduced the Chesapeake Bay TMDL. The TMDL established pollution load reductions for nitrogen, phosphorus, and sediment that are required to achieve established water quality objectives for the Chesapeake Bay. The load reductions are allocated to tributaries to the Bay, and for each tributary to source sectors, including agriculture. The scientific foundation for the TMDL is a suite of water quality models developed through a USEPA-led environmental science initiative to determine the effects of human activity on the Bay's waters and living resources and to provide a scientific foundation for measures to restore the Bay beginning in the 1970s (NRC 2012). Models in the suite include:

- The Chesapeake Bay Watershed Model quantifies the impacts of land use and point source discharges on nutrient and sediment loads to the Bay.

[15] BMPs are generally defined as practices to reduce agricultural pollution loads compared to "conventional" practices. Further discussion is presented in Chap. 4.

- The Chesapeake Bay Airshed Model estimates atmospheric deposition of pollutants.
- The Chesapeake Bay Estuary Model estimates the impacts of nutrient loads on the water quality in the Bay.
- The Chesapeake Bay Scenario Builder is a planning tool that can be used to estimate pollution loads under alternative scenarios about land use, population, and other drivers. The model is intended to help state-level environmental agencies responsible for meeting the load limits develop and evaluate plans.

Similar in concept but not in scope, scale, and detail is the BALTSEM model developed by the Baltic Nest Institute in Sweden to calculate and allocate nutrient load reductions to the Baltic Sea between different basins and countries on the Baltic coast. These reductions are transformed into country-specific Maximum Allowable Inputs for the Baltic Sea Action Plan. Unlike the Chesapeake Bay TMDL load allocations to tributaries and sectors, the Baltic Sea Action Plan allocations are not mandated.

A third illustrative application of interest is the Overseer model in New Zealand (Overseer 2020). The model was developed for pollution planning but is particularly notable for its use to quantify nitrogen load reductions from farms in the Lake Taupo watershed. The watershed is the location of the only cap-and-trade nutrient trading program applied to agricultural nonpoint sources (Chap. 6). The program imposes a cap on nitrogen loads from agriculture and uses pollution trading to allocate load reductions across farms. The Overseer model is used to quantify load reductions.

3.7.2 Biophysical Model Uncertainty and Complexity

While biophysical models are essential for effective and efficient agricultural pollution management, it is important to recognize their limitations. We highlight two concerns especially relevant to their use for policy design. One is uncertainty. Models are needed because nonpoint pollution involves highly complex relationships between land use practices and water quality outcomes. If models could predict the water quality consequences of field- and farm-specific controls with certainty, the nonpoint management problem would be greatly simplified as the significance of individual nonpoint sources to pollution loads would be known with certainty. Specifically, the effects of changes in farming practices could be predicted without error. But this is not the case. Prediction errors from water quality models are substantial (Arabi et al. 2012; Beven 2013, 2016; Loucks and van Beek 2017; NRC 2001; Osmond et al. 2012; Testa et al. 2017). Even the most sophisticated water quality models are simplified representations of the complex processes of pollution formation, transport, fate, and all are limited by data availability and quality. Further, all are based on theories of these processes. Uncertain conceptual knowledge about these processes, imperfections in the representation of conceptual in models, and data limitations will result in prediction errors in the best of circumstances. Implementation and interpretation errors can be another source of prediction error (Beven 2005, 2006; Moriasi et al. 2016).

In general, simple models are beneficial for regulatory purposes because they reduce regulatory burdens and costs related to data requirements and acquisition, computation, modeling expertise, and translation of complex relationships into policy design and administration. The appeal of more sophisticated complex models is the expectation of greater accuracy and reliability, greater likelihood of success in achieving water quality targets, and therefore greater program and policy credibility. However, the presumption that more sophisticated models are more accurate is not always warranted (Loucks and van Beek 2017). As we proceed through the book, we develop criteria for models that are not perfect but useful in that they facilitate cost-effective solutions without excessive regulator costs.

3.7.3 Economic Models and Integrated Assessment Models

Like biophysical models, economic models are used at multiple scales for planning, policy design, and evaluation. And like biophysical models, useful model types depend on the question. When economic models are linked to biophysical models, the combination is an integrated assessment model. As described in Chap. 2, economic models may be used for *ex ante* or *ex post* policy analysis.

A common use of models is to estimate the costs of pollution control activities. The simplest models are accounting or budgeting frameworks to estimate the costs of individual BMPs based on technical specifications, installation costs, operating and maintenance costs, and depreciation rates.

More complex models are developed for determining the costs of managing nutrients, sediments, and pests at field and farm scales, models for determining the costs of whole-farm plans, and models for determining the costs of watershed plans the cover multiple sources.

Models used for evaluating costs at small to large scales are sometimes simple accounting models which feed data on agricultural practice scenarios into equations that calculate costs. An example is the Chesapeake Assessment Scenario Tool (CAST). The model enables analysts to estimate the cost and water quality impacts of individual or suites of BMPs implemented within defined planning areas. Models of this type are essentially spreadsheets. More sophisticated are models that can select suites of practices to minimize the costs of achieving management targets. Models of this type are implemented using optimization routines (i.e., computer programs that select the best combinations from a suite of alternatives subject to environmental performance requirements). For example, a field-scale optimization model would select field-scale management practices to minimize the costs of an edge-of-field target load. A whole-farm model would choose practices for each field in a farm to minimize the cost of a limit on edge-of-field emissions of a farm. Watershed-scale models would select farming practices distributed spatially across a watershed to minimize the cost of watershed load targets. Example of models of this type are presented in the Appendix to Chap. 5. The applications examine watershed management at multiple scales for reducing pollution loads to the Chesapeake Bay.

Economic and integrated models can alternatively be used to model behavioral responses to water quality policy interventions. Behavioral models are designed to estimate how a producer or set of producers will respond to various types of interventions. Simple illustrative models of this type are presented in Chap. 4 and subsequent chapters. If the scale of policy intervention is large enough that the agricultural response will affect the prices of agricultural products or inputs like land, economic models should include models of the affected markets to determine price changes and their effects on resource allocation. An illustration of an integrated economic-biophysical model of this type is presented in Chap. 5. The application considers the costs of achieving targeted nutrient reductions to the Gulf of Mexico from the very large Mississippi watershed in the US. The region is a large share of the agricultural commodity supplies in some markets, with the result that prices in these markets are affected by the changes in agricultural production.

Accounting and optimization models are typically "representative farm models." Representative farm models are models of hypothetical producers who are assumed to be profit maximizers. The options for farm outputs (crops, livestock) farming practices (rotations, tillage, nutrient, and pest management practices, livestock feeding), land and capital resource, and marketing practices are selected by the modeler to be typical or representative of farming practices in the location of the application. Technical information is often derived from the results of agricultural researchers. We use simple representative farm models to develop concepts of farm decision making in Chap. 4. While this approach can be very useful for estimating social costs and gaining insights about the cost and benefits to producers from policy interventions, representative farm models cannot capture the true complexity and heterogeneity of farms and producer behavior (e.g., Fleming et al. 2018). As we discussed briefly in Chap. 2 in relation to *ex ante* versus *ex post* policy analysis, differences in predicted and actual outcomes of pollution policy interventions are routine.

Indicative of the difficulty of predictive modeling of complex behavioral responses to pollution control interventions in agriculture is the fact that economic models for farms and watersheds are almost always special-built for the specific problem. This is in contrast to biophysical models where there are standard off-the-shelf modeling options (e.g., USLE, SWAT, and other models listed in Table 4.1). Economic complexity and heterogeneity are simply too large. State-of-the-art model applications explicitly recognize uncertainty and use various techniques to understand the implications of data limitations, and uncertainty about technology, prices, behavioral objectives, and other determinants of economic choices.

Kling et al. (2017) provide a comprehensive review of existing integrated models applied for food, energy, and water systems. They also list avenues for improving the accuracy and usefulness of the models. For the biophysical parts, they emphasize the modelling of loading in tile drains, sedimentation during the riverine transport and wetlands and the impacts of active conservation practices on pollutants. There are also certain crops, like bioenergy crops, for which basic modelling (nutrient responses, loading characteristics) needs to be improved. The accuracy of the impacts of conservation practices on crop yields could be improved. The economic representation of farm choices should be improved from simple static behavior into more refined behavioral models. In addition to increasing the accuracy of individual model components, they emphasize the need to improve the integration of economic decision-making models with crop yield and water quality models.

Various methods exist to learn about how actual as opposed to hypothetical producers behave and to develop corresponding predictive models. One technique involves the use of experimental economics in which policy experiments are used to learn about behavior. Experiments place subjects, who are often university students but are sometimes producers, in decision-making frameworks that mimic the decision problems that would be created by a policy intervention. Some relevant applications of experimental economics are to develop messaging to increase producer participation in conservation programs (Chap. 5), to test how they respond to complex pollution control incentives (Chap. 5), and to test how auction formats affect producer behavior in conservation or water quality auctions (Chap. 7). Interviews and sample surveys of producers can be used to learn about choices made in response to actual interventions or about how choices might be made in response to anticipated interventions. There is extensive research examining determinants of farm conservation practice adoption based on survey research. Data sets on farm practices developed through survey are also used to conduct statistical analyses and develop predictive models.

3.8 Summary

This chapter describes the primary types of water quality problems that result from agriculture land, the significance of these problems, the pollutants that cause them, and the farming practices that produce them. It describes macro-scale economic drivers that determine the scale and composition of agricultural output, and the methods of agricultural production. Though agricultural problems are pervasive, solutions are optimally place-based. The chapter introduced an essential ingredient of place-based solutions, which is a watershed-based approach. Essential to implementation of this approach are data and models for characterizing problems, their sources, and analyzing policy interventions. The chapter provides a brief introduction to model types, purposes, and limitations.

The chapter provides foundational material for subsequent chapters. Water pollutant problems are complex and the impacts of agriculture are multi-dimensional. For policy making we must be able to quantify the links between the actions and the outcomes. This means that all parts of the process need modelling of some precision: economic decision making, crop yields, loading from the farms, delivery processes, water quality impacts, and the economic significance of these impacts. This is the arena in which agricultural policy making needs to operate. The following chapters will describe in greater depth economic decision making, the basic tools and instruments that can be used to influence these choices and novel instruments such as water quality trading and auctions. To assure economic efficiency and ecological integrity any instrument must be evaluated and designed keeping in mind the biophysical operational environment of agriculture.

References

Ahlvik, L. and Iho, A. 2018. Optimal geoengineering experiments. *Journal of Environmental Economics and Management* 92: 148–168.

Ahtiainen, H., J. Artell, M. Czajkowski, B. Hasler, L. Hasselström, K. Hyytiäinen, J. Meyerhoff, J.C. Smart, T. Söderqvist, K. Zimmer, and J. Khaleeva. 2013. Public preferences regarding use and condition of the Baltic Sea—An international comparison informing marine policy. *Marine Policy* 42: 20–30.

Alexander, R.B., R.A. Smith, G.E. Schwarz, E.W. Boyer, J.V. Nolan, and J.W. Brakebill. 2008. Differences in phosphorus and nitrogen delivery to the Gulf of Mexico from the Mississippi River Basin. *Environmental Science & Technology* 42 (3): 822–830.

Aloe, A.K., F. Bouraoui, B. Grizzetti, G. Bidoglio, and A. Pistocchi. 2014. Managing nitrogen and phosphorus loads to water bodies: Characterisation and solutions. *Towards macro-regional integrated nutrient management.* Joint Research Centre, JRC-Ispra.

Arabi, M., Meals, D.W. and Hoag, D.L.K. 2012. Watershed modelling: National institute of food and agriculture – conservation effects assessment project. In D.L. Osmond, D.W. Meals, D.L.K. Hoag, and M. Arabi, eds., *How to build better agricultural conservation programs to protect water quality: National Institute of Food and Agriculture conservation effects assessment project* (pp. 84–119). Ankeny, IA: Soil Conservation Society of America.

Arnold, J.G., R. Srinivasan, R.S. Muttiah, and J.R. Williams. 1998. Large area hydrologic modeling and assessment part I: model development 1. *JAWRA Journal of the American Water Resources Association 34* (1): 73–89.

Ator, Scott W., John W. Brakebill, and Joel D. Blomquist. 2011. *Sources, fate, and transport of nitrogen and phosphorus in the Chesapeake Bay watershed: An empirical model.* Vol. 5167. US Department of the Interior, US Geological Survey.

Baker, D.B., Confesor, R., Ewing, D.E., Johnson, L.T., Kramer, J.W. and Merryfield, B.J. 2014. Phosphorus loading to Lake Erie from the Maumee, Sandusky and Cuyahoga rivers: The importance of bioavailability. *Journal of Great Lakes Research* 40 (3): 502–517.

Barlow, P.M., and E.G. Reichard. 2010. Saltwater intrusion in coastal regions of North America. *Hydrogeology Journal* 18 (1): 247–260.
Beven, K. 2005. On the concept of model structural error. *Water Science and Technology* 52 (6): 167–175.
———. 2006. A manifesto for the equifinality thesis. *Journal of Hydrology* 320 (1–2): 18–36.
———. 2016. Facets of uncertainty: Epistemic uncertainty, non-stationarity, likelihood, hypothesis testing, and communication. *Hydrological Sciences Journal* 61 (9): 1652–1665.
Bicknell, B.R., J.C. Imhoff, J.L. Kittle Jr, A.S. Donigian Jr, and R.C. Johanson. 1993. Hydrologic simulation program-FORTRAN (HSPF): User's Manual for Release 10. Rep. No. EPA/600/R-93, 174.
Billen, G., J. Garnier, and L. Lassaletta. 2013. The nitrogen cascade from agricultural soils to the sea: Modelling nitrogen transfers at regional watershed and global scales. *Philosophical Transactions of the Royal Society B: Biological Sciences* 368 (1621): 20130123.
Bouwman, L., K.K. Goldewijk, K.W. Van Der Hoek, A.H. Beusen, D.P. Van Vuuren, J. Willems, M.C. Rufino, and E. Stehfest. 2013. Exploring global changes in nitrogen and phosphorus cycles in agriculture induced by livestock production over the 1900–2050 period. *Proceedings of the National Academy of Sciences* 110 (52): 20882–20887.
BP. 2020. Statistical review of world energy 2020. 69th edn. Available at https://www.bp.com/content/dam/bp/business-sites/en/global/corporate/pdfs/energy-economics/statistical-review/bp-stats-review-2020-full-report.pdf
Breitburg, D., Levin, L.A., Oschlies, A., Grégoire, M., Chavez, F.P., Conley, D.J., Garçon, V., Gilbert, D., Gutiérrez, D., Isensee, K. and Jacinto, G.S. 2018. Declining oxygen in the global ocean and coastal waters. *Science* 359 (6371).
Brownell, H., and S.D. Eaton. 1975. The Colorado River salinity problem with Mexico. *The American Journal of International Law* 69 (2): 255–271.
Canfield, D., E. Kristensen, and B. Thamdrup. 2005. *Aquatic geomicrobiology*. Elsevier.
Cao, P., C.C. Lu, and Z. Yu. 2018. Historical nitrogen fertilizer use in agricultural ecosystems of the contiguous United States during 1850–2015: Application rate, timing, and fertilizer types. *Earth System Science Data Discussion* 10: 969.
Carpenter, S.R. 2008. Phosphorus control is critical to mitigating eutrophication. *Proceedings of the National Academy of Sciences* 105 (32): 11039–11040.
Carpenter, S.R., and E.M. Bennett. 2011. Reconsideration of the planetary boundary for phosphorus. *Environmental Research Letters* 6 (1): 014009.
Carpenter, S.R., J.J. Cole, J.R. Hodgson, J.F. Kitchell, M.L. Pace, D. Bade, K.L. Cottingham, T.E. Essington, J.N. Houser, and D.E. Schindler. 2001. Trophic cascades, nutrients, and lake productivity: Whole-lake experiments. *Ecological Monographs* 71 (2): 163–186.

Ceballos, G., and P.R. Ehrlich. 2002. Mammal population losses and the extinction crisis. *Science* 296 (5569): 904–907.

Champetier, A., and D.A. Sumner. 2019. Marginal costs and likely supply elasticities for pollination and honey. *American Journal of Agricultural Economics* 101 (5): 1373–1385.

Chesapeake Bay Program. 2017. Water quality standards attainment and monitoring 2020. Available at https://www.chesapeakeprogress.com/clean-water/water-quality. Retrieved August 18, 2020.

Chesapeake Bay Program (CBP). 2020. Water Quality Standards Attainment and Monitoring. Published online at Chesapeake Progress. Retrieved from: https://www.chesapeakeprogress.com/?/clean-water/water-quality [online resource]

Clearwater, R.L., T. Martin, and T. Hoppe, eds. 2016. *Environmental sustainability of Canadian agriculture: Agri-environmental indicator report series—Report #4.* Ottawa: Agriculture and Agri-Food Canada.

Conley, D.J. 1999. Biogeochemical nutrient cycles and nutrient management strategies. In *Man and River Systems*, 87–96. Dordrecht: Springer.

De Vries, W., J. Kros, C. Kroeze, and S.P. Seitzinger. 2013. Assessing planetary and regional nitrogen boundaries related to food security and adverse environmental impacts. *Current Opinion in Environmental Sustainability* 5 (3–4): 392–402.

Deliman, P.N., R.H. Glick, and C.E. Ruiz. 1999. *Review of watershed water quality models. TR W-99-1. U.S.* Washington, DC: Army Corps of Engineers.

Diaz, R.J., and R. Rosenberg. 2008. Spreading dead zones and consequences for marine ecosystems. *Science* 321 (5891): 926–929.

Dodd, R.J. and Sharpley, A.N. 2016. Conservation practice effectiveness and adoption: Unintended consequences and implications for sustainable phosphorus management. *Nutrient cycling in agroecosystems* 104 (3): 373–392.

Dodd, Jason R., and Antonio P. Mallarino. 2005. Soil-test phosphorus and crop grain yield responses to long-term phosphorus fertilization for corn-soybean rotations. *Soil Science Society of America Journal* 69 (4): 1118–1128.

Dodds, W.K., W.W. Bouska, J.L. Eitzmann, T.J. Pilger, K.L. Pitts, A.J. Riley, J.T. Schloesser, and D.J. Thornbrugh. 2009. Eutrophication of US freshwaters: Analysis of potential economic damages. *Environmental Science & Technology* 43 (1): 12–19.

Döll, P. 2002. Impact of climate change and variability on irrigation requirements: A global perspective. *Climatic Change* 54 (3): 269–293.

Dudley, N., and S. Alexander. 2017. Agriculture and biodiversity: A review. *Biodiversity* 18 (2–3): 45–49.

Dunlap, C.R., K.S. Sklenar, and L.J. Blake. 2015. A costly endeavor: Addressing algae problems in a water supply. *Journal-American Water Works Association* 107 (5): E255–E262.

Einarsson, R., D. Pitulia, and C. Cederberg. 2020. Subnational nutrient budgets to monitor environmental risks in EU agriculture: Calculating phosphorus budgets for 243 EU28 regions using public data. *Nutrient Cycling in Agroecosystems*: 1–15.

Ekholm, P. and Lehtoranta, J., 2012. Does control of soil erosion inhibit aquatic eutrophication? *Journal of environmental management* 93 (1): 140–146.

European Commission. 2012. A blueprint to safeguard Europe's water resources. Communication from the Commission to the European Parliament, the Council, the European Economic and Social Committee and the Committee of the Regions.

FAO. 2014. Area equipped for irrigation (infographic). AQUASTAT. Database. Rome. http://www.fao.org/nr/water/aquastat/infographics/Irrigation_eng.pdf. Retrieved September 2020.

Faunt, C.C., M. Sneed, J. Traum, and J.T. Brandt. 2016. Water availability and land subsidence in the Central Valley, California, USA. *Hydrogeology Journal* 24 (3): 675–684.

Flanagan, D.C., and M.A. Nearing. 1995. USDA-Water Erosion Prediction Project: Hillslope profile and watershed model documentation. *Nserl Rep* 10: 1–123.

Fleming, Patrick, Erik Lichtenberg, and David A. Newburn. 2018. Evaluating impacts of agricultural cost sharing on water quality: Additionality, crowding In, and slippage. *Journal of Environmental Economics and Management* 92: 1–19.

Forrest, R.D. 1982. Early history of wound treatment. *Journal of the Royal Society of Medicine* 75 (3): 198.

Gallai, N., J.M. Salles, J. Settele, and B.E. Vaissière. 2009. Economic valuation of the vulnerability of world agriculture confronted with pollinator decline. *Ecological Economics* 68 (3): 810–821.

Geissen, V., H. Mol, E. Klumpp, G. Umlauf, M. Nadal, M. van der Ploeg, S.E. van de Zee, and C.J. Ritsema. 2015. Emerging pollutants in the environment: A challenge for water resource management. *International Soil and Water Conservation Research* 3 (1): 57–65.

Goldhamer, D.A., and E. Fereres. 2017. Establishing an almond water production function for California using long-term yield response to variable irrigation. *Irrigation Science* 35 (3): 169–179.

Goldstein, R.A. 2001. *Watershed analysis risk management framework (WARMF): Update one*. Topical Report. EPRI, Palo Alto.

Goulson, D. 2014. Pesticides linked to bird declines. *Nature* 511 (7509): 295–296.

Grafton, R.Q., G.D. Libecap, E.C. Edwards, R.J. O'Brien, and C. Landry. 2012. Comparative assessment of water markets: Insights from the Murray–Darling Basin of Australia and the Western USA. *Water Policy* 14 (2): 175–193.

Gren, M. 2001. International versus national actions against nitrogen pollution of the Baltic Sea. *Environmental and Resource Economics* 20 (1): 41–59.

Gunnell, D., and M. Eddleston. 2003. Suicide by intentional ingestion of pesticides: A continuing tragedy in developing countries. *International Journal of Epidemiology* 32 (6): 902–909.

Haith, D.A., and L.L. Shoemaker. 1987. Generalized watershed loading functions for stream flow nutrients 1. *JAWRA Journal of the American Water Resources Association 23* (3): 471–478.

Hamdan, M.N., M.J. Post, M.A. Ramli, and A.R. Mustafa. 2018. Cultured meat an Islamic perspective. *Journal of Religion & Health 57* (6): 2193–2206.

Hart, B.T., P. Bailey, R. Edwards, K. Hortle, K. James, A. McMahon, C. Meredith, and K. Swadling. 1991. A review of the salt sensitivity of the Australian freshwater biota. *Hydrobiologia* 210 (1–2): 105–144.

Harter, T., and J. Lund. 2012. *Addressing nitrate in California's drinking water: Executive summary.* Report, UC Davis. Available at http://groundwaternitrate.ucdavis.edu/files/138958.pdf

Harvey, J.W. 2016. Hydrologic exchange flows and their ecological consequences in river corridors. In *Stream ecosystems in a changing environment*, 1–83. Academic Press.

Helander, B., A. Bignert, and L. Asplund. 2008. Using raptors as environmental sentinels: Monitoring the white-tailed sea eagle Haliaeetus albicilla in Sweden. *Ambio: A Journal of the Human Environment* 37 (6): 425–431.

HELCOM. 2018. HELCOM Thematic assessment of eutrophication 2011–2016. Baltic Sea Environment Proceedings No. 156. Available at http://stateofthebalticsea.helcom.fi/pressures-and-their-status/eutrophication/#core-indicators-on-direct-effects

Henley, W.F., M.A. Patterson, R.J. Neves, and A.D. Lemly. 2000. Effects of sedimentation and turbidity on lotic food webs: A concise review for natural resource managers. *Reviews in Fisheries Science* 8 (2): 125–139.

Ho, J.C., and A.M. Michalak. 2017. Phytoplankton blooms in Lake Erie impacted by both long-term and springtime phosphorus loading. *Journal of Great Lakes Research* 43 (3): 221–228.

Howarth, R.W. 1988. Nutrient limitation of net primary production in marine ecosystems. *Annual Review of Ecology and Systematics* 19 (1): 89–110.

Huffman, W.E., and R.E. Evenson. 2001. Structural and productivity change in US agriculture, 1950–1982. *Agricultural Economics* 24 (2): 127–147.

Hyytiäinen, K., B. Hasler, S. Ericsdotter, M. Nekoro, K. Blyh, J. Artell, L. Ahlvik, and H. Ahtiainen. 2013. Worth it: Benefits outweigh costs in reducing eutrophication in the Baltic. In *BalticSTERN summary report for HELCOM 2013 ministerial meeting.* Helsinki: Helsinki Commission. [Online]: http://helcom.fi/Documents/Ministerial2013/Associated%20documents/Background/BalticSternReport.pdf.

Ireland, R.L. 1983. Land subsidence in the San Joaquin Valley, California, as of 1983. Water-resources investigations report 85-4196. U.S. Geological Survey.

Available at https://pubs.usgs.gov/wri/1985/4196/report.pdf. Retrieved September 2020.

Jarvie, H.P., L.T. Johnson, A.N. Sharpley, D.R. Smith, D.B. Baker, T.W. Bruulsema, and R. Confesor. 2017. Increased soluble phosphorus loads to Lake Erie: Unintended consequences of conservation practices? *Journal of Environmental Quality* 46 (1): 123–132.

Kadim, I.T., O. Mahgoub, S. Baqir, B. Faye, and R. Purchas. 2015. Cultured meat from muscle stem cells: A review of challenges and prospects. *Journal of Integrative Agriculture* 14 (2): 222–233.

Kahru, M., R. Elmgren, J. Kaiser, N. Wasmund, and O. Savchuk. 2020. Cyanobacterial blooms in the Baltic Sea: Correlations with environmental factors. *Harmful Algae* 92: 101739.

Keeler, B.L., J.D. Gourevitch, S. Polasky, F. Isbell, C.W. Tessum, J.D. Hill, and J.D. Marshall. 2016. The social costs of nitrogen. *Science Advances* 2 (10): e1600219.

Kirkby, M.J., and N.J. Cox. 1995. A climatic index for soil erosion potential (CSEP) including seasonal and vegetation factors. *Catena* 25 (1–4): 333–352.

Kling, C.L., R.W. Arritt, G. Calhoun, and D.A. Keiser. 2017. Integrated assessment models of the food, energy, and water nexus: A review and an outline of research needs. *Annual Review of Resource Economics* 9: 143–163.

Knisel, W.G. 1980. CREAMS: A field scale model for chemicals, runoff, and erosion from agricultural management systems (No. 26). Department of Agriculture, Science and Education Administration.

———. 1991. CREAMS/GLEAMS: A development overview. Proceedings of the CREAMS/GLEAMS Symposium, Publication (Vol. 4, pp. 9–17).

Kosenius, A.K. 2010. Heterogeneous preferences for water quality attributes: The case of eutrophication in the Gulf of Finland, the Baltic Sea. *Ecological Economics* 69 (3): 528–538.

Lake Erie LaMP. 2019. Environment and Climate Change Canada and the U.S. Environmental Protection Agency. Lake Erie Lakewide Action and Management Plan, 2019–2023.

Lee, J.A., and T.E. Gill. 2015. Multiple causes of wind erosion in the Dust Bowl. *Aeolian Research* 19: 15–36.

Lekagul, A., V. Tangcharoensathien, and S. Yeung. 2019. Patterns of antibiotic use in global pig production: A systematic review. *Veterinary and Animal Science* 7: 100058.

Levin, M.D. 1983. Value of bee pollination to US agriculture. *American Entomologist* 29 (4): 50–51.

Loucks, D.P. and van Beek, E. 2017. Water resources planning and management: An overview. Water Resource Systems Planning and Management, pp.1–49.

Ludena, C.E., T.W. Hertel, P.V. Preckel, K. Foster, and A. Nin. 2007. Productivity growth and convergence in crop, ruminant, and nonruminant production: Measurement and forecasts. *Agricultural Economics* 37 (1): 1–17.

Maccoux, M.J., A. Dove, S.M. Backus, and D.M. Dolan. 2016. Total and soluble reactive phosphorus loadings to Lake Erie: A detailed accounting by year, basin, country, and tributary. *Journal of Great Lakes Research* 42 (6): 1151–1165.

MacDonald, J.M. 2020. Tracking the consolidation of U.S. agriculture. *Applied Economic Perspectives and Policy* 00 (00): 1–19.

Mahmoud, M.R., and L.A. Garcia. 2000. Comparison of different multicriteria evaluation methods for the red bluff diversion dam. *Environmental Modelling & Software* 15 (5): 471–478.

Mateo-Sagasta, J., and J. Burke. 2010. *State of land and water (SOLAW). Background report on water quality and agriculture interactions, a global overview*. Rome: FAO.

Mateo-Sagasta, J., S.M. Zadeh, and H. Turral, eds. 2018. *More people, more food, worse water? A global review of water pollution from agriculture*. Rome/Colombo: FAO/International Water Management Institute (IWMI). CGIAR Research Program on Water, Land and Ecosystems (WLE).

Micklin, P., N.V. Aladin, T. Chida, N. Boroffka, I.S. Plotnikov, S. Krivonogov, and K. White. 2020. The Aral Sea: A story of devastation and partial recovery of a large Lake. In *Large Asian Lakes in a Changing World*, 109–141. Cham: Springer.

Miller, S.N., D.J. Semmens, D.C. Goodrich, M. Hernandez, R.C. Miller, W.G. Kepner, and D.P. Guertin. 2007. The automated geospatial watershed assessment tool. *Environmental Modelling & Software 22* (3): 365–377.

Moore, C., D. Guignet, K. Maguire, C. Dockins, and N. Simon. 2015. *A stated preference study of the Chesapeake Bay and Watershed Lakes* (No. 2168-2018-8163).

Morgan, R.P.C., J.N. Quinton, R.E. Smith, G. Govers, J.W.A. Poesen, G. Chisci, and D. Torri. 1998. The EUROSEM model. In *Modelling Soil Erosion by Water*, 389–398. Berlin, Heidelberg: Springer.

Moriasi, D.N., King, K.W., Bosch, D.D., Bjorneberg, D.L., Teet, S., Guzman, J.A. and Williams, M.R. 2016. Framework to parameterize and validate APEX to support deployment of the nutrient tracking tool. *Agricultural Water Management* 177: 146–164.

Mottet, A., C. de Haan, A. Falcucci, G. Tempio, C. Opio, and P. Gerber. 2017. Livestock: On our plates or eating at our table? A new analysis of the feed/food debate. *Global Food Security* 14: 1–8.

Mund, M.D., U.H. Khan, U. Tahir, B.E. Mustafa, and A. Fayyaz. 2017. Antimicrobial drug residues in poultry products and implications on public health: A review. *International Journal of Food Properties* 20 (7): 1433–1446.

National Research Council (NRC). 2001. *Assessing the TMDL approach to water quality management*. National Academies Press.

———. 2012. *National Research Council, 2012. Science for environmental protection: the road ahead*. National Academies Press.

Niskanen, O., A. Iho, and L. Kalliovirta. 2020. Scenario for structural development of livestock production in the Baltic littoral countries. *Agricultural Systems* 179: 102771.

OECD. 2013. *OECD compendium of Agri-environmental indicators*. Paris: OECD Publishing. https://doi.org/10.1787/9789264186217-en.

———. 2017. *Diffuse pollution, Degraded waters: Emerging policy solutions. OECD studies on water*. Paris: OECD Publishing. https://doi.org/10.1787/9789264269064-en.

Ohio, E.P.A. 2010. *Ohio Lake Erie phosphorus task force final report*. Ohio EPA OH Task Force.

Ollikainen, M., and J. Honkatukia. 2001. Towards efficient pollution control in the Baltic Sea: An anatomy of current failure with suggestions for change. *Ambio: A Journal of the Human Environment* 30 (4): 245–253.

Osmond, D., D. Meals, D. Hoag, M. Arabi, and A. Luloff. 2012. Improving conservation practices program to protect water quality in agricultural watersheds: Lessons learned from the National Institute of Food and Agriculture–Conservation Effects Assessment Project. *Journal of Soil and Water Conservation* 67 (5): 122A–127A.

Overseer. 2020. Overseer scientific model. Available at https://www.overseer.org.nz/our-model. Retrieved September 2020.

Paerl, H.W. 2009. Controlling eutrophication along the freshwater–marine continuum: Dual nutrient (N and P) reductions are essential. *Estuaries and Coasts* 32 (4): 593–601.

Pavela, R. 2014. Limitation of plant biopesticides. In *Advances in plant biopesticides*, 347–359. New Delhi: Springer.

Playán, E., and L. Mateos. 2006. Modernization and optimization of irrigation systems to increase water productivity. *Agricultural Water Management* 80 (1–3): 100–116.

Post, M.J. 2012. Review: Cultured meat from stem cells: Challenges and prospects. *Meat Science* 92 (3): 297–301.

Pote, D.H., Daniel, T.C., Moore Jr, P.A., Nichols, D.J., Sharpley, A.N. and Edwards, D.R. 1996. Relating extractable soil phosphorus to phosphorus losses in runoff. *Soil Science Society of America Journal* 60 (3): 855–859.

Pretty, J.N., C.F. Mason, D.B. Nedwell, R.E. Hine, S. Leaf, and R. Dils. 2003. Environmental costs of freshwater eutrophication in England and Wales. *Environmental Science & Technology* 37: 201–208.

Pufahl, P.K., and L.A. Groat. 2017. Sedimentary and igneous phosphate deposits: Formation and exploration: An invited paper. *Economic Geology* 112 (3): 483–516.

Ritchie, H. 2017. Meat and dairy production. Published online at OurWorldInData.org. Retrieved from: https://ourworldindata.org/meat-production [online resource].

———. 2019. "Half of the world's habitable land is used for agriculture". Published online at OurWorldInData.org. Retrieved from: https://ourworldindata.org/global-land-for-agriculture [Online Resource]

Roberson, J.A. 2014. A tough anniversary year. *Journal-American Water Works Association* 106 (11): 18–23.

Rockström, J., W. Steffen, K. Noone, Å. Persson, F.S. Chapin III, E. Lambin, T.M. Lenton, M. Scheffer, C. Folke, H.J. Schellnhuber, and B. Nykvist. 2009. Planetary boundaries: Exploring the safe operating space for humanity. *Ecology and Society* 14 (2).

Rolland, N.C.M., C.R. Markus, and M.J. Post. 2020. The effect of information content on acceptance of cultured meat in a tasting context. *PLOS ONE 15* (4): e0231176.

Ronquillo, M.G., and J.C.A. Hernandez. 2017. Antibiotic and synthetic growth promoters in animal diets: Review of impact and analytical methods. *Food Control* 72: 255–267.

Roser, M., and H. Ritchie. 2013 Hunger and undernourishment. Published online at OurWorldInData.org. Retrieved from: https://ourworldindata.org/hunger-and-undernourishment [online resource].

Savchuk, O.P. 2018. Large-scale nutrient dynamics in the Baltic Sea, 1970–2016. *Frontiers in Marine Science* 5: 95.

Schechner, V., E. Temkin, S. Harbarth, Y. Carmeli, and M.J. Schwaber. 2013. Epidemiological interpretation of studies examining the effect of antibiotic usage on resistance. *Clinical Microbiology Reviews* 26 (2): 289–307.

Schmidtko, S., L. Stramma, and M. Visbeck. 2017. Decline in global oceanic oxygen content during the past five decades. *Nature* 542 (7641): 335–339.

Selman, M., and S. Greenhalgh. 2010. Eutrophication: Sources and drivers of nutrient pollution. *Renewable Resources Journal* 26 (4): 19–26.

Seré, C., H. Steinfeld, and J. Groenewold. 1996. *World livestock production systems*. Food and Agriculture Organization of the United Nations.

Sharpley, Andrew. 2000. Phosphorus availability. *Handbook of Soil Science*: D18–D30.

Sharpley, A., H.P. Jarvie, A. Buda, L. May, B. Spears, and P. Kleinman. 2013. Phosphorus legacy: Overcoming the effects of past management practices to mitigate future water quality impairment. *Journal of Environmental Quality* 42 (5): 1308–1326.

Shoemaker, L., Dai, T., Koenig, J. and Hantush, M. 2005. *TMDL model evaluation and research needs*. Cincinnati OH: National Risk Management Research Laboratory, US Environmental Protection Agency.

Shortle, J.S., and R.C. Griffin. 2001. Economic perspectives on irrigated agriculture and the environment. In *Irrigated agriculture, and the environment*, ed. J.S. Shortle and R.C. Griffin. Edward Elgar Publishing Ltd.

Shortle, J. and Horan, R.D. 2017. Nutrient pollution: A wicked challenge for economic instruments. *Water Economics and Policy* 3 (02): p.1650033.

Shortle, J.S., Mihelcic, J.R., Zhang, Q. and Arabi, M. 2020. Nutrient control in water bodies: A systems approach. *Journal of Environmental Quality* 49 (3): 517–533.

Smolders, A., and J.G.M. Roelofs. 1993. Sulphate-mediated iron limitation and eutrophication in aquatic ecosystems. *Aquatic Botany* 46 (3–4): 247–253.

Southwick, E.E., and L. Southwick Jr. 1992. Estimating the economic value of honeybees (Hymenoptera: Apidae) as agricultural pollinators in the United States. *Journal of Economic Entomology* 85 (3): 621–633.

Steffen, W., K. Richardson, J. Rockström, S.E. Cornell, I. Fetzer, E.M. Bennett, R. Biggs, S.R. Carpenter, W. De Vries, C.A. De Wit, and C. Folke. 2015. Planetary boundaries: Guiding human development on a changing planet. *Science* 347 (6223).

Stone, W.W., R.J. Gilliom, and K.R. Ryberg. 2014. Pesticides in US streams and rivers: Occurrence and trends during 1992–2011. *Environ. Sci. Technol.* 48 (19): 11025–11030.

Sutton, M.A., A. Bleeker, C.M. Howard, J.W. Erisman, Y.P. Abrol, M. Bekunda, A. Datta, E. Davidson, W. De Vries, O. Oenema, and F.S. Zhang. 2013. *Our nutrient world, The challenge to produce more food & energy with less pollution*. Centre for Ecology & Hydrology.

Testa, J.M., Clark, J.B., Dennison, W.C., Donovan, E.C., Fisher, A.W., Ni, W., Parker, M., Scavia, D., Spitzer, S.E., Waldrop, A.M. and Vargas, V.M. 2017. Ecological forecasting and the science of hypoxia in Chesapeake Bay. *BioScience* 67(7): 614–626.

Toledo. 2020. *A watertight plan*. Ohio: City of Toledo. Available at http://toledoh2o.com/the-plan/. Retrieved September 2020.

Tuomisto, H., and M.J.T. Teixeira de Mattos. 2011. Environmental impacts of cultured meat production. *Environmental Science and Technology* 46 (14): 6117–6123.

United Nations. 2019. *World population prospects 2019*. Department of Economic and Social Affairs. Available at https://population.un.org/wpp/Graphs/DemographicProfiles/Line/900.

USBR. 2015. *Shasta Lake water resources investigation, feasibility report*. U.S. Department of the Interior, Bureau of Reclamation. Available at https://www.usbr.gov/mp/ncao/slwri/docs/feasability/slwri-final-fr-full.pdf. Retrieved September 2020.

USDA. 2017. Cost of pollination. Available at https://downloads.usda.library.cornell.edu/usda-esmis/files/d504rk335/ht24wn48h/zg64tp76q/CostPoll-12-21-2017.pdf. Retrieved September 2020.

———. 2019. USDA ERS data. *Fertilizer Use and Price*. Available at https://www.ers.usda.gov/data-products/fertilizer-use-and-price.aspx. Retrieved August 2020.

———. 2020. USDA ERS. U.S. bioenergy statistics. Available at https://www.ers.usda.gov/data-products/us-bioenergy-statistics/. Retrieved August 2020.

USEPA, 2016. *National Rivers and Streams Assessment 2008–2009: A Collaborative Survey*. Washington, DC: U.S. Environmental Protection Agency, Office of Water and Office of Research and Development, EPA/841/R-16/007, March.

USGS, 2020. U.S. Geological Survey, Mineral Commodity Summaries. 2020. Phosphate Rock. Available at https://pubs.usgs.gov/periodicals/mcs2020/mcs2020-phosphate.pdf. Retrieved August 2020.

van Dijk, K.C., J.P. Lesschen, and O. Oenema. 2016. Phosphorus flows and balances of the European Union member states. *Science of the Total Environment* 542: 1078–1093.

van Ittersum, M.K., and K.G. Cassman. 2013. Yield gap analysis—Rationale, methods, and applications—Introduction to the special issue. *Field Crops Research* 143: 1–3.

van Kauwenbergh, S.J., M. Stewart, and R. Mikkelsen. 2013. World reserves of phosphate rock… a dynamic and unfolding story. *Better Crops* 97 (3): 18–20.

van Weert, F., J. Van der Gun, and J. Reckman. 2009. *Global overview of saline groundwater occurrence and genesis*. International Groundwater Resources Assessment Centre.

Vieno, N., P. Hallgren, P. Wallberg, M. Pyhälä, S. Zandaryaa, and Baltic Marine Environment Protection Commission. 2017. *Pharmaceuticals in the aquatic environment of the Baltic Sea region: A status report*. Vol. 1. UNESCO Publishing.

von Liebig, J.F. 1859. *Letters on modern agriculture*. J. Wiley.

Wainger, L.A., D.H. Secor, C. Gurbisz, W.M. Kemp, P.M. Glibert, E.D. Houde, J. Richkus, and M.C. Barber. 2017. Resilience indicators support valuation of estuarine ecosystem restoration under climate change. *Ecosystem Health and Sustainability 3* (4): e01268.

Wheeler, D.M. 2015. OVERSEER®TechnicalManualTechnical Manual for the description of the OVERSEER®Nutrient Budgets engine. ISSN: 2253-461X.

Wheeler, S., A. Loch, A. Zuo, and H. Bjornlund. 2014. Reviewing the adoption and impact of water markets in the Murray–Darling Basin, Australia. *Journal of Hydrology* 518: 28–41.

Whittemore, R.C. 1998. The BASINS model. *Water Environment & Technology 10* (12): 57–61.

Williams, J.R. 1985. The physical components of the EPIC model. In *Soil Erosion and conservation*, ed. S.A. El-Swaify, W.C. Moldenhauer, and A. Lo, 272–284. Ankeny: Soil Conservation Society of America.

Wischmeier, W.H., and D.D. Smith 1978. Predicting rainfall erosion losses: A guide to conservation planning (No. 537). Department of Agriculture, Science and Education Administration.

Young, R.A., C.A. Onstad, D.D. Bosch, and W.P. Anderson. 1987. *AGNPS, Agricultural Non-Point Source Pollution Model: A Watershed Analysis Tool. USDA Conservation Report 35*. Washington, DC: USDA-ARS.

Yuan, L., T. Sinshaw, and K.J. Forshay. 2020. Review of Watershed-Scale Water Quality and Nonpoint Source Pollution Models. *Geosciences 10* (1): 25.

Zaffiri, L., J. Gardner, and L.H. Toledo-Pereyra. 2012. History of antibiotics. From salvarsan to cephalosporins. *Journal of Investigative Surgery* 25 (2): 67–77.

Zhang, W., T.H. Ricketts, C. Kremen, K. Carney, and S.M. Swinton. 2007. Ecosystem services and dis-services to agriculture. *Ecological Economics* 64 (2): 253–260.

CHAPTER 4

Decision Making at the Farm Level

4.1 Introduction

Policy research offers many examples of agri-environmental policies that did not deliver expected outcomes or that had adverse outcomes because behavioral responses of producers to the policies were not adequately considered. Policy research also offers many examples of agri-environmental policies that cost society far more than required to achieve policy objectives because of inadequate attention to the economics of agricultural production and farm decision making. If water pollution goals for agriculture are to be met, and to be met without an undue burden of social cost, then farm decision making as it affects water quality outcomes and responses to policy interventions must be understood and taken into account.

This chapter introduces economic concepts and tools for analyzing farm decision making on key choices affecting water quality outcomes. It begins with a brief introduction to some farming-related terms we use routinely in this and subsequent chapters, and to choices and objectives in farm decision making. We then introduce standard economic models used to explain producers' use of productive inputs that directly affect pollution (e.g., fertilizer), crops that influence the polluting input intensity of production, and the spatial distribution of agricultural activities in watersheds. These models are used to illustrate how market-driven agricultural

J. Shortle et al., *Water Quality and Agriculture*, Palgrave Studies in Agricultural Economics and Food Policy,
https://doi.org/10.1007/978-3-030-47087-6_4

landscapes differ from Pareto Efficient landscapes given water pollution externalities. Subsequent sections introduce additional concepts and tools relevant to farm decision making on polluting inputs, crops, the spatial structure of production, and the use of best management practices.

4.2 Some Basics

Conventional crop production entails planting and subsequently harvesting land areas. The basic economic facility of agricultural production is the farm, essentially a defined land area managed for crop or livestock production. The "producer" is the entity making decisions about what and how to produce. The producer may be an individual, a family, a business partnership of multiple unrelated individuals, or a corporation. The producer may own the land, rent the land, or use some combination of both. The arable land of a crop farm is typically subdivided into fields, which are the basic production units of the farm. A field is a defined area that is managed (mostly) as a single unit.[1] Farms that include livestock may allocate land to pasture. The producer organizes activities across the collection of fields or pastures included in the farm as an integrated economic entity for profit making and possibly other objectives.

A field may be planted and harvested annually, as with corn, or planted in a perennial, like hay. In some farming systems, fields may be planted with multiple crops in a year. An example is planting corn in the spring, harvesting the corn in the fall, replanting the field after the corn harvest in a cover (or catch) crop like winter wheat, which is then harvested in the spring. Different fields within a farm may be planted in the same crops or in different crops. Crop rotations are multi-year sequences of crop types, for example, corn-soybeans or corn-hay.

Farming activities within fields promote crop growth by active management. Plowing or other types of tillage prepare land for planting and may serve other purposes like pest management. Plant nutrients are

[1] Field boundaries are defined by various factors including physical features, legal boundaries, and historical practice. We say a field is mostly operated as a single unit as the area of the field will be planted to the same crop, or crop rotation, and subjected to the same field operations (tillage, planning, fertilization, harvesting) at more or less the same time. But advances in information technology enable "precision farming" that allows for spatial fine-tuning of certain farming practices like fertilizer applications within fields that would be otherwise managed as homogenous units. Within-field variations can result from variations in soils and other factors.

applied in the form of mineral fertilizers, or manure from farm livestock or from other natural occurring organic sources (e.g. bat guano). Weed, insect, and disease management is conducted in the growing season using mechanical (cultivation), chemical (pesticides), and biological controls (e.g., the use of insect predators to control insect populations). Crop rotations also are used for managing nutrients and pests. In arid environments, irrigation is used to manage soil moisture. At the end of the season crops are harvested.

Key determinants of water quality outcomes are the use and management of inputs applied to the land (nutrients, pesticides) that can be harmful to water quality, practices that extract potential pollutants (sediment, salts) that are naturally present in soils, and practices that influence pollutant removal processes (e.g., leaching, runoff, volatilization).

A fundamental consideration for systematic explanation of observed choices and for policy analysis is producers' objectives. What is it that they seek to achieve? When viewing farms as conventional businesses, the standard economic paradigm calls for modeling farm decisions as motivated by profit. This is the approach we take in this chapter. It is a useful approach given the substantial research demonstrating that profitability is a key driver of farm decisions regarding what crops to produce and how to produce them, and that producers are highly responsive to economic incentives (Cattaneo 2003; Reichelderfer and Boggess 1988; Roberts and Lubowski 2007; Segerson 2013; Suter et al. 2008; Pannell and Claassen 2020).

A broader view and one that we acknowledge recognizes that producers generally have multiple objectives. Important within the context of this book, producers to varying degree have environmental motives, value stewardship, and are willing to trade profit in service of their environmental preferences (Chouinard et al. 2008; Pannell et al. 2006; Weaver 1996; Kosenius and Ollikainen 2019). That this is the case implies that there are private benefits to producers from environmental protection and stewardship activities. But we know from the large contribution of agriculture to water quality outcomes that in aggregate these private benefits from environmental protection do not significantly offset the private benefits from polluting activities. The private benefits from water quality protection however large they may be do not negate water quality externalities.[2]

[2] Water quality protection measures undertaken by producers motivated by their own environmental preferences entails private provision of a public good. Were producers the sole cause of a problem and their private benefits sufficiently great, private provision could achieve

Given that environmental preferences appear comparatively weak and are likely heterogeneous, it is understanding and managing the private benefits from polluting activities that is of greatest utility for policy design. This calls for a focus on profit-seeking behaviors. Key questions are what to produce, how to produce, and where to produce to maximize profit.

4.3 Private Production Decisions: Input Intensities and Land Allocation

Producers make decisions on varying timescales. Decisions about durable assets (e.g., tractors and other machinery, tillage and irrigation equipment, barns) consider the returns received from the investments over multi-year periods. These decisions result in the *fixed inputs* in farm production that are not varied on short timescales. Decisions about perennial crops and herd sizes are also long-term decisions. Decisions about annual crops or crop rotations of a few years' duration are made on shorter timescales. Correspondingly, decisions about the use of *variable inputs* used to increase farm output (e.g., fertilizer, irrigation, animal feeds) or to protect output or the quality of output from pests or diseases (e.g., pesticides, pharmaceuticals) are made annually or even intra-annually. Short-term decisions regarding crops or variable inputs are generally constrained by long-term choices (fixed inputs) defining feasible short-term options.[3]

This section presents two interrelated basic economic models used to analyze questions of what crops to produce, where to produce, and the use of productive inputs that contribute directly to pollution. One is a standard model from the economics of production for understanding and predicting the use of variable inputs. The second is a standard model for understanding and predicting arable land use, in our context the allocation of crops to fields and the spatial distribution of agricultural production in a watershed.

We begin by considering a hypothetical crop farm with several fields and the producer's decision about which crops to plant in each field and the level of variable inputs to use in crop production. For simplicity we assume that crop decisions about what and where to produce are made

the Pareto Optimum. While there may be localized instances where this is the case, it is clearly not the general condition.

[3]Maximizing short-term profits takes the farm capital as given. Farm capital comprises items like machinery, drainage of fields, and buildings. In the long run, producers also choose farm capital.

independently from one year to the next. Decisions about crop rotations are analytically similar but operate on a sequence of interlinked crops. Rotations and spatial dependencies between fields are discussed subsequently. Further, we limit the variable input to nitrogen fertilizer. The producer's decisions are represented as a two-step optimization procedure. In the first step, the producer considers economically plausible[4] candidate crops for each field on the farm. For each candidate crop and field, the producer determines the nitrogen use that maximizes the net return from the crop. This first step essentially produces a list for each field of the net return for each candidate crop when input use is optimized for the crop and location. In the second step, the producer compares the net returns for each candidate crop for each field and selects for each field the candidate crop with the highest net return. The result is an allocation of crops, possibly the same or different, across the farm fields.

4.3.1 Nitrogen Fertilization Use

The net return from producing a crop is the revenue from selling the crop less the production costs. Revenue is the harvested crop yield multiplied by the crop price. Total production costs are the costs of variable and fixed inputs. As fixed inputs are fixed, it is variable inputs that control crop yield. The information needed to determine the net revenue-maximizing use of variable inputs is technical knowledge of how crop yield increases with the application of inputs, the crop price when harvested, and the unit cost of the inputs.[5]

In our illustration, the variable input is nitrogen and the variable cost is the cost of its application. Other costs include those of field operations (tillage, planting, pesticide applications, harvesting) requiring fuel, chemicals, labor, and machinery. These are largely proportional to the land area of the field and can be reasonably treated as fixed for the field. For reasons explained below, we exclude land costs from the calculation of net return in this first step. The net return is then the revenue from crop sales less the variable nitrogen fertilizer cost and the fixed costs of the field operations.

[4] Climate and soils, for example, impose large constraints on economically viable crops. Some climates and soils are suitable to fruits or vegetable or wine grapes and some are not. Fixed inputs (farm labor, machinery, structures) also impose constraints on economically viable crops in the short run.

[5] In the general economic theory of production, the technical relationship between input levels, both fixed and variable, and output is referred to as a production function.

The technical relationship between yield-increasing input use and crop yield for an input is called the *yield response function*. The trade-off between revenues and costs from fertilizer applications is crucially dependent on this relationship. Yield response functions are mathematical equations that describe how the expected crop yield increases with the application of inputs to the land area.[6] At the plant scale, research indicates that crop yield is controlled by the availability of the required plant nutrient that is most scarce relative to the needs of the plant. This principle is known as von Liebig's Law of the Minimum.[7] If nitrogen is abundant and phosphorus scarce, it is phosphorus that controls growth. If phosphorus is abundant and nitrogen is scarce, it is nitrogen that controls growth. Irrigation is used to increase yields when soil moisture is the limiting factor. Our model focusing on the nitrogen decision is consistent with a situation in which phosphorus is abundant and nitrogen is limiting.

Figure 4.1 describes the shape of a general yield response function, denoted $y = f(n)$, that indicates how crop yield per hectare (y) increases along the vertical axis as the use of nitrogen fertilizer input (n) is increased along the horizontal axis. $f(n)$ is notation representing the algebraic equation that describes the relationship between yield and fertilizer input. As there is no single equation that describes yield response for all crops and locations, the specific algebraic form of $f(n)$ will depend on the crop, location, and possibly other variables.

Figure 4.1 shows yield increasing with the amount of nitrogen fertilizer but at a decreasing rate. This response shows up in the curve flattening with greater use of the input, implying that the additional yield resulting from additional fertilizer decreases as more is applied. This flattening is the classical finding that agriculture is subject to diminishing marginal returns. The yield in Fig. 4.1 is shown to increase indefinitely with further application of the input. This is a property of the so-called Mitscherlich response function which is used in some studies to estimate response. The yield function in Fig. 4.1 illustrates barley yields on average boreal lands, where the typical fertilizer application rates range from 80 to 140 kg/ha.

[6] In the general economic theory of production, yield response functions are total product functions for crop production inputs.

[7] The Law of the Minimum implies at the plant level that yield responds linearly to fertilizer inputs up to a maximum at which some other input becomes limiting. Producers, however, do not manage individual plants but populations of plants in fields. Research demonstrates that when aggregating many plants over heterogeneous land areas, empirical response functions for land management units (i.e., fields) are nonlinear and display diminishing marginal returns (Berck and Helfand 1990).

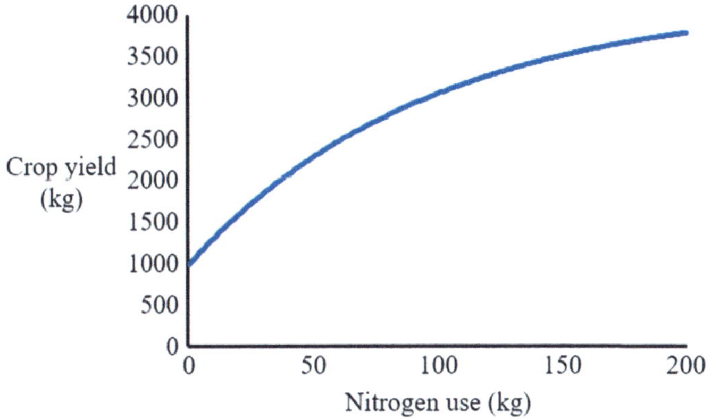

Fig. 4.1 Yield response function. (Source: Authors' creation)

A common alternative used in research is a quadratic response function, which exhibits decreasing yields if the producer uses too high fertilization rates (Dillon and Anderson 2012; Heady and Dillon 1988; Rasmussen 2011).

Given the yield response function applicable to the field, the net return obtained from production can be expressed as a function of the nitrogen application rate given the price of the crop, the cost of nitrogen, and the fixed costs. Revenues are the per unit price of the crop multiplied by the crop yield, which depends on the fertilizer rate. Variable costs are the costs of fertilizer application. Net return is written as follows:

$$\pi = pf\left(n\right) - cn - M \tag{4.1}$$

where π denotes the net return, p is the crop price per unit (e.g., bushel or tons), c is the unit cost of fertilizer application, M is the fixed cost, and n and $f(n)$ are, as defined above, the nitrogen application and yield response function. The first term on the right side of the equation is the price multiplied by yield and is therefore an expression for revenues dependent on the fertilizer rate. The second term, the product of the unit cost of fertilizer and the amount of fertilizer, is the total cost of the fertilizer application.

Given that M is fixed, maximizing net return requires selecting the fertilizer application n to maximize the difference between crop revenues and fertilizer costs. As described in Chap. 2, economics uses marginal analysis to analyze trade-offs and identify conditions for optimal choices. A standard result of production economics is that the quantity of an input that maximizes net returns is the quantity that equalizes the ***marginal revenue*** generated from the use of an input with the unit cost of the input. The marginal revenue from additional fertilizer is the ***marginal product*** of fertilizer, denoted by *MP*, multiplied by the price of the crop. The marginal product is the incremental increase in yield that results from an incremental increase in fertilizer. Marginal revenue is the marginal product of the input multiplied by the crop price giving the incremental increase in revenue that results from an incremental increase in fertilizer use. With this, the marginal condition for the net return-maximizing fertilization application is written

$$pMP = c. \tag{4.2}$$

This condition indicates that the net return-maximizing fertilizer rate is the rate at which the marginal increase in revenues, pMP, from the last unit applied fertilizer is equal to its unit cost.

The fertilizer rate at which this condition is satisfied depends on the shape of the yield response function. For fertilization to be profitable it must be the case that $pMP > c$ for some fertilizer levels. And when this is the case, the incremental gain exceeds the incremental cost so that it pays to fertilize more. The reverse is true when $pMP < c$. However, given diminishing marginal returns, *MP* decreases with the level of fertilizer use. This means that the difference between pMP and c shrinks as fertilization increases until eventually the difference is zero. At this point net return is maximized. Additional fertilization would increase costs more than revenues.

This optimization result is illustrated in Figs. 4.2 and 4.3 for the application of fertilizer to barley in Finland. In Fig. 4.2, the horizontal axis is the fertilizer rate and the vertical axis is revenues and costs. The revenue curve is obtained by multiplying the barley crop yield at each fertilization rate, as given by the response function in Fig. 4.1, by the barley price. The cost line indicates the fertilizer costs as a function of fertilizer rate. For clarity, the figure omits the fixed cost, set at $M = 250$. This cost is incurred regardless of the fertilizer level.

Figure 4.2 describes Eq. (4.1). In the figure, the net revenue-maximizing fertilizer rate is the amount which makes the vertical difference between revenue curve and variable cost line the greatest. In the figure the optimal fertilization is about140 kg/ha giving 3440 kg as the yield and 268.5 €/ha as the net revenue.

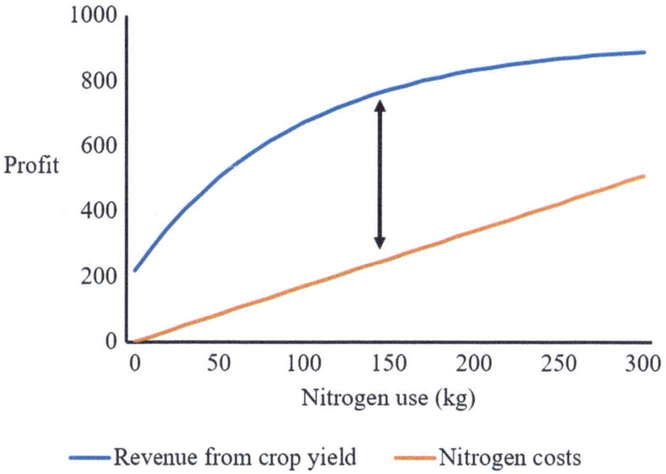

Fig. 4.2 Profit-maximizing choice of fertilizer application. (Source: Authors' creation)

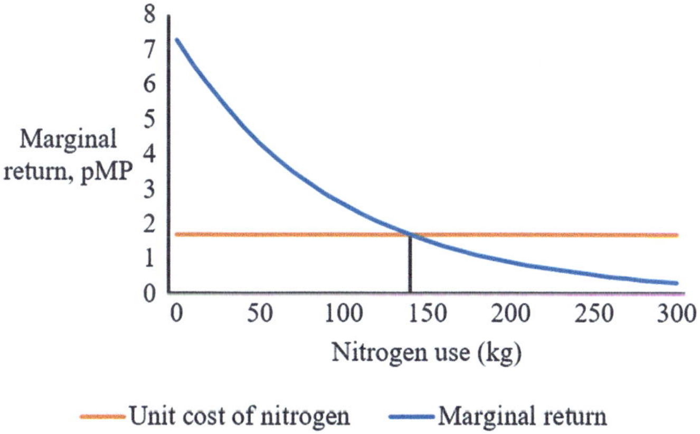

Fig. 4.3 Profit-maximizing choice of fertilizer application: marginal analysis. (Source: Authors' creation)

The same result is illustrated graphically using the marginal condition (4.2) in Fig. 4.3. The horizontal axis is again the fertilizer application rate. The vertical axis is marginal revenue and marginal cost. The marginal revenue curve (*pMP*) decreases to the right due to diminishing marginal returns, and the unit price of nitrogen fertilizer is constant. The point where the graphs cross defines the optimal nitrogen fertilizer use, which is here 140 kg/ha. Figure 4.3 illustrates that using nitrogen fertilizer increases revenue from selling the crop but this increase takes place at a diminishing rate. As long as the marginal revenue is greater than the unit cost of nitrogen, profits increase. In the figure a shift from 139 kg to 140 kg increases revenue as much as costs, at higher application levels profits increase less than costs, so it is not economically wise to increase nitrogen fertilization beyond that point.

Box 4.1 Parametric Illustration of Fertilizer Optimization

To illustrate the net return-maximizing nitrogen rate consider the following parametric example. To make calculation easy, suppose that the nitrogen yield response function for a crop is quadratic, $y = A + a_1 n - 0.5 a_2 n^2$, where A is a constant, and a_1 and a_2 are coefficients that determine the shape of the function. The corresponding profit equation is $\pi = p(A + a_1 n - 0.5 a_2 n^2) - cn - M$. The optimality condition ($pMP - c = 0$) for nitrogen application is given by $p(a_1 - a_2 n) - c = 0$. Solving this equation for the value of n that satisfies the equality gives the following optimal nitrogen rate, $n^0 = (pa_1 - c)/a_2 p$. This condition defines nitrogen demand as a function of crop price and fertilization costs. From the optimality conditions we see that if fertilizer price (c) increases, the optimal fertilization rate decreases and if the crop price (p) increases, the optimal fertilization rate increases.

Given that water quality damages are related to fertilizer use, it is of interest to ask how crop and fertilizer prices affect use. Consider a crop price increase. A higher barley price would shift the revenue curve in Fig. 4.2 and the marginal profit curve in Fig. 4.3 upward, while the costs would be unchanged. The outcome would be an increase in the fertilization rate. Other things equal, this implies that market conditions or public policies that lead to higher (lower) prices of nitrogen-intensive crops lead to higher (lower) fertilization rates and potentially increased (decreased) water quality problems. Alternatively, consider an increase in the nitrogen cost. The cost increase would rotate the cost line in Fig. 4.2 upward reducing the profits from fertilization at every fertilization rate and also reducing the fertilization rate at which net revenues are maximized. In Fig. 4.3, the cost increase shifts the marginal cost line upward, shifting the point at which marginal revenues equal to marginal cost to the left. This implies that the cost of nitrogen fertilizer, whether policy or market induced, affects stresses on water quality.

Similarly, it is of interest to ask how land quality influences land use. Assuming the same crop is planted, the yield from a more fertile field will be greater than that from a less fertile given the same fertilizer application rate. The marginal returns will therefore be greater resulting in greater fertilizer use per unit of land area on the higher than lower-quality soils. This does not, however, imply that higher-quality soils will contribute more to nitrogen loads than low-quality soils in the same crop. One consideration is that greater plant growth on the high-quality soil means greater nitrogen utilization and less nitrogen available for removal to water resources. Other factors are location of land relative to water resources and other aspects of the farming system affecting nitrogen loads.

We note that we have greatly simplified farm nitrogen management to focus on the fundamental question of how much nitrogen a producer applies when making fertilization decisions. There are details, including the effects of manure application history, rotation history, soil testing application methods and timing. It remains, nevertheless, that the primary drivers are marginal revenues and marginal costs.

4.3.2 *Optimal Land Allocation on the Farm*

The second step in the two-step optimization uses the results from the first step to determine which crop to place in each farm field. The procedure is simple under the assumptions we have made. The first step provides a list of optimized net returns for each candidate crop for each field. In the second step the candidate crop that provides the highest net return for a field is the crop assigned to the field. With each field in its highest value use, total net returns to the farm are maximized. If the fields are of the same quality, the same crop will be assigned to all. If quality varies, farm fields may be put into the same crop or a mix of crops.

The model we have described is the Ricardian model of agricultural land use. The model was introduced by its namesake David Ricardo's work on land rents and land use in the early 1800s. Modern interpretations are due to Hochman and Zilberman (1977), and Lichtenberg (1989, 2002). The Ricardian model predicts that land will be allocated to the use which provides the largest *Ricardian Rent*. Ricardian rent is the net return from land exclusive of payments or opportunity costs of the land use. It is equivalently the net returns from production calculated in the two-step model. The first step determined the intensity of production for candidate land uses that would maximize the Ricardian rent if land were put to that use. The second step selected land uses that maximize Ricardian rent. The theory recognizes that soils vary in their suitability for different crops and that there are variations in productivity within suitability classes, often called "land qualities." Land quality generally refers to soil organic matter content, which is an increasing indicator of three important components of soil productivity: ability to hold water, fertility, and tilth (Lichtenberg 1989, p. 190). For example, in boreal agriculture, oats and rapeseed perform better in low-quality lands but wheat performs better in high-quality lands (Lankoski and Ollikainen 2003). Land quality differences along with crop prices and production costs determine the net returns from different candidate crops on different fields. The allocation of crops to fields is illustrated in the next section which considers land use at the watershed scale where the same basic principles apply.

4.3.3 Land Use at Watershed Scales

The Ricardian theory can be extended to land use at multiple scales. At the farm scale, it explains land use across the farm. At larger scales the theory can be used to explain the overall pattern agricultural land use as a function of the same variables that determine Ricardian rents at the farm level (prices, costs, land qualities) but also Ricardian rents from competing land uses, such as forestry or urban and at the rural–urban fringe. Within rural lands, rents from agricultural land uses versus less developed land uses, such as forestry, determine the allocation among rural land uses. Within each land use category, the spatial pattern of use is explained by the spatial pattern of rents. For water quality protection we are especially interested in land use patterns within watersheds.

The range of land qualities found on a farm will generally be smaller than the range of qualities found across a larger area within which it is located. A greater diversity of crops may therefore be expected in the broader area than on individual farms. Figure 4.4 illustrates the choice between two hypothetical crops over a large set of farms. The horizontal

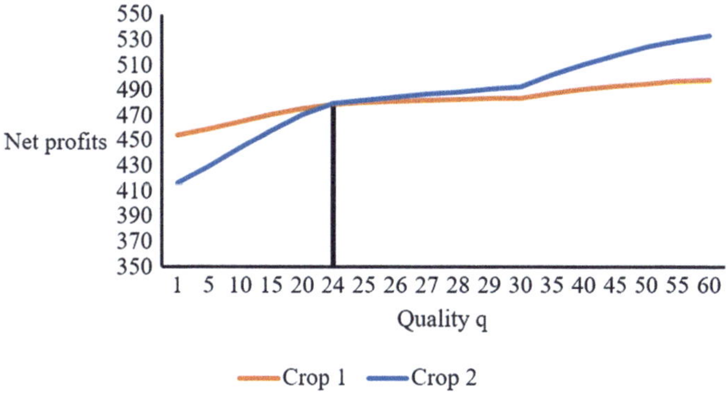

Fig. 4.4 Land allocation between crops: private optimum. (Source: Adapted from Lankoski and Ollikainen 2003)

axis describes a range of land qualities of fields ranked from 1 to 60 from the lowest quality on the left to the highest quality on the right. The vertical axis is the net returns per hectare for each crop under optimal variable input management. The yellow curve is the net return for hypothetical Crop 1 and the blue curve is the net return for hypothetical Crop 2. For both crops, net returns are higher when planted in fields with higher land quality. However, the relative net returns vary across land qualities. Crop 1 provides the greater net return on lands below quality 24 while Crop 2 provides the greater net return on lands above quality 24. Accordingly, Crop 1 would be allocated to fields with land quality values less than 24, and Crop 2 would be allocated to fields with land quality greater than 24.

The results are contingent on crop prices. A higher price for Crop 1 relative to Crop 2, for instance, would increase fertilization use and net profits for Crop 1 for each land quality (see Figs. 4.3 and 4.4). The consequent change in land allocation between crops would, however, take place only in few land qualities around the breakeven point. The breakeven land quality would be less than 24 so that more land would be allocated to Crop 1. A similar analysis can be provided for input prices. An increase in fertilizer price would decrease fertilizer intensity and profits for both crops but this decrease would more for Crop 2 and more land would be shifted to Crop 1.

The lowest quality land in agricultural production in a watershed defines the ***entry/exit margin*** for agriculture in the watershed. The margin is not a geographic concept but an economic one defining the land quality level that divides land allocated to agricultural production and land allocated to lower-intensity uses, such as forestry, or left idle. The margin depends on crop and input prices, agricultural technology, and agricultural and land use policies affecting land rents and land use.

This margin is important for the effects of agricultural policies, whether for water quality protection or other purposes, on both the total land in agricultural production and the change in the land use pattern between agriculture and other land use forms. Policies that increase land rents shift the margin, resulting in increased agricultural land use. The water quality effects of agricultural policies (or any policies that affect land rents) thus have two ways of impacting water quality. First, they impact input use and possibly green infrastructure investments such as buffer strips and wetlands. These impacts tend to increase pollution from agriculture. Second, they impact watershed-level land use by making farming more profitable compared to other land uses. If these other land uses are less polluting than agriculture – as they tend to be – this effect increases the pollution loads. This effect is often referred to as the external margin effect of agricultural policies.

For instance, Lankoski and Ollikainen (2013) found that the introduction of EU area payments in the 1990s led to increased agricultural land rents relative to forestry in Finland resulting in conversion of forest land to crop production. Because the per hectare nutrient pollution is higher in agriculture than in forestry, this land use change led to a corresponding increase in the risk of nutrient loading (Lankoski and Ollikainen 2013). Another example of a policy increasing cropland rents leading to expansion of cropland at the extensive margin with adverse impacts on water quality is the 2007 US Renewable Fuel Standard (RFS) discussed in Chap. 2.

4.4 Pareto Efficient Production Intensities and Land Allocation

Pareto Efficient agriculture optimizes the social net benefits from agricultural production. Whereas the market economy will maximize Ricardian rents from agriculture in a watershed, a Pareto Efficient allocation maximizes Ricardian rents less the social costs of water pollution. Except for the differences in the information used, the optimization can be done in two steps as above. The first step is to optimize a set of candidate farming systems for fields. The second step selects from the optimized candidates the set that maximize social net returns.

4.4.1 Pareto Optimal Production Intensity

To illustrate we again consider the first of the two optimization steps: choosing the nitrogen fertilization for a single crop. To include damage costs in the optimization problem we need a nitrogen loading function that describes the relationship between water quality outcomes and the fertilization choice. The loading function in this case would consist of a field-scale model describing the fraction of nitrogen applied that is not removed by the crop but delivered to the edge-of-field, and a transport model that describes the fraction of the edge-of-field nitrogen that contributes to ambient nitrogen pollution.

In addition to loading and transport models, we also need a damage cost function indicating the social cost of ambient nitrogen pollution. Underlying the damage cost function are biophysical relationships introduced in Chap. 3 that describe the effects of pollution on water quality conditions and the resulting costs of water quality degradation, and the effects of water quality degradation on the use and nonuse values derived from the impaired water. As in Chap. 2, the marginal damage cost of ambient pollution at the receptor is denoted *MDC*. The marginal damage cost of nitrogen applied to the crop is denoted MDC_n. The difference between *MDC* and MDC_n is due to various processes that affect the fraction of nitrogen applied that becomes edge-of-field nitrogen emissions, and the fraction of edge-of-field emissions that reach the receptor. Given that nitrogen delivered is a fraction of the nitrogen applied, it is the case that $MDC_n < MDC$. Factors affecting the difference include soil properties affecting leaching and runoff, and characteristics of the flow path from the field to the receptor.

Pareto Efficient management in this simple case maximizes the social net returns from production. This is the Ricardian land rent, π, less the damage cost. As described above, due to a diminishing marginal product, the land rent increases at a decreasing rate with the fertilizer application. Damage cost increases with fertilization. Where the amount of nitrogen removed by crop growth is proportional to yield, diminishing marginal product implies nitrogen lost to the environment increases at an increasing rate. In this case, other things equal, the marginal damage cost of applied nitrogen increases at an increasing rate.

Following the principles of marginal analysis, the optimal fertilizer rate sets the marginal revenue from nitrogen equal to the cost of applying nitrogen plus the marginal damage cost of applied nitrogen:

$$pMP = c + MDC_n \tag{4.3}$$

The sum $c + MDC_n$ is the social cost of using nitrogen in cultivation. Because $MDC_n > 0$, accounting for the marginal damage from nutrient loads reduces the use of nitrogen input.

The optimization result is illustrated in Figs. 4.5 and 4.6. Figure 4.5 shows the total revenue from crop sales, and the total costs as a sum of fertilizer costs and damage cost from nitrogen runoff. The difference between revenues and costs is the social net benefit. The vertical distance is greatest at 95 kg/ha, making it the social net benefit maximizing level. The figure also shows the nitrogen rate that maximizes net revenues from crop production excluding the damage cost. This occurs at 149 kg/ha where the vertical distance between revenue and fertilizer cost is greatest. The example shows that accounting for the damages reduces the socially efficient nitrogen application rate relative to the rate that maximizes net revenues.

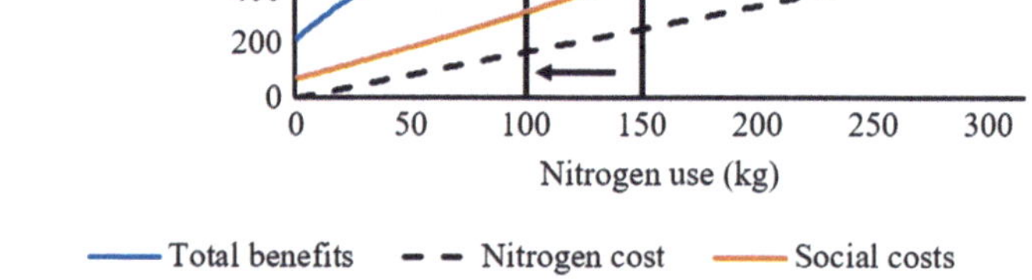

Fig. 4.5 Social net benefit maximizing choice of fertilizer application. (Source: Authors' creation)

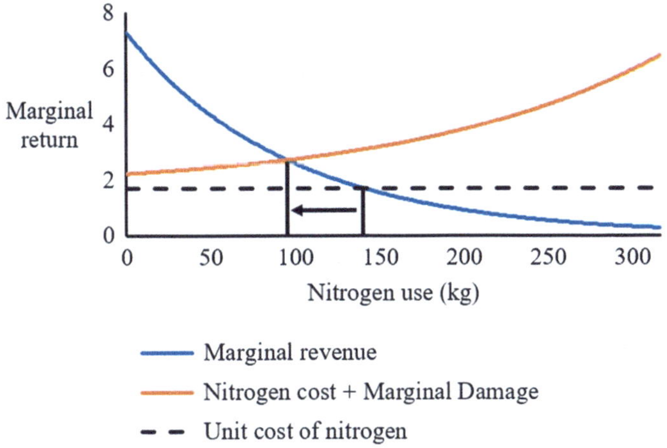

Fig. 4.6 Social net benefit maximizing choice of fertilizer application: marginal analysis. (Source: Authors' creation)

The same outcome is shown in Fig. 4.6 but this time using marginal analysis. As in Fig. 4.3, the marginal benefit of fertilizer is given by the decreasing marginal revenue curve (*pMP*). The marginal cost of fertilizer is now the sum of the increasing marginal damage and the constant unit cost of nitrogen fertilizer. The intersection of marginal revenue and the social marginal cost is to the left of the intersection of marginal revenue and the unit cost of fertilizer. The socially optimal solution requires a lower fertilization rate than the rate that maximizes private returns.

How do changes in market prices affect socially optimal fertilization? A higher crop price increases the monetary benefits from fertilizer relative to the damage from water quality and therefore increases fertilizer application but less than in the private solution. An increase in fertilizer cost has the opposite effect. Suppose that a technological innovation increases the marginal product of fertilizer at every application rate. This would shift up the marginal revenue curve. Following the analysis in Fig. 4.6 this innovation would increase the amount of fertilizers applied. An increase in the marginal productivity thus increases fertilizer application. Finally, the higher the marginal damage cost of applied fertilizer (MDC_n), the lower the fertilizer application. Variations in MDC_n may be due to variations in land characteristics affecting the edge-of-field emissions from nitrogen applications, factors affecting the delivery ratio of edge-of-field emissions to the receptor, and the marginal damage cost at the receptor.

Our analysis thus far has focused on the farm level. A key consideration at the farm level is that prices are exogenous to the farmer. Box 4.2 gives a concrete example of the nature, scope, and distribution of the social costs of reducing pollution from agriculture in a large watershed with many producers. Prices are exogenous to individual farmers but costs and responses are affected by policy-induced price changes.

Box 4.2 The Social Cost of Pollution Control in Agriculture: Gulf of Mexico Case Study

Our focus on farm-level decision making in this chapter results in an emphasis on the farm-level costs of changes in agricultural production to reduce water pollution. As we have described, the farm-level cost is the loss of economic rents from agricultural production. Excluding the transaction costs of policy development and administration, the social cost of changes in farm production will be limited to the loss of economic rents if the prices of inputs and outputs used in agriculture are unaffected by policy interventions. This will be the case if the farms subject to pollution controls are too small a share of supply in the outputs or demand in the input markets to significantly affect the markets. The social costs of policy interventions affecting a small part of large agricultural markets will be reduction in economic rents from agricultural land use. This cost may be borne by the farmland owners or it may be transferred to the public if landowners are compensated.

If the farms subject to pollution controls are a large enough share to affect the market supply of agricultural products, the social cost will include a loss of consumer surplus. Increased production costs from undertaking pollution controls will reduce market supplies. The supply reductions will lead to new market equilibria with higher prices and lower production. Increased prices will reduce consumer welfare. As was illustrated in Fig. 2.2, the social cost to consumers of internalizing externality is reduction in consumer surplus. Higher prices will mitigate the costs to producers to some degree.

The large seasonal Gulf of Mexico hypoxic zone, commonly referred to as the Gulf of Mexico Dead Zone (see Chap. 3), is driven largely by nutrient losses from the Mississippi/Atchafalaya River Basin (MARB). This very large region contains about 70% of US cropland and is a large share of the supply of major agricultural commodities (e.g., corn, wheat). The Mississippi River-Gulf of Mexico Watershed Nutrient Task Force, comprised of State and Federal agencies and led

(*continued*)

Box 4.2 (continued)

by the US Environmental Protection Agency (US EPA), established a goal for reduction in the Dead Zone that would require nutrient load reductions to the Gulf of Mexico by 45% or more.

The US Department of Agriculture's Economic Research Service conducted a study to estimate the costs of implementing the proposed 45% reduction from agriculture under two scenarios (Marshall et al. 2018). One scenario meets the goal with regionally targeted pollution reductions based on differences in regional abatement costs within the MARB. The second scenario requires the 45% goal to be met in all regions regardless of cost. The first scenario is by design more cost-effective than the second and will have the lower social cost.

Given the magnitude of the reduction and the importance of the region to agricultural commodity supplies, significant economic costs to producers and consumers could result if the reduction was implemented. The study estimated that cropland in production in the region would decline by 4.4% if the goal were achieved by the targeted reductions and 10.9% if the reductions were not regionally targeted (the baseline year for these changes is 2007). Prices of most crops would rise. Prices for some major crops were estimated to increase by up to 10% in the costlier scenario. Higher prices partially offset producers' pollution control costs. Higher prices also pass some of the cost on to consumers of agricultural products, including livestock producers, food consumers, and biofuel feedstock users.

Under the targeted scenario, US consumer surplus is reduced by an estimated 2.5% (1.9 billion USD). Under the costlier nontargeted scenario, consumer welfare drops an estimated 4.4% (3.3 billion USD). Under the targeted scenario, producer net returns in the MARB basin increase by 1.3% (847 USD). The increase in producer net returns occurs because commodity prices more than compensate for increased production costs. The increase in the total, however, masks gains to some producer and losses to others depending on crop and location within the region. In the nontargeted scenario, net returns decrease by 0.4% (264 million USD). The higher prices of the crops do not offset the higher production costs. In both scenarios, social costs estimated by the study are almost entirely costs to consumers due to increased production costs. The estimates of producer and consumer costs in the study do not net out environmental benefits they may gain.

4.4.2 Pareto Efficient Land Allocation in a Watershed

A Pareto Efficient configuration of land uses in a watershed optimizes land use considering all uses. Accordingly, in the second step of the two-step societal optimization, optimized social net returns for competing land uses must be considered along with the optimized agricultural social net returns in allocating fields and other land parcels to uses. The first step of optimizing private net returns less pollution damage costs is performed for all plausible candidate uses for all parcels. The relative social net returns of these candidates will depend on the size of the private net returns from the commercial use relative to the social costs. Some uses may not be for the provision of private goods but for the provision of public good (e.g., buffers, wetlands, bioretention ponds, public parks). For those uses, societal returns from the environmental benefits provided must be compared to social opportunity costs. Total and marginal pollution damage costs for fields and other parcels are generally contingent on the overall all level of emissions from rural and urban sources. In consequence, net social returns for candidate uses on any given parcel depend on the candidate uses considered for others. The social optimization determines the amount of land and the spatial pattern of land use in agricultural and competing nonagricultural uses.

We emphasize that this implies that optimal pollution controls are being applied in nonagricultural as well as agricultural uses. The Pareto Efficient watershed requires efficiency in pollution control in all uses. The Pareto Efficient allocation trades off the social costs of reducing pollution from agricultural production and other land uses for the social benefits of reduced water pollution.

Within the land allocated to agriculture, the candidate crop that provides the highest net social return is assigned to each field. With each field in its highest valued use to society, total social net benefits from agriculture are maximized. If the fields are of the same quality, as defined previously to reflect agronomic productivity, the same crop will be assigned to all. If quality varies, farm fields may be assigned to different crops. Figure 4.7 repeats Fig. 4.4 replacing the net private returns on the vertical axis with the net social returns from the crops when placed on lands of differing quality. In constructing Fig. 4.7 we assume that differences in pollution from different lands depend only on the fertilization rates when accounting for pollution damage costs in calculating social net returns. This simplifying assumption highlights the effect of crop choice on net social returns as it affects nitrogen

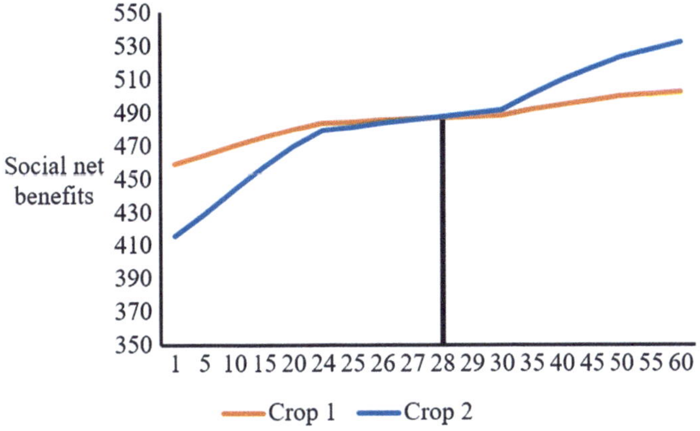

Fig. 4.7 The socially optimal land allocation between crops. (Source: Adapted from Lankoski and Ollikainen 2003)

intensity and pollution damage costs. The horizontal axis describes the land qualities and the vertical axis the social net returns for each quality for both crops. For each crop, social net benefits are greater on the higher-quality lands. As with private net returns in Fig. 4.4, Crop 1 (orange) also provides higher social net returns on lower-quality lands and Crop 2 (blue) on higher-quality lands. Pollution damage costs reduce social net returns relative to private net returns. The optimal land use shifts in favor of Crop 1, which is now cultivated on higher land qualities. The breakeven land quality is $q = 28$ when social costs are included in contrast to $q = 24$ when only private net returns were considered. The map of agricultural land use would be determined by the spatial pattern of land qualities to which the crops are allocated.

The analysis of the preceding paragraph removes the effect of characteristic of land, their location in space, and the nitrogen efficiency of crops on net social returns from alternative crops in alternative locations. Depending on soil properties and topography, some lands produce greater edge-of-field pollution than others (see Box 4.3). Location and flow paths from fields to impaired waters result in differences in the proportion of edge-of-field pollution that is delivered to impaired waters. In consequence, marginal damage costs will differ across fields in a watershed when placed in the same crop and fertilized at the same rate. The optimal spatial pattern of agricultural production will therefore reflect properties of soils and their locations affecting pollution damages

Box 4.3 Pollution Propensities – Critical Source Areas

We use the term pollution propensities to capture the existence of significant differences in the likelihood that particular locations will contribute to pollution loads based on characteristic of the land and landscape rather than management characteristics. High-propensity areas are called critical source areas or variable source areas in hydrology research on nonpoint pollution. Not all agricultural land in a watershed contributes equally to downstream water quality problems and would not even if identically managed. Disproportionality is often the case with a small fraction of the land area contributing a large fraction of the pollution loads (e.g., Diebel et al. 2008; Qiu 2009; Nowak et al. 2006; Pionke et al. 2000; Rao et al. 2009; Reaney et al. 2019; Ribaudo 1989; Shore et al. 2014; White et al. 2009; White et al. 2014; Xu et al. 2019). In the Mississippi River Basin, for example, 10% of cropland is estimated to contribute 30% of the entire nitrogen load from cultivated cropland to the Gulf of Mexico (White et al. 2014). At a smaller scale, White et al. (2009) examine land sources of phosphorus and sediment within six watersheds in the US state of Oklahoma. Three watersheds were agricultural dominated with cultivated land accounting for more than 40% of the total land use. The others were grassland and forestland dominated. In the agriculturally dominated watersheds, the most polluting 5% of the lands in the watersheds yielded between 24% and 30% of phosphorus load and between 33% and 29% of sediment loads. The most polluting 5% of the agricultural lands were on average responsible for 22% of phosphorus and sediment loads of agriculture's total pollution. Pollution from these critical source areas was more than five times higher than the average load from the remaining agricultural land. In an extensively studied watershed in Pennsylvania located within the Chesapeake Bay Basin, 90% of the algal-available phosphorus came from 10% of the land (Pionke et al. 2000).

The key implication of disproportionality is that for effectiveness and efficiency, pollution control activities need to be targeted to the

(*continued*)

Box 4.3 (continued)

locations where the pollutant or pollutants causing water quality impairments are originating. Critical source areas are produced by the combination of high source risk and strong connected pathways to affected waterbodies (Reaney et al. 2019). Climate, topography, and geology are important determinants of hydrological activity. Artificial features, like drainage systems can also affect the generation and delivery of pollution loads. Critical source areas within watershed will not be identical for different pollutants. Transport mechanisms and processes for pollutants also are important. For example, sediment and sediment-bound pollutants move overland in runoff on steep slopes and under intense storms. Soluble forms of pollutants may move more in shallow subsurface flow or may be caught on vegetation during slow overland flow. In well-vegetated humid regions with steep landscapes, shallow, permeable soils underlain by a restricting layer, runoff is the dominant mechanism of phosphorus transport (Rao et al. 2009). Smaller fractions of watersheds nearer to streams tend to contribute to a larger fraction of the total delivered phosphorus loads, while nitrogen sources will be distributed over a larger area (Heathwaite et al. 2000; Pionke et al. 2000).

Watersheds will differ in the amount of lands in critical source areas for different pollutants and the size of the areas. Highly variable landscapes will be more heterogenous in terms of the loading propensity (White et al. 2009). Some areas vary in size, increasing or decreasing seasonally. These are referred to as variable source areas. Climate change will change the location and importance of critical sources areas in watersheds (Wagena and Easton 2018).

Locations of critical source areas within watersheds is not routine knowledge. Accordingly, a key task for effective nonpoint pollution management is identification of critical source areas. Various methods are being refined through research to produce tools for spatially targeting pollution controls (Easton et al. 2020; Reaney et al. 2019).

as well as yields and nitrogen responses. To illustrate, consider Fig. 4.7. Suppose first that pollution propensity, for reasons of location or other considerations, is positively correlated with land quality. The social net return of Crop 2 is reduced relative to Crop 1 and more land is allocated to Crop 1. The opposite will be true if, for reasons of location or other considerations, pollution propensity is negatively correlated with land quality. In any case, positive correlation, negative correlation, or a completely random distribution of both, the map of agricultural land use in the watershed would depend now on the spatial pattern of land qualities and pollution propensities. Box 4.4 provides a simple parametric illustration of interactions between land qualities and pollution propensities as affected by delivery ratios using the simple models developed in this chapter. Box 4.5 provides a concrete illustration of changes in crop mix, production practices, and production locations that would be required to reduce agricultural nutrient loads from the MARB to the Gulf of Mexico at least cost.

The amount and pattern of agricultural land use in the Pareto Efficient watershed will depend on net social returns not only from alternative agricultural uses but also from nonagricultural land uses. There is a huge difference in nutrient use intensity between agricultural and other rural land uses (e.g., crop production vs. forestry), and also between uses within agriculture (e.g., barley production vs. low density pasture). The gradient of nutrient intensity in land use extends from intensive agriculture, such as crop production, to extensive agriculture such as pasture lands and nonagricultural land uses such as urban development or forestry.

To illustrate, suppose that Crop 1 in Fig. 4.7 is a forest land use. The private net revenues from forestry would be as in Fig. 4.5: lower-quality lands would be more profitable in forest than crops. Nutrient loads per hectare from forestry are generally lower than from crops. Therefore, taking damages into account would shift the land use from the more polluting activity, agriculture, into the less polluting one, forestry. The entry-in/exit-from margin for agriculture in a Ricardian model based social net benefits shifts to reflect the effect of pollution damage costs.

4.4.3 *Summing Up*

There are several messages we wish to make with this discussion. One is that for any given set of crop and input prices, the intensity of polluting input use will be greater in a market-driven agriculture with externalities

Box 4.4 Socially Optimal Fertilization Choice with Varying Nutrient Delivery Rates

The concepts discussed above are illustrated numerically using the simple parametric model introduced to reflect the socially optimal solution for two homogeneous fields. One field is located near the waterway, and the other at some distance from it. The nitrogen yield response function for both fields are as in Sect. 4.3.2: $y = A + a_1 n - 0.5a_2 n^2$. Let d denote the (constant) marginal damage of nitrogen loads at the receptor. This load is identical at the field edge in both fields and given by nitrogen runoff function $r_N = \sigma_1 n + 0.5\sigma_2 n^2$. We label fields as i (near the receptor) and j. The delivery ratio for field i is $\beta_i = 1$ and for field j is $0 < \beta_j < 1$. The nitrogen load to the receiving water body is the sum of loads: $Z_N = \beta_i r_N^i + \beta_j r_N^j$, and the damage from nutrient loads is dZ_N. The social net benefit function subtracts the damage from the land rents. The total net social benefits from the two field parcels is therefore $W = \pi^i + \pi^j - dZ_N$.

The rent function for both parcels is identical, $\pi = p(a_1 n - 0.5a_2 n^2) - cn - M$. The condition for net social benefits maximizing nitrogen fertilization for field i is $p(a_1 - a_2 n) - c - d\beta_i(\sigma_1 + \sigma_2 n) = 0$. Recalling that $\beta_i = 1$ we obtain as the socially optimal nitrogen fertilization for field i: $n_i^* = \left(pa_1 - c - d\sigma_1\right) / \left(a_2 p + d\sigma_2\right)$. The field j delivery ratio is $\beta_j < 1$. By the same procedure, the optimal nitrogen intensity is given by $n_j^* = \left(pa_1 - c - \beta_j d\sigma_1\right) / \left(a_2 p + \beta_j d\sigma_2\right)$.

Looking at the optimal n_i^* and n_j^* leads to two conclusions. First, relative to the private optimum ($n^0 = (pa_1 - c)/a_2 p$ for both), the social optimum leads to a lower level of nitrogen application, as a response to damage from water pollution. Technically, this shows up as a smaller numerator and higher denominator relative to private optimum (which is obtained by setting the marginal damage, $d = 0$). Second, the optimal nitrogen application is higher in the field parcel located at a greater distance from the receiving water body than in the parcel near to the water body. Given that fields are otherwise identical, this reflects the effect of the different delivery ratios.

Box 4.5 The Structure and Scope of Cost-Effective Portfolios: Gulf of Mexico Case Study

Pareto Efficient changes in the use of polluting inputs, use of best management practices, allocation of crops to land, and changes in land use are necessarily social cost minimizing. The scope and structure of cost-effective changes in agriculture for cost-effective control are illustrated by the United States Department of Agriculture study estimates of the reducing nitrogen (N) and phosphorus (P) loading to the Gulf of Mexico from the Mississippi-Atchafalaya (MARB) basin by 45% (Marshall et al. 2018) (see Box 4.2).

Working-Land Best Management Practices (BMPs): Meeting the 45% reduction goal at least cost would require 26.6 million acres (10.8 million hectares) of working lands to implement nutrient management BMPs and 2.8 million acres (1.1 million hectares) to be planted with cover crops over 2007 baseline levels. Another 5.7 million acres (2.31 million hectares) would be required to implement drainage water management BMPs. Erosion control BMPs individually or in a portfolio of BMPs (e.g., nutrient management, cover crop, and drain water management) would be required on 8.35 million acres (3.38 million hectares). In total, additional nutrient-reduction measures would be required on about 35.4 million cropland acres (14.3 million hectares). The nutrient-reduction BMPs on working lands do not include reduced tillage as these are not a preferred practice for addressing nutrient loads to the Gulf of Mexico. Almost all (96%) of the required increase in use of BMPs is in the Ohio River Basin, Lower Mississippi, and Upper Mississippi sub-basins (see Box 3.5). Little change is required in the Missouri and Arkansas-White-Red sub-basins as the costs of delivered nutrient reductions to the Gulf are comparatively high, due in part to lower nutrient delivery ratios from those basins.

Green Infrastructure BMPs: BMPs on working lands in a cost-minimizing portfolio are not sufficient to achieve the 45% reduction target. The cost-effective portfolio also requires conversion of 1.01 million acres (0.41 million hectares) of cropland to riparian buffers and 0.34 million acres (0.14 million hectares) of cropland to wetlands. Most of the buffer expansion would occur in the Upper

(*continued*)

Box 4.5 (continued)

Mississippi, Lower Mississippi, and Arkansas-White-Red regions, with lesser amounts in the Ohio sub-basin. Most of the wetland expansion would occur in the Upper Mississippi and Ohio River sub-basins where much of the cropland was formerly wetlands and currently requires drainage to remain in production. Smaller areas of wetland expansion across all the other MARB sub-basins except the Missouri, which does not restore any wetland area.

Cropland Area: Cost-effective control to achieve the 45% reduction target is estimated to reduce the cropland area in the MARB by 4.3%. The reductions are from conversion of cropland to buffers and wetlands, from converting environmentally sensitive cropland to conservation uses under the US Conservation Reserve Program (CRP) (see Chap. 5), and from idling of cropland. Cropland removed from production are predominantly highly erodible land in the Ohio, Tennessee, and Upper/Lower Mississippi sub-basins, where cropland losses were concentrated. These reductions in highly erodible acreage are offset by increased erodible land acreages in the Missouri sub-basin, leaving the MARB-wide proportion of production on highly erodible acreage (29%) roughly the same.

Irrigated Versus Non-Irrigated Cropland: Irrigated cropland, which accounts for about 12.3% of the working-land acreage in the MARB, accounts for 18.6% of the cropland reduction. This result is driven almost entirely by acreage shifts in the Lower Mississippi where 40% of the acreage is irrigated in the baseline, 49% of the decline in acreage under the Gulf Constraint scenario comes from irrigated production. Irrigated land is increased in some regions.

Crops: Across the MARB, hectares in corn would be reduced by 2%, hay by 3.5%, soybeans by 5.6%, and wheat 3.8%. The reductions reflect the regional distribution of production as it affects delivery, and the relative costs of reducing nutrient reductions from the crops. Increased crop prices with the MARB supply reductions lead to increased production of the crops outside of the MARB so that the net supply reduction is smaller than the MARB reductions.

than in Pareto Efficient agriculture. We have illustrated this for nitrogen fertilizer, but the conclusion generalizes. When inputs increase external costs, their use will be excessive from a societal perspective without policy intervention. The private net benefit of their use is maximized by a higher quantity than that which maximizes the social net benefit. The degree will depend on a variety of factors including crop and input prices that determine input use. A second message that we have not developed explicitly but that is a corollary of the first message is that too little use will be made of inputs that reduce pollution damages as the private net benefits of their use will be less than the social net benefits of their use in the presence of pollution externalities. We takeup this topic further in a subsequent section on best management practices.

A third message is that the amount of land in agricultural use and the pattern of use in watersheds will differ between market-driven land use with externalities and Pareto Efficient land use patterns. In both cases, the net returns from competing land uses matter. In the market-driven landscape with externalities, it is private net returns from private uses that determine overall patterns and amounts. In the Pareto Efficient landscape, social net returns from both public and private uses determine overall patterns and amounts.[8] Social net returns take into account both the private net benefits and the pollution damage costs of land uses as affected by the interactions between land use, land characteristics, location, and the social costs of water quality impairments. In general, we expect the Pareto Efficient watershed to have less land area in pollution-intensive crop and livestock activities. Some agricultural land may be shifted from intensive crop production into green infrastructure that captures and filters pollutants such as riparian buffers and constructed wetlands (Sect. 4.4.6). Some may be shifted into low-intensity commercial uses like pasture or perhaps forest. Some may be even by shifted to developed uses in which significant

[8] As a caveat we note that land allocation and land use change is a considerably more complex topic than we have represented. The basic messages regarding the amount and pattern hold true, but a comprehensive analysis of market-driven and socially optimal landscapes requires consideration of many additional issues. There are many externalities in land use other than water quality that affect optimal patterns of use. There can be significant adjustment costs in land use change that affect market and Pareto Efficient allocations. Some land use changes are economically or ecologically irreversible so that the history of land use constrains subsequent use.

pollution controls are applied (Wrenn et al. 2019). In addition to shifts in overall land use, we would see changes in the spatial pattern of use reflecting the effects of location on external costs. Pollution-intensive agriculture would be removed from high-propensity locations to be replaced by lower-intensity agricultural or nonagricultural uses.

In addition to these messages the analysis also prompts thinking about scope of information required for watershed planning and the utility of empirical models for policy analysis (see Box 4.6). This is a theme we continue to develop through the book.

Box 4.6 Empirical Simulation Models and the Ricardian Agricultural Land Use Model

The structure of agricultural production that reduces nutrient loads from the Mississippi-Atchafalaya basin to the Gulf by 45% described in Boxes 4.2 and 4.4 is computed using the US Department of Agriculture's Economic Research Service Regional Environment and Agriculture Programming (REAP) model. REAP is a large and sophisticated example of an integrated economic-biophysical model for *ex ante* policy analysis.

REAP was developed to represent agricultural production and its environmental impacts for 273 production regions across the United States (Johansson et al. 2007; Marshall et al. 2018). It is essentially a numerical implementation of the Ricardian land use model where the production regions replace the farms and fields that we described as the spatial production units. For each production region the model includes a set of candidate crop rotations and candidate farming practices for producing them. It also includes regional livestock production activities. The Ricardian model in our presentation assumes that market-driven agriculture maximizes Ricardian rents. However, in our conceptual model, prices of agricultural commodities are exogenous. Because the REAP model represents a very large agricultural economy that can affect commodity prices, REAP allocates production acreage among candidate crop rotations and farming practices available in each region to maximize the sum of

(*continued*)

Box 4.6 (continued)

consumers' and producers' surpluses from production and consumption. As described in Chap. 2, competitive markets maximize this joint surplus. The REAP model therefore determines market equilibrium commodity prices, crop supplies, cropland allocations, rotations, etc. However, consistent with our presentation of the Ricardian model, the structure of agricultural production in the REAP model maximizes Ricardian rents given the market equilibrium commodity prices.

The application of the REAP model to the Gulf Hypoxia solves for the market equilibrium subject to restrictions on nutrient delivery to the Gulf of Mexico. In the scenario where nutrient reductions are not predefined for production regions but are allocated according to relative costs of delivered load reductions, maximizing economic surpluses subject to the restriction minimizes the social cost of the nutrient reduction from agriculture (ignoring other agricultural externalities).

Integrated economic-biophysical models consistent with the Ricardian theory such as REAP are commonly used for *ex ante* analysis of agricultural water pollution controls in agriculture. REAP is unusual in the size of the agricultural economy that it represents. Economic research makes extensive use of smaller-scale models for fields, farms, and small watersheds. Trade-offs exist between the spatial scale captured by a model and fine scale management details and representation physical processes. Trade-offs also exist between the degree to which models represent complex physical processes and complex economic decision making and behavior (Kling et al. 2017).

4.5 Generalizing the Ricardian Model

Sections 4.3 and 4.4 introduced basic economic models for explaining producers' selection of crops, the location of production activities, and the nitrogen intensity of production. The same models were used to introduce the concept of Pareto Efficient agricultural production at farm and watershed scales and to demonstrate that market-driven agriculture is not Pareto Efficient when agriculture produces water pollution externalities. The models we introduced were sufficient to develop the key messages summarized above but do not include some variables and topics important to farm decision making and the spatial structure of agriculture relevant to water quality.[9] We discuss these in this and subsequent sections.

4.5.1 Phosphorus Fertilization

Yield response to soil phosphorus is analogous to nitrogen. It increases with the availability of phosphorus in the soil and displays diminishing marginal returns. However, the economics of phosphorus management is different. Nitrogen behavior in soils is highly dynamic and weather dependent. Nitrogen applied to the soils that is not removed by plant growth does not remain in the soil. It is removed by various processes (leaching, surface runoff, denitrification, ammonia volatilization) on fairly short timescales. In contrast, phosphorus that is not removed by plant growth generally remains in the soil. Soil phosphorus increases if applications exceed crop use. In soils in which soil phosphorus has been maintained at adequate levels, nitrogen is typically the limiting nutrient. Producers must apply some nitrogen annually to achieve profitable yields. Crop response to annual phosphorus applications is typically small. Phosphorus applications are aimed at maintaining soil phosphorus reserves. Farms with sufficiently large dairy or swine operations generally need not apply purchased phosphorus as cattle and swine manure is rich in phosphorus. Manure applied to cropland supplies more phosphorus than needed for crop growth.

In terms of economic decision making, optimal phosphorus management is a time-dependent problem. In equilibrium, phosphorus is applied to offset removal to maintain the soil phosphorus at the optimal level from year to year. If the pool of soil phosphorus is below or above the target level, fertilization is used to elevate or deplete the pool at the rate dictated by economic considerations and the soil processes (Iho and Laukkanen 2012). Soil phosphorus processes therefore play a key role in optimal fertilization choices.

From a societal perspective, phosphorus applications and soil phosphorus levels in intensively farmed agricultural areas are excessive. As with nitrogen, this is because producers do not account for the social costs of phosphorus pollution. Unlike nitrogen, the environmental problems and

[9] The models we have presented are bare bones versions to introduce key concepts and messages with minimal complexity. Models used in economic research are considerably more complex in the variables and relationships they include.

management solutions have a strong time dimension in two ways. First, phosphorus in agricultural soils can continue to flow from farms to water resources for many years after reductions in phosphorus inputs. Second, phosphorus accumulates in the sediments of water bodies where it can contribute to water quality problems long after loads from agricultural land have decreased. In consequence, very long lags, decades or more, can exist between reductions in phosphorous use in agricultural production reduced phosphorous pollution.

4.5.2 *Manure and Livestock*

Manure management is a significant feature of livestock operations. Producers value manure as a source of nutrients and organic matter that promote soil health. There are many dimensions to handling and the application of manure that affects its utility as a fertilizer and the cost of its use. These include the methods by which manure is collected from animal facilities (e.g., barns, barnyards), the ways in which animals are kept (e.g., concentrated housing vs. free range), whether the manure is stored and how it is stored, whether the manure is treated in some way (e.g., fed to a manure digester), the timing and methods of application, and where it is spread on the farm. Also important is the volume of manure relative to the land area of the farm. Manure can shift from being a useful resource for a farm to a waste disposal problem if the volume of manure produced is sufficiently large relative to the land area of the farm.

The simple economic model of nitrogen fertilization presented above can be used to gain some insight into the economics of manure use as a fertilizer. With livestock on the farm, nitrogen can be supplied to crops in some combination of purchased fertilizers and livestock manure. The cost of the mineral fertilizer denoted c above, is determined by the price and application cost. The cost of manure is determined by the application cost. Denote this cost as c_m per unit. The equipment used for purchased fertilizer and manure are not the same. For this and other reasons, application costs differ between the two. If the cost of providing manure nitrogen to the crop is less than the cost of purchased fertilizer ($c_m < c$), manure is clearly economically preferred to purchased nitrogen. If the nitrogen supply in manure from farm livestock is less than the amount that would maximize net return, it is then economically advantageous to increase the overall application with purchased fertilizer. In this case, the supplemental purchased fertilizer application must again equalize marginal revenue and marginal cost (i.e., $pMP = c$) to maximize net return. The overall nitrogen application would be the same as without manure. Manure from the farm provides a share of the total that depends on the manure supply relative to the optimal application.

If manure supplies more nitrogen than the rate at which the purchased fertilizer optimization condition is satisfied (i.e., $pMP < c$ if all the manure is applied), supplemental purchased fertilizer is uneconomic.[10] However, if available, additional manure application is economic up to the point at which marginal revenue is equal the marginal manure application costs ($pMP = c_m$). Beyond this point additional manure application reduces net returns. If the supply of manure exceeds the net return-maximizing level, the excess manure is a waste management problem. The economic problem then is the cheapest way to handle excess manure. Choices include applying the manure and suffering the loss of net returns, paying for off-farm disposal, the use of crops and tillage systems that accommodate larger manure applications, and adjusting the scale of the animal operation relative to the farm size.[11] The loss of net returns from excess manure handling is a cost of the livestock operation. There is therefore a trade-off between the net returns from additional animals and the costs of handling excess manure.[12]

Moving manure from barns or storage facilities is costly. It requires time, labor, fuel, and equipment costs. Essentially, the cost of manure application on a farm, denoted by c_m above, increases with the distance from the manure storage. The profit-maximizing solution has manure applied only up to a certain threshold distance. This critical radius depends on hauling costs but also on the alternative sources of nutrients, availability of cropland, willingness of crop farm to have the manure applied on their fields and the agronomic needs of the cultivated crops (Fleming et al. 1998). If mineral fertilizer costs increase, manure is spread over longer

[10] Even with ample manure producers may use some purchased fertilizer as a starter to initiate plant growth. Fertilizer management can use split applications. A starter is applied early in the season. Additional fertilizer may be applied after planting based on soil test indicating need.

[11] Limits on manure application to fields may exist for reasons other than diseconomies of fertilizer application. For example, large manure application volumes may interfere with some field operations.

[12] In some locations there may be a demand for manure sufficient to support markets in which excess manure can be sold as a fertilizer, soil conditioner, animal bedding (after treatment), or for other purposes (Parker 2004). The ability to sell manure can reduce the cost of excess manure when available.

distances (Innes 2000; Schnitkey and Miranda 1993; Kaplan et al. 2004). Also, the application rates within the critical radius decrease with distance. The application rate at the critical distance is lower than the rate on fields closer to the facility (Lötjönen et al. 2020). The use of the farmland area, herd size, crop choices, diet and manure application rates are jointly optimized to maximize net returns from crop and livestock production. The optimal solution may also have manure exported off-farm, particularly if regulations restrict manure application on the farm's fields.

Concentrating phosphorus applications close to facilities leads to high soil phosphorus concentrations in those locations. Phosphorus applications are typically excessive even though nitrogen would be applied exactly to crop agronomic needs. For almost all livestock types, the ratio of manure phosphorus and nitrogen is higher than the ratio of agronomic needs of typical crops. And if nitrogen is applied excessively, phosphorus application will be even more so. Long-term manure applications are a prominent reason for legacy phosphorus problems in intensive production areas, such as Delmarva peninsula by the Chesapeake Bay (Kleinman et al. 2019).

From a societal perspective, privately optimal manure applications are excessive. Accounting for the environmental damage from the nutrient loading implies that manure is hauled for longer distances, herd sizes are smaller, more land is allocated to silage (Lötjönen et al. 2020). What the optimal management choice for one farm would be, depends on the current herd size, the livestock type, the availability of land for manure applications on and near the farm, and the costs of treatment technologies.

4.5.3 Irrigation

Analogous to nutrient response functions are water response functions that describe how crop yields increase with water inputs. Like nutrient response functions, water response functions show diminishing marginal returns. Unless a crop is grown in standing water like rice, crop yields fall if too much water is applied.

Economically, the water input decision is like the fertilization decision (Boggess et al. 1993; Caswell and Zilberman 1986; Pfeiffer and Lin 2014; Green and Sunding 1997; Green et al. 1996; Griffin 2016). Water inputs increase output and revenues but at a cost. The cost of the water input depends on the price of water if water is purchased, the cost of pumping water if it is taken from aquifer, and the cost of applying the water to fields. The rule for net revenue maximization is to equalize marginal revenues with the application cost. The decision is dependent on the crop type, which determines net revenues from irrigation and the marginal cost of water.

Crops vary in their water requirements. Some are water intensive, like corn. Some are less so, like wheat. Irrigation technologies vary greatly in the efficiency in delivering water to crops. Water-saving irrigation systems are capital intensive. Producers face a trade-off between cost savings from lower overall water usage versus the capital costs of the water-efficient irrigation technologies. This trade-off is strongly influenced by land quality as it affects land allocation to crops that vary in their water needs and compatibility with alternative irrigation systems. Water cost savings is only one of a variety of factors influencing the economics of irrigation technology adoption (Pfeiffer and Lin 2014; Green et al. 1996; Green and Sunding 1997).

Irrigation water applied to crops can be taken up by the crop, other vegetation (e.g., weeds), evaporate, runoff in irrigation return flows, or percolate through the soil to groundwater and eventually to surface water. Water moving through soils captures nutrients, pesticides, and salts. The volume of irrigation water is a key determinant of water quality outcomes. As in the case of fertilizers, when irrigation practices result in water quality or other externalities (e.g., subsidence), producers do not bear the external cost of their irrigation decisions. The result will be socially excessive water use and excessive allocation of land to water-intensive crops (Dwyer et al. 2006).

Water conservation policies can amplify this outcome. For instance, in an example of the rebound effect, Pfeiffer and Lin (2014) have found that policies to encourage the adoption of more water-saving irrigation technologies in the US state of Kansas increased the profitability of producing higher-value crops on previously marginal land, resulting in a shift away from dryland crops to irrigated crops. It may also induce the planting of more water-intensive crops on already irrigated land as more efficient irrigation increases the amount of water the crop receives per unit applied.

Amplifying this outcome is water pricing. The cost of water to producers is rarely market determined as water markets are rare. Some producers access water directly from groundwater or surface water supplies. When this is the case, producers' costs are determined by the cost of private acquisition. Because of the economies of scale in their development, irrigation systems are typically developed by governments (Griffin 2016). Various methods for allocating water are used, with water pricing in some form being common (Johansson et al. 2002). Water subsidies or other pricing defects that result in producers paying less for water than the social cost of the water encourage water use, magnifying the distortion from missing markets for pollution or other externalities associated with irrigation.

4.5.4 *Pesticides*

The economics of pesticides differs from nutrient and irrigation water in fundamental ways. Whereas nutrients and water have a direct yield-increasing effect on output, the effect of pesticides is indirect. Pesticides are a damage control input (Lichtenberg and Zilberman 1986). Damage control inputs do not increase potential output but contribute to production by reducing yield or quality losses (Sexton et al. 2007). Animal immunizations and antibiotics are other examples of damage control inputs.

Like fertilizers and irrigation decisions, producers' decisions regarding the use of pesticides involves a comparison of their private benefits and costs. And like fertilizer and irrigation decisions, crop types, land qualities, and climate strongly influence net returns. To a greater degree than fertilizers and water, optimal pesticide management is information intensive. It requires information on pest pressures, how they are affected by farm conditions, the effects of infestations on yields or quality, the suite of chemical and other methods available for pest management, and their effectiveness (Waterfield and Zilberman 2012). The information intensity and costs of finely tuned pest management is a barrier to adoption (Waterfield and Zilberman 2012).

Some use of pesticides in agriculture is prophylactic, analogous to the use of immunization of livestock. The dominant economic model for pesticide decision making is the economic threshold model in which pesticide use is in response to the presence of a pest and dependent on the pest density. At low densities pest damages are too small to warrant application. When densities pass an economic threshold, pesticides are applied to limit damages to yield and/or crop quality. The economic threshold is the density at which the cost of applying pesticides equals the value of the avoided damage. Factoring into pest management decisions is the development of pest resistance. The damage control approach leads to increased use of an agent with the development of resistance until the resistance is so severe that the agent is no longer cost-effective (Sexton et al. 2007).

Like fertilizers and irrigation water, externalities associated with pesticides result in socially inefficient use. Unlike other agricultural inputs, pesticide use is often heavily regulated. Most regulations are of the "command and control" type, controlling which pesticides can be placed on the market and the conditions of use (Waterfield and Zilberman 2012). An important but controversial innovation for pesticide use is the development of pest-resistant Genetically Modified (GM) crops. Advances in Genetically Modified Organisms (GMO) may also have other beneficial environmental effects (see Box 4.7).

Box 4.7 Genetically Modified Organisms

Technological innovation is of tremendous importance to managing agriculture's impacts on the environment given the increased stresses that will come with demand growth and climate change. Innovations in integrated pest management, precision agriculture, and other farming practices are one important domain of technological change. Another important domain of innovation is the development of new products that reduce the demand for environmentally harmful inputs and practices. We discussed innovations in industrial processes to supply proteins in Chap. 3 (Box 3.4). Another important area of technological development is Genetically Modified Organisms (GMOs).

The crops and livestock that are produced today are mostly the result of thousands of years of human selection. In the beginning, selection was passive. Farmers observed differences in the performance of different plants and saved seeds from plant with preferred traits for subsequent planting. With time humans learned to actively develop plants or animals with desirable traits by selective breeding. Individuals within a species are selected for mating to produce progeny with desired characteristics. Genetic engineering (GE) emerged after the 1970s as a method to produce plants and animals with desired traits using recombinant genetic technologies. The technology generally increases the speed and precision of producing new organisms, referred to as GMOs. It enables the production of organisms that could not be produced by traditional breeding methods.

The technology has appeal as a means to rapidly increase overall agricultural productivity, reducing the costs of food and other agricultural products, and reduce environmental harms from agricultural

(continued)

Box 4.7 (continued)

production, reducing the social cost of agriculture (Bhullar and Bhullar 2010; NRC 2010; Zilberman et al. 2018). GE applications to crops have evolved over time. Initially GMO development focused on traits related to pest control, either by improving plant resistance to pests, which can reduce the use of herbicides, or plant tolerance to herbicides. This was followed by developments focused on enhancement of other traits affecting, for example, nutritional value, shelf life, drought tolerance. Recent developments focus on traits that allow new uses of plants, e.g., to produce proteins and chemicals used in pharmaceuticals, industrial products such as silk, or substitutes for plastic or fossil fuels. There is much interest in the possibility of using GE to increase the nitrogen use efficiency of cereal grains and other crops and therefore as mechanism to help reduce fertilization and nutrient pollution.

Klümper and Qaim (2014) indicate that the benefits from using genetically modified crops were the highest for the poorest farmers in developing countries. This promise has great appeal given the growing demands for agricultural production and the increased pressures this will place on land and water resources. This GE promise also has great appeal given the challenges that climate change poses to meeting future food demands and providing food security in developing countries (Baldos and Hertel 2014; Wheeler and Von Braun 2013).

The spread of GMOs has been constrained by regulations emphasizing precautionary approaches and reflecting public skepticism. Labeling GMOs has been made mandatory by the European Commission and some countries have banned the growing of GMO products. Many other countries have imposed other restrictions on their use. There are fears of unintended risks of GMOs, making it difficult to weigh the potential benefits and risks of the new technology. Essentially, it is not possible to prove—with any new technology—that risks are completely absent (Zilberman et al. 2018). Therefore, regulation of new technology is eventually a social choice that must be made without full scientific certainty.

4.5.5 *Spatial Interdependencies at the Farm Level*

In the Ricardian model presented in Sects. 4.3 and 4.4 the net returns from alternative land uses are independent. The crop choice for one field does not affect the net returns from other crops in other fields. This is a useful simplification for introducing the model and deriving some general conclusions, but it is not the general case. One source of interdependencies is scarce farm resources that must be allocated across fields. For example, operator labor must be allocated across fields and livestock activities. Demands on operator time can be a major factor in farm decision making. Another source of interdependencies is joint production technologies. Crops and livestock when produced on the same farm are jointly produced. Crops are selected in combination with purchased feeds to meet the dietary needs of the livestock enterprise at the lowest cost. Livestock manure provides nutrients for the crops. Land use choices across farms and rotations are used for managing pests, soil fertility, and soil erosivity.

4.5.6 *Spatial Interdependencies at the Watershed Level*

Economic and technological interdependencies also exist across farms affecting spatial patterns of crops and farming activities in watersheds (Belcher et al. 2005; Lee et al. 2019; Lewis et al. 2008; Parker and Munroe 2007; Rucker et al. 2012). Organic farming, coexistence of genetically and nongenetically modified crops, pollination, and invasive species management are examples of agricultural land uses that interact, affecting net returns and influencing land use decisions. Interdependencies between farms result in externalities and economically inefficient landscapes for reasons other than water quality externalities when Coasian bargaining fails to coordinate farm decision making (Lewis et al. 2008; Parker and Munroe 2007).

4.5.7 *Co-benefits, Rural Amenities and Disamenities*

A final set of considerations when thinking comprehensively about the economic efficiency of agricultural land use patterns are co-benefits and amenities and disamenities. The term co-benefits is commonly used to describe environmental public goods that are jointly produced by conservation and water quality protection practices in agriculture. Carbon sequestration, wildlife habitat, and biodiversity are examples. Co-benefits are an important consideration for pollution policy design in agriculture. We describe co-benefits from specific agricultural practices in a subsequent section. In Chap. 8 we describe policies for integrating co-benefits into pollution policies for agriculture. We should note that co-produced environmental effects are not always benefits. They are sometimes costs. For example, riparian buffers installed along streams to filter nutrient and sediment in runoff can harbor weeds and provide habitat for plant disease carrying insects. Generally, any action done on the farm may influence the entire matrix of environmental attributes.

Rural amenities and disamenities are respectively public goods and bads that agricultural landscapes provide to people who live or work in proximity to agriculture or visit farming areas. Many people find agricultural landscapes aesthetically pleasing and culturally significant. Land use policies are often used to limit conversion of agricultural land to developed uses. Maintaining economically viable agriculture in a major objective of EU agricultural policy. Common agricultural disamenities are odors and flies associated with livestock operations, and dust from tillage operations, and smoke from burning fields.

4.6 Agricultural Best Management Practices

We have focused to this point on decisions regarding productive inputs (land, water, and fertilizer) and damage control inputs (pesticides, pharmaceuticals) that have direct effects on water pollution. We turn now to Best Management Practices (BMPs), which are not individual inputs but technologies used to conserve soils, manage pests, and reduce water polluting emissions (Arabi et al. 2007; Giri et al. 2014). Some BMPs involve routine dimensions of agricultural production that serve agricultural productivity, but the economics of their use are not well-described by the productive input decision paradigm we have developed thus far. This is because BMP implementation often involves discrete choices with indirect effects on output rather than the continuous variables with direct effects on output we have considered in previous sections. BMPs are also often long-lived compared to the annual decisions we have described for fertilizer and crops allocations. Some BMPs do not serve agricultural productivity and would not be implemented by producers unless they are provided incentives to adopt or required for regulatory compliance.

4.6.1 BMP Types and Economic Aspects

BMPs can be categorized in a variety of ways. One is by environmental function. Common types within this category are BMPs for animal and crop nutrient management, manure management, pest management, erosion and sediment control, and water conservation. Another way to categorize is by production system. Common within this category are BMPs for crops, livestock, and pastures. Still another is by physical characteristics. Structural BMPs involve the construction of "gray" or "green" infrastructure. Gray infrastructure typically uses concrete and steel. An example is a manure storage structure. Green infrastructure uses vegetation and soils. Examples are terraces, sediment basins, riparian buffers, and grassed waterways. Most agricultural BMPs are nonstructural. Examples are conservation tillage practices, contour tillage, and strip cropping for erosion and runoff management, soil testing to inform fertilization decisions, crop rotations to manage soil nutrients, erosion, and pests, and cover (catch crops) to capture nitrogen.

BMP implementation entails costs. Almost all BMPs require some planning to fit BMPs into the physical circumstances of a farm. Producers may do the planning on their own, use planning services (technical assistance) provided by government agencies, or hire farm consultants. Structural BMPs involve construction or establishment costs and annual maintenance costs. And, when land must be removed from production for the facility, there is the opportunity cost of forgone income from the land. Some nonstructural BMPs may require capital investments in equipment (tillage equipment, manure application equipment) or annual outlays for materials or services (e.g., soil tests).

The economic benefits of BMPs are commonly divided into private and public benefits (Pannell 2008). Private benefits are those that accrue to the producer or landowner. Contributions to farm profits stand out but research on adoption recognizes the producers to varying degree derive other benefits from BMP adoption. These include the satisfaction they receive from environmental stewardship and standing in their communities (e.g., Chouinard et al. 2008; Pannell et al. 2006; Pannell 2008; Pannell and Claassen 2020). Public benefits are the resulting external benefits, such as improvements in water quality. BMPs for water quality protection can also generate a variety of co-benefits to the public.

When making decisions about BMPs producers or landowners weigh their private benefits against their private costs. These will often accrue over years, so the decision is often an investment decision (see Box 4.8).

Box 4.8 BMP Decisions for Private Net Benefits

Economic decisions about variable agricultural inputs entail weighing the effects on farm revenues and costs in specific production periods. Decisions that involve major equipment purchases, earthworks, planting of long-lived plants, or removal of land from production for long periods of time require considering effects on revenues and costs over the life of the decisions.

A key concept when evaluating the economics of investments is the ***opportunity cost of capital***. To illustrate, consider a producer who can invest 10,000 USD today in new farm equipment or place the money in a financial investment that pays an annual dividend. The new equipment would increase annual farm profits by 1000 USD for 15 years. The annual net rate of return from the equipment is 10%. If the rate of return from the annual dividend from the financial investment is greater than 10%, the financial investment is the better economic choice. If the annual dividend is less than 10% the new equipment is the better economic choice. The opportunity cost of capital for private investment decisions is generally the forgone return from the investment that provides the highest net returns.

Alternative farm investments can be ranked by comparing their annual rates of return if the annual net returns are constant over time, as in the example. It is more often the case that there is variability over time. For example, agricultural BMPs may have large initial costs and then annually recurring costs of operation, maintenance, or opportunity costs from removing land from crop production over the life of the BMP. When subsidies are available, these costs may be offset by public payments. The private benefits from the BMP, if any, may vary from year to year. For example, the annual returns from BMPs that benefit producers by reducing variable costs (e.g., labor and fuel from reduced tillage practices, water from irrigation BMPs, pesticides from IPM) will vary with input and crop prices.

A standard method for evaluating investments with differing time net return profiles is to compare their ***net present values***. Net present values convert sums received at different points in time into units of equal value. Fundamental to the concept is differences in the ***time value of money***. For various reasons, a USD, EUR, or other currency received in the present is worth more than the same amount in the

(*continued*)

Box 4.8 (continued)

future. One reason for this is that money invested earns a return. In our example, if financial investment provides the producer with a 15% annual return, then in the second year the 10,000 USD has grown to value to 11,500 USD, i.e., the initial investment plus the first-year return of 1500 USD. If the producer can earn 15% in a year on the 10,000 invested today, then the producer as an investor would not be willing to trade 10,000 USD today for any less than 11,500 USD in a year. 10,000 USD is worth less a year from now than it is today. The equivalent value next year to 10,000 USD today is 11,500 USD. This is not due to inflation but due to the opportunity to earn a return on money invested. Consumers as well as producers display time preference when making purchasing decisions.

Calculating present values involves two steps. The first step is to convert future net returns in each period into present values. This step is done by ***discounting*** the future values to the present using a ***discount rate***. The discount rate for private investment decisions is the opportunity cost of capital. To illustrate, suppose when considering a BMP investment that the producer can earn an interest of r% on money invested. The general formula for the present value of an amount received or expended t years from the present, denoted NR_t, is $PV_t = NR_t/(1 + r\%)^t$. For example, if $t = 1$ (i.e., "next year"), $NR_t = 1{,}000$, and r% = 0.15, then $PV_1 = \frac{1{,}000}{1.15} = 869.57$. The second step adds the discounted sums over time. The formula for a project with a life of T years a constant discount rate or r% is $PV = \sum_{t=0}^{T} PV_t = \sum_{t=0}^{T} NR_t / (1 + r\%)^t$. From this formula it is apparent that the present value of a flow of net returns is less than the nominal sum of the flow. It is also apparent that present values vary inversely with the discount rate. Public subsidies toward costs would be included as a reduction in costs in the calculation of present values in the time periods in which they are received.

There are some important implications of discounting. First, the weight placed on sums expended or received decreases exponentially over time. This creates an economic bias against BMPs or other projects with high upfront costs and low near-term benefits. A second is that discount rate has a very strong impact on present value. Small increases in the discount rate can produce large decreases in present value.

Because producers do not consider all public benefits, the decisions they make will be socially inefficient. They may adopt some practices that are both privately and publicly beneficial but not to the extent that is socially optimal. They cannot be expected to adopt practices with large social benefits but negative private net benefits. Public information programs, technical assistance programs, and subsidies for BMP adoption that are introduced in Chap. 5 are based on the idea that BMPs provide public and private benefits and intended to reduce producers' costs so as to increase their private net benefits.

4.6.2 BMPs for Crop Production

Tillage Practices Planting and tilling are routine functions in crop production and directly affect profits. And as described in Chap. 3, tillage has a significant impact on erosion and runoff. Conservation tillage practices are an important set of agricultural BMPs that limit use of tillage or the soil disruption from tillage. No-till systems are emphasized. Private benefits can be substantial as no-till conserves soils and improves soil health, representing an investment in farm's natural capital, and adding to farm income by saving fuel, time, and machinery costs (Soane et al. 2012). Private benefits from conservation tillage has led to significant adoption (Caswell et al. 2001; Friedrich et al. 2012). However, it eliminates the weed control achieved by traditional ploughing and hence increases costs from prophylactic pesticide treatments (Caswell et al. 2001). In consequence, no-till is little used in organic agricultural production systems (Mirsky et al. 2012; Zikeli and Gruber 2017). It may also provide lower yields during wet growing periods (Soane et al. 2012).

Potential public benefits include reduced sedimentation and reduced nitrogen and total phosphorus loads. However, no-till tends to elevate the loading of dissolved phosphorus and therefore careful considerations should be made to balance the pros and cons related to its environmental effects (Sharpley and Smith 1994; Dodd and Sharpley 2016; Jarvie et al. 2017).

There are various other conservation practices related to tillage, each with varying energy, labor, and machinery costs as well as indirect costs from, say, increased need for chemical weed control. Common practices include chiseling and subsoiling, which minimize the mixing of surface soil but improve aeration and infiltration of water; contour tillage for steeply sloped fields; and strip cropping, which alternates, say, strips of grass and tilled soil to mitigate erosion on one hand and to obtain the benefits of conventional tillage on the other.

Cover Crops Cover crops (grasses, legumes) are planted with or after the harvest of a primary crop—winter wheat, for example, after a corn harvest. They are used to provide a vegetative cover to protect soil after the harvest of the primary crop and to capture nitrogen. There are multiple potential private benefits associated with the use of cover crops. Cover crops increase water infiltration, reduce erosion, increase soil carbon, improve the natural capital of the farm, and contribute to weed control (Miller et al. 1994). However, there are some disadvantages, including an increase in plant disease risks (Hoorman 2009). Public benefits include reducing nitrogen and sediment loading. However, the practice has been shown to increase the loading of dissolved phosphorus (Bechmann et al. 2005; Smith et al. 2017).

Integrated Pest Management (IPM) IPM refers to a suite of practices that include alternatives to the use of synthetic pesticides and that increase the cost-effectiveness of pesticides when used. A key component of IPM is continuous monitoring and identification of pests to assess risks (Waterfield and Zilberman 2012). Pesticide use is minimized and triggered after some threshold in pest quantity is crossed. Other actions include biological controls (use of natural enemies of the pests) and use of pest inhibiting farming practices (e.g., rotations). Specific IPM practices vary regionally with variations in crops, field size, farming practices, climate, and landscape structure.

IPM can provide significant private and public benefits. In a study of the economic impacts of IPM use to control soybean aphid, Song and Swinton (2009) estimate gains to consumers and soybean growers for the worth of $1.3 billion over 15 years. Kim et al. (2008) find similar benefit ranges for soybean producers in the US. However, private adoption rates appear to be slowed by the information intensity and complexity of IPM (Waterfield and Zilberman 2012).

IPM is enhanced by models for forecasting the timing and location of pest infestation, economic damages from infestations, and the effectiveness of control practices. The large role of information and forecasting in IPM and the constant evolution of pest risk make for a significant role of research and public information systems in pest management.

Green Infrastructure This set of BMPs focuses on land use. Riparian buffer strips are areas covered with permanent vegetation adjacent to ditches or streams. They capture sediment and nutrients in runoff from fields before they reach water bodies. The effectiveness of buffers as nutrient sinks is controversial. Like any other permanent cover, they tend to enrich the topsoil with phosphorus and thereby increase the loading of dissolved phosphorus (Hoffmann et al. 2009). They may sometimes act as habitat for species that transmit plant diseases such as in the case of Pierce's disease (Brown et al. 2002). Wetlands are areas where the soil is covered by water at least most of the time. Constructed wetlands are intended to slow the runoff and allow the sediments settle and nutrients undergo biogeochemical processes that reduce their eventual loading. They serve as sinks for nutrients but also provide biodiversity benefits (Van der Valk et al. 1978; Nichols 1983).

BMPs that require removing land from production to create buffers or wetlands entail significant costs and do not contribute to farm productivity via improved soil capital or significant income. The costs are those of establishing and maintaining the buffer of wetland, and the opportunity cost of removing land from crop production. The net private benefits are distinctly negative with the result that private adoption without public subsidies is rare.

Precision farming Precision farming refers to input-management practices where input use, for example fertilization rates, follow the variability in growing conditions within the field scale. GIS techniques are combined with measurement data to detect and utilize the variability. The spatial scale can be very small, up to 4000 grids per hectare (Tisseyre et al. 2018). If the variability is measured with soil sampling instead of remote sensing, the scale is much larger.

Soil phosphorus, for instance, exhibits strong spatial distribution (Kleinman et al. 2011; Uusitalo et al. 2007). Its measurement is based on manual sampling and is therefore done on a coarser scale. Because crops respond to the potentially plant-available phosphorus, optimal phosphorus fertilization rates are dependent on the indices. By knowing the soil phosphorus status, producers may save phosphorus application costs (if the levels are excessively high). This also helps mitigate phosphorus loading.

Improving the fertilizer use efficiency with precision farming is based on obtaining and utilizing a more accurate understanding on soil fertility and hence on crop response. Fertilizing an entire field at the same rate will result in overfertilization of some areas and underfertilization of others. Both result in a loss of profit. The benefit of precision farming is to reduce the quantity of inputs required to achieve desired yields. On the costs side, precision farming requires investment on measurement devises and data management. The adoption of precision farming technologies is driven by farm's capabilities to handle the investment costs. Large farms are more likely adopters. Human capital is decisive, particularly the familiarity with computers (Khanna 2001; Pierpaoli et al. 2013).

Improved nutrient use efficiency also reduces nutrient loads, given there are no land use changes. There are thus social benefits to precision farming as well. However, this does not remove the difference between socially and privately optimal fertilization rates as described in Chap. 2 and earlier in this chapter. Even though the fertilization rates would be lower in privately optimal farming with precision farming than in socially optimal traditional farming, the socially optimal fertilization rates with precision farming would be even lower.

Phytoextraction and Soil Amendments Phytoextraction is a method for depleting the soil phosphorus accumulated in the past. It involves planting grasses with high phosphorus uptake to extract soil phosphorus. High nitrogen fertilization rates are used to maximize growth and phosphorus removal. Van der Salm et al. (2009) found 30% to 90% reductions on dissolved phosphorus concentrations in the solution of the topmost soil layer after five years of intensive phytoextraction. Dodd et al. (2014) found comparable results for two New Zealand soil types, emphasizing the potential benefits of accelerating the soil phosphorus depletion by promoting growth with higher nitrogen fertilization.

Phytoextraction offers some interesting trade-offs. To maximize phosphorus removal requires producer to apply nitrogen at *higher* rates than privately optimal, reducing farm profits. The social benefit-cost reasoning is that in addition to the private net revenues from nitrogen use there is a benefit to society: the rate with which the additional nitrogen application unit accelerates soil phosphorus depletion and therefore phosphorus loading. However, the overall social benefit-cost calculus must also consider the external costs of increased nitrogen, raising an interesting question of balancing damages from phosphorus and nitrogen loading.

Soil amendments are added to affect the chemical composition of soil. Alternatives include liming, structural liming, and gypsum. The liming and structural liming provide private benefits by improving the soil structure and thereby the yield potential. Public benefits result from lowered nutrient loading. The use of gypsum as a soil amendment is a relatively new practice. Gypsum is obtained either as a by-product of phosphorus fertilizer manufacturing or from the flue gas desulfurization processes. It may be amended on soils or used as filtering ditches (Ekholm et al. 2012; Bryant et al. 2012).

4.6.3 BMPs for Livestock Production

BMPs for livestock production deal mostly with the storage, handling, and application of manure, and with livestock management practices to prevent or limit livestock access to waterways (Wilson et al. 2014).

Livestock Exclusion Conventional farming practices allow animals in pastures to access streams for watering because keeping them out is costly. Livestock in streams discharge nutrients, pathogens, and pharmaceuticals in urine and feces. Livestock access to streams also causes damage to stream banks that increase bank erosion, sedimentation, and reduces stream health. Applicable BMPs include fencing along streams, alternative water sources, and protected stream crossing. The net private benefits without public subsidies are negative. Like green infrastructure, private adoption in this category of BMPs is very rare without public subsidies.

Manure Management Plans Manure management plans are designed to counter manure application practices that concentrate manure applications close to animal facilities and apply manure nutrients substantially in excess of crop removal rates. They may also address other externalities from manure handling (odors, flies, ammonia emissions). Nutrient Management Plans are required in several US states and are mandatory under the Clean Water Act for large, confined animal operations (CAFOs).

As noted above, increasing nutrient use efficiency provides private benefits to producers. Nutrient management plans may help, but nutrient application standards that serve the environment may not serve profits. When water quality damages are due to phosphorus, water quality protection, the source of public benefits, is best served by application standards for phosphorus. But research demonstrates that restrictions on manure applications to balance inputs with nutrient removal can impose significant private costs dependent on farm size, availability of land for manure application, animal type, and producers' willingness to substitute chemical fertilizers for manure (Ribaudo et al. 2004; Ribaudo and Agapoff 2005).

Feed Amendments and Manure Treatment Chemicals can be added to feed to reduce potential pollution. Feed enzymes have been used to improve livestock productivity since the 1950s (Adeola and Cowieson 2011). Phytase, for instance, may improve digestion of feed and thereby reduce production costs and provide some private benefits (Selle and Ravindran 2007; Ponnuvel et al. 2014). Feeding animals phytase has shown to be effective in reducing phosphorus concentration of swine, dairy, and poultry excretions (Smith et al. 2004; Knowlton et al. 2007; Bosch et al. 1997). This, in turn, reduces the excess of phosphorus in manure applications. It can thus also be a source of public benefits. Phytase reduces phosphorus in manure and helps deplete soil phosphorus levels and reduce dissolved phosphorus loads in water. Phytase is also a comparatively low-cost method for reducing phosphorus loads (Ribaudo et al. 2003). However, animal feeding is a complex combination of multiple feed and enzyme components and clear-cut effects of changes in individual enzymes is hard to determine (Adeola and Cowieson 2011).

Manure Treatment Manure treatments, for instance with aluminum or iron chloride or other chemical compounds, can inhibit the release of phosphorus from manure (Smith et al. 2004; Zhang and Lei 1998). Manure can also be treated physically by separating the dry matter and the liquid fractions, which makes the product more favorably a substitute for chemical fertilizers by decreasing the hauling costs. The costs of such practices are machinery and operation costs. In each of these, there are no private benefits to be obtained from manure treatment. However, as they may be effective in reducing nutrient loading, getting producers to adopt such practices could be socially beneficial.

Manure Storage and Application Methods Manure storage enables producers to manage the timing of manure applications to better serve crops and to reduce environmental damages (Sharpley et al. 2006). In the absence of storage, applications to fields must be done frequently. Manure nutrients are more likely to be lost to the environment than made available to crops. Consequently, there are private and public benefits. Different storage systems have different capacities to preserve nutrients. Uncovered lagoon storage, for instance, loses more than half of its nitrogen content compared to covered lagoons (Aillery et al. 2005). Producers may utilize this volatilization as means to manage nitrogen if they are facing expensive nitrogen limits in terms of manure applications on land but no restrictions regarding ammonia emissions to air (Sweeten et al. 2000). Manure storage choices also have a direct effect on manure spills (Innes 2000). Breached storage facilities can cause significant harm to aquatic life if the spill flows into streams or lakes (Armstrong et al. 2010).

The timing and methods of manure application affect the environmental fate of manure nutrients. Nutrients in manure are readily removed in runoff when manure is spread on the soil surface, especially on frozen ground. Surface runoff can be reduced by storing manure during winter months to avoid the necessity of spreading on frozen ground and injecting manure into soils when applying. Injecting liquid manure into soils reduces immediate surface losses and helps preserve the nutrients. It does not, however, alleviate the problem of overapplication of nutrients and all the short- and long-term problems it causes to surface and groundwaters.

4.6.4 Multiple Pathways, Multiple Receptors, Complex Trade-offs

BMP installation is often presented as the solution to water quality problems caused by agriculture. This "technology is the solution" approach ignores the economic drivers of producers' choices and the private and public economics of BMPs. It also ignores significant environmental trade-offs in the application of BMPs. As described in Chap. 3, nutrients applied to farm fields can move along multiple pathways from fields to impaired surface of groundwater receptors. The existence of multiple pathways of nutrient transport is important because the location, timing, and nature of environmental outcomes resulting from nutrient emissions from a place will depend on the pathways followed from that place and because these pathways can be influenced by management choices (Shortle and Horan 2017). BMPs to mitigate nutrient movements through one pathway can increase movement through another, shifting the type, location, and timing of aquatic ecosystem damages and other environmental harms, and implying trade-offs between the types of water quality outcomes and the cost-effectiveness of alternative control strategies. We have noted previously how management practices affect the dissolved and particulate phosphorus fractions creating a trade-off between erosion control for sediment and particulate phosphorus control and dissolved phosphorus control. Another is that manure management practices designed to curtail emissions of ammonia-N, which can be a significant cause of particulate air pollution in animal-intensive production regions, can exacerbate leaching of nitrate-N (Aillery et al. 2005).

4.6.5 *Social Net Benefits*

The societal benefits of BMPs installed for water quality protection are the private benefits if any, plus the water quality benefits, plus the net value of co-benefits (costs). BMPs provide net social benefits when the benefits exceed the expenses of installation and maintenance, whether borne publicly or privately, plus the opportunity costs of lost net returns from crop production if any. Society receives beneficial externalities when BMPs provide sufficient net private benefits so that producers will adopt them without public intervention. No-till is an example. But this is not the case for many BMPs. There is no social efficiency loss if BMPs with negative net private benefits also have negative net social benefits. There is if BMPs with negative net private benefits provide positive net social benefits.

The social benefits of BMPs for water quality protection depend on their effectiveness in reducing pollution loads and the economic benefits of the water quality protection. Effectiveness varies between BMP types and where they are located within fields and watersheds. Water quality benefits vary with the type and cost of the pollution problem. Because water quality benefits are often unknown or highly uncertain, economic assessments of BMPs often focus on their cost-effectiveness. Table 4.1 provides estimates of the cost-effectiveness of various BMPs for reducing nutrient loads to the Gulf of Mexico. The Appendix: Measuring Social Cost-Effectiveness: Present Values and Annualized Costs to this chapter further describes the concept of cost-effectiveness and its calculation.

4.7 Conclusions

This chapter examined the economics of production decisions in agriculture affecting water quality. The chapter presented analytical frameworks for explaining the choices that producers make and the choices that would maximize the social net benefits of agricultural production when agriculture is a source of water quality externalities.

Our analysis leads to a useful taxonomy of production and land use choices: variable inputs that are pollution increasing (fertilizers, pesticides, pharmaceuticals) or pollution decreasing (e.g., gypsum, phytase), management practices that influence pollution loads (tillage, nutrient and pest management practices, livestock feeding, and manure management practices), and land allocation to green infrastructure and to crops. In the presence of externalities but in the absence of policy interventions, economic theory and research predict a market-driven agriculture will allocate land on farms and within watersheds to the crops and/or livestock that provide the greatest net returns to producers, maximizing Ricardian rents. Producers will use inputs and farming practices as required to maximize Ricardian rents. The map of crop/livestock production in a watershed will reflect spatial patterns in land qualities that determine the relative net returns from alternative crops and/or livestock. The amount of land in agriculture in a watershed will be influenced by competing land uses and by the range of land qualities. Depending on input and output prices some land may be of too low a quality to support profitable production. Land that provides greater rents in developed uses will be unavailable to agriculture.

Pareto Efficient agricultural landscapes consider the external costs and benefits of land use in agriculture and in competing land uses. Given water quality externalities, Pareto Efficient agricultural land use maximizes the Ricardian rents from agricultural production less the water quality damage costs. Each of the production and land use choice types would be affected. Compared to market-driven agriculture, we would expect in general less-pollution-intensive crops, less use of variable inputs that are pollution increasing, greater use of variable inputs that are pollution reducing, greater use of management practices that reduce pollution loads, and allocation of cropland to green infrastructures instead of crop production. The map of agricultural production would change to reflect the relative net social returns from alternative crops in alternative locations that vary in the agronomic qualities and pollution propensities. The amount of land in agriculture in a watershed will again be influenced by competing land uses and the range of qualities. Land that was profitable for crop production might no longer be when external costs are deducted and converted to alternative uses.

The policy challenge that emerges is how to shift from the market-driven crops, input intensities, pest, nutrient, and erosion and runoff management practices, and landscape patterns to the Pareto Efficient driven crops, input intensities, pest, nutrient, and erosion and runoff management practices, and landscape patterns. Or at least, to one that better balances social benefit and costs. One common suggestion for the use of economic incentives is to tax polluting inputs and subsidize pollution-reducing inputs. Another is the common approach of subsidizing the use of best management practices. The taxonomy of choice we have developed and the description of market-driven versus Pareto Efficient landscapes indicate these two approaches, individually or in combination, would be at best partial solutions as they do not address the complete set of choices. A comprehensive policy requires attention to the complete set of choices, adding crops and land use to the mix. We begin our exploration of the options in the next chapter.

Appendix: Measuring Social Cost-Effectiveness: Present Values and Annualized Costs

The concept of cost-effectiveness is used in many contexts to compare the cost-efficiency of alternatives methods for achieving an end. Examples that people often encounter as consumers are comparisons of the cost-effectiveness of alternative energy and water conservation technologies. Cost-effectiveness is often measured as the cost of an option divided by a measure of the physical outcome of the option (e.g., Boardman et al. 2017). This is an average but very common and useful measure when comparing alternative technologies. The issue in such comparisons is not to evaluate incremental changes but differences between discrete choices. Dividing the cost of a pollution control practice by the emissions or ambient pollution reduction it provides is an example.

Because costs and pollution reductions typically take place over time, cost-effectiveness measures must account for the time profiles of costs and effects. Costs incurred over time are typically measured in one of two ways. One cost measurement is the *present value* of the flow of costs over time. The use of present values for evaluating BMPs as farm investments is introduced in Box 4.4. The same concept and method apply to public investments. Like individual producers and consumers, there are social time preferences. Indeed, social time preferences emerge from the fact that people discount the future. As with private discounting, social discounting converts social net benefits received at different times into common units of value using ***social discount rates***. A social discount rate can be thought of as an exchange rate for exchanging values over time. A discount rate higher than zero means that a future USD or EUR is worth less than a current USD or EUR.

The proper choice of social discount rates is significant and complex topic in the economics of social benefit-cost analysis and beyond the scope of this book (see e.g., Gollier and Weitzman 2010 and Gollier 2013). Government agencies typically provide rules or guidelines for the use of discounting in public project and policy analyses, including the choice of the discount rate. Typical discount rates for public policy and project evaluation are between 3% and 7%. The general formula for the present value of a cost incurred t years from the present, C_t, is $PV_t = C_t/(1 + r_s\%)^t$, where $r_s\%$ is the social discount rate. For example, if $t = 1$, $C_t = \$10$, and $r_s\% = 7\%$, then $PV_1 = \frac{\$10}{1.07} = \$9.35$. The second step adds the discounted costs incurred over time. The formula for a project with a life of T years is $PV = \sum_{t=0}^{T} PV_t = \sum_{t=0}^{T} C_t / (1 + r_s\%)^t$. From this formula it is apparent that the present value of a flow of costs is less than the sum of the undiscounted costs. If a set of options has equivalent environmental outcomes, the cost-effective option is the one with the lowest present value cost. As with private discounting, the weight placed on costs decreases exponentially over time. Other things equal, this creates an economic bias against BMPs or other projects with high upfront costs. A second is that the social discount rate has a very strong impact on present value. Small increases in the discount rate can produce large decreases in present value.

The second method of cost measurement uses ***annualized*** costs. Annualized costs are useful when evaluating projects, or constructing portfolios of projects, that have differing lifetimes and effects on target variables. Annualized cost is the ***annual equivalent cost*** of a project. The annualized cost of an investment is the cost that, if it were to occur equally in every year over the lifetime of the investment, would give the same present value cost as the actual investment plus subsequent operations, maintenance costs, and opportunity costs. The formula for the annualized cost of a project that last T years with present value PV and discount rate $r_s\%$ is $AC = PV * r_s\% /[1 - (1 + r_s\%)^{-T}]$. If T is very large, for example a permanent riparian buffer installation, the annualized cost is approximately $PV * r_s\%$. For example, if a very long-lived project has a PV = \$10,000 at a discount rate of 7%, the annualized cost is approximately \$700.

Cost-effectiveness assessments of agricultural pollution control technologies or plans typically compute the cost-effectiveness as the annualized cost divided by the long-term annual average reduction in emissions or ambient pollution expected from the practice or plan (see the Appendix for Chap. 5). For example, to aide planning for pollution controls to meet nutrient pollution control targets for the Chesapeake Bay, the US Environmental Protection Agency's Chesapeake Bay Program provides estimates of the cost-effectiveness for nitrogen and phosphorus control for 247 Best Management Practices for agriculture, urban, and other land uses in the Chesapeake Bay watershed. The average cost-effectiveness of agricultural BMPs for reducing nitrogen is \$586 per kilogram of reduction and for phosphorus is \$1144 per kilogram of reduction (2018 USD) (CBP 2020). The comparable numbers for reductions from developed land (e.g., urban stormwater) are for nitrogen \$11,482 and for phosphorus \$90,683. The ranges of BMP cost-effectiveness across BMP types for agriculture are also very narrow compared to those for developed land. It is differences in cost-effectiveness of this kind that indicate the substantial inefficiency that results from policies that rely heavily on point sources to meet water quality goals.

Another example of cost-effective measures comes from the Marshall et al. (2018) study or reducing pollution loads to the Gulf of Mexico from the MARB. Table 4.1 shows cost-effectiveness estimates for seven alternative BMPs implemented in different sub-basins of the MARB. The estimates show substantial differences in cost-effectiveness based on both BMP and location. For example, wetland and riparian buffers are comparatively low-cost BMPs in all region. Regions with relatively high delivery ratios (e.g., Lower Mississippi Basin) for any practice are more cost-effective than regions with relatively low delivery ratios (Missouri River).

While common and useful, several limitations of conventional measures of cost-effectiveness should be recognized. One limitation pertains to the timing of environmental outcomes. The timing of pollution reductions can vary greatly between options. Reductions in nitrogen fertilizer use for example can provide contemporaneous reductions in potential emissions

Table 4.1 Cost-effectiveness of selected management practices for reducing N delivered to the Gulf of Mexico by practice and MARB sub-basin

Practice	*Sub-basin*	*Cost per lb. N reduction to the Gulf of Mexico (USD)*
Cover crops	Lower Mississippi	5.29
	Tennessee River	6.89
	Upper Mississippi River	23.40
	Missouri River	253.27
Cover crops with erosion controls	Lower Mississippi	3.96
	Tennessee River	4.44
	Ohio River	5.58
	Upper Mississippi	9.64
	Arkansas-White-Red River Basin	123.46
Nutrient management	Ohio River	6.03
	Upper Mississippi River	8.31
	Arkansas-White-Red River Basin	32.78
Nutrient management with erosion controls	Tennessee River	4.50
	Upper Mississippi	12.26
	Arkansas-White-Red River Basin	55.48
	Missouri River	357.81
Drainage water management with erosion controls	Lower Mississippi	4.42
	Tennessee	8.03
	Ohio	8.46
	Upper Mississippi	18.91
	Arkansas-White-Red River Basin	22.55
	Missouri	392.47
Riparian buffers	Lower Mississippi	2.36
	Tennessee	3.21
	Arkansas-White-Red River Basin	4.93
	Upper Mississippi	12.27
	Ohio	16.61

(*continued*)

Table 4.1 (continued)

Practice	*Sub-basin*	*Cost per lb. N reduction to the Gulf of Mexico (USD)*
Wetlands	Lower Mississippi	0.47
	Tennessee	0.49
	Arkansas-White-Red River Basin	0.89
	Ohio	1.28
	Missouri	40.51

Source: Adapted from Table 3.2 in E. Marshall et al. 2018. ***Reducing nutrient losses from cropland in the Mississippi/Atchafalaya River Basin: Cost efficiency and regional distribution***, ERR-258, US Department of Agriculture, Economic Research Service, September 2018

while riparian buffers take years to establish. Differences in timing are masked using long-term annual average pollution reductions for measuring environmental performance.

A second limitation pertains to differences in environmental risk between options due to uncertainty about their effectiveness or reliability. Such risk is also masked by using long-term annual average reductions (McSweeny and Shortle 1990; Shortle 1990). Accordingly, when comparing practices, plans, or policies, differences in the time profiles of their effects and their risk should be considered. For example, two practices with equivalent cost-effectiveness measured in the standard way may differ substantially in timing and risk.

A third limitation pertains to options that affect multiple targeted pollutants, such as nitrogen and phosphorus. When this is the case, an option will have cost-effectiveness measures for each pollutant. Projects that are relatively cost-effective for one may not be for another. When this is the case, additional information is again required to rank options. One method is an Environmental Benefits Index (EBI) (see Chap. 7) that develops a single performance metric based on a weighted average of the pollution reductions. If options have environmental benefits that are not targeted the cost measure should be adjusted to include the social benefits. For example, if an option being evaluated for nitrogen reduction also reduces greenhouse gases, the cost estimate for the option should deduct the benefits of the greenhouse gas reduction (Lötjönen and Ollikainen 2019).

Finally, we note that benefit-cost comparisons may similarly use present values or annualized values. Like costs, benefits are realized over time. Discounting is used to convert into common units. Present values add discounted benefits received over time to provide a single benefit value. The annualized cost of a project or policy is the benefit that, if it were to occur equally in every year over the lifetime of the project or policy, would give the same present value benefit-cost as the actual nominal flow of benefits. The Appendix to Chap. 5 illustrates the use of annualized costs, benefits, and benefit-cost ratios.

References

Adeola, O., and A.J. Cowieson. 2011. Board-invited review: Opportunities and challenges in using exogenous enzymes to improve nonruminant animal production. *Journal of Animal Science* 89 (10): 3189–3218.

Aillery, M.P., N.R. Gollehon, R.C. Johansson, J.D. Kaplan, N.D. Key, and M. Ribaudo. 2005. *Managing manure to improve air and water quality* (No. 1477-2016-121076).

Arabi, M., J.R. Frankenberger, B.A. Engel, and J.G. Arnold. 2007. Representation of agricultural conservation practices with SWAT. *Hydrological Processes* 22 (16): 3042–3055.

Armstrong, S.D., D.R. Smith, P.R. Owens, B. Joern, and C. Williams. 2010. Manure spills and remediation methods to improve water quality. In *Genetic engineering, biofertilisation*, Soil quality and organic farming, 201–215. Dordrecht: Springer.

Baldos, U.L.C., and T.W. Hertel. 2014. Global food security in 2050: The role of agricultural productivity and climate change. *Australian Journal of Agricultural and Resource Economics* 58 (4): 554–570.

Bechmann, M.E., P.J. Kleinman, A.N. Sharpley, and L.S. Saporito. 2005. Freeze–thaw effects on phosphorus loss in runoff from manured and catch-cropped soils. *Journal of Environmental Quality* 34 (6): 2301–2309.

Belcher, K., J. Nolan, and P.W. Phillips. 2005. Genetically modified crops and agricultural landscapes: Spatial patterns of contamination. *Ecological Economics* 53 (3): 387–401.

Berck, P., and G. Helfand. 1990. Reconciling the von Liebig and differentiable crop production functions. *American Journal of Agricultural Economics* 72 (4): 985–996.

Bhullar, G.S., and N.K. Bhullar. 2010. Acceptance of GMOs worldwide: A consumer and producer perspective. In *Advances in biotechnology*, 279.

Boardman, A.E., D.H. Greenberg, A.R. Vining, and D.L. Weimer. 2017. *Cost-benefit analysis: Concepts and practice*. Cambridge University Press.

Boggess, W., Lacewell, R. and Zilberman, D., 1993. Economics of water use in agriculture. Agricultural and environmental resource economics.

Bosch, D.J., M. Zhu, and E.T. Kornegay. 1997. Economic returns from reducing poultry litter phosphorus with microbial phytase. *Journal of Agricultural and Applied Economics* 29 (1379-2016-113460): 255–266.

Brown, C., L. Lynch, and D. Zilberman. 2002. The economics of controlling insect-transmitted plant diseases. *American Journal of Agricultural Economics* 84 (2): 279–291.

Bryant, R.B., A.R. Buda, P.J. Kleinman, C.D. Church, L.S. Saporito, G.J. Folmar, S. Bose, and A.L. Allen. 2012. Using flue gas desulfurization gypsum to remove dissolved phosphorus from agricultural drainage waters. *Journal of Environmental Quality* 41 (3): 664–671.

Caswell, M.F., and D. Zilberman. 1986. The effects of well depth and land quality on the choice of irrigation technology. *American Journal of Agricultural Economics* 68 (4): 798–811.

Caswell, M., K.O. Fuglie, C. Ingram, S. Jans, and C. Kascak. 2001. *Adoption of agricultural production practices: Lessons learned from the US Department of Agriculture Area Studies Project* (No. 1473-2016-120785).

Cattaneo, A. 2003. The pursuit of efficiency and its unintended consequences: Contract withdrawals in the environmental quality incentives program. *Review of Agricultural Economics* 25 (2): 449–469.

Chesapeake Bay Program (CBP). 2020. Chesapeake Bay scenario assessment tool. https://cast.chesapeakebay.net/. Last accessed 15 Oct 2020.

Chouinard, H.H., T. Paterson, P.R. Wandschneider, and A.M. Ohler. 2008. Will farmers trade profits for stewardship? Heterogeneous motivations for farm practice selection. *Land Economics* 84 (1): 66–82.

Diebel, M.W., J.T. Maxted, P.J. Nowak, and M.J. Vander Zanden. 2008. Landscape planning for agricultural nonpoint source pollution reduction I: A geographical allocation framework. *Environmental Management* 42 (5): 789–802.

Dillon, J.L., and J.G. Anderson. 2012. *The analysis of response in crop and livestock production*. Elsevier.

Dodd, R.J., and A.N. Sharpley. 2016. Conservation practice effectiveness and adoption: Unintended consequences and implications for sustainable phosphorus management. *Nutrient Cycling in Agroecosystems* 104 (3): 373–392.

Dodd, R.J., R.W. McDowell, and L.M. Condron. 2014. Manipulation of fertiliser regimes in phosphorus enriched soils can reduce phosphorus loss to leachate through an increase in pasture and microbial biomass production. *Agriculture, Ecosystems & Environment* 185: 65–76.

Dwyer, G., R. Douglas, D.C. Peterson, J. Chong, and K. Madden. 2006. *Irrigation externalities: Pricing and charges*. Available at SSRN 903719.

Easton, Z.M., K. Stephenson, A. Collick, P.M. Fleming, E. Kellner, J. Martin, M. Ribaudo, and G. Shenk. 2020. Increasing effectiveness and reducing the cost of non-point source best management practice implementation: Is targeting the answer? Chesapeake Bay Science and Technical Advisory Committee Publication Number 20-002.

Ekholm, P., P. Valkama, E. Jaakkola, M. Kiirikki, K. Lahti, and L. Pietola. 2012. Gypsum amendment of soils reduces phosphorus losses in an agricultural catchment. *Agricultural and Food Science* 21 (3): 279–291.

Fleming, R.A., B. Babcock, and E. Wang. 1998. Resource or waste? The economics of swine manure storage and management. *Review of Agricultural Economics* 20 (1): 96–113.

Friedrich, T., Derpsch, R. and Kassam, A., 2012. Overview of the global spread of conservation agriculture. Field Actions Science Reports. The journal of field actions, (Special Issue 6).

Giri, S., A.P. Nejadhashemi, S. Woznicki, and Z. Zhang. 2014. Analysis of best management practice effectiveness and spatiotemporal variability based on different targeting strategies. *Hydrological Processes* 28 (3): 431–445.

Gollier, C. 2013. The debate on discounting: Reconciling positivists and ethicists. *Chicago Journal of International Law* 13 (2): 551–564.

Gollier, C., and M. Weitzman. 2010. How should the distant future be discounted when discount rates are uncertain? *Economics Letters* 107 (3): 350–353.

Green, G.P., and D.L. Sunding. 1997. Land allocation, soil quality, and the demand for irrigation technology. *Journal of Agricultural and Resource Economics*: 367–375.

Green, G., D. Sunding, D. Zilberman, and D. Parker. 1996. Explaining irrigation technology choices: A microparameter approach. *American Journal of Agricultural Economics* 78 (4): 1064–1072.

Griffin, R.C. 2016. *Water resource economics: The analysis of scarcity, policies, and projects.* MIT Press.

Heady, E.O., and J.L. Dillon. 1988. *Agricultural production functions.* New Delhi: Kaliany Publishers.

Heathwaite, L., A. Sharpley, and W. Gburek. 2000. A conceptual approach for integrating phosphorus and nitrogen management at watershed scales. *Journal of Environmental Quality* 29 (1): 158–166.

Hochman, E., and D. Zilberman. 1977. Examination of environmental policies using production and pollution microparameter distributions. *Econometrica* 46 (4): 739–759.

Hoffmann, C.C., C. Kjaergaard, J. Uusi-Kämppä, H.C.B. Hansen, and B. Kronvang. 2009. Phosphorus retention in riparian buffers: Review of their efficiency. *Journal of Environmental Quality* 38 (5): 1942–1955.

Hoorman, J.J. 2009. *Using cover crops to improve soil and water quality. Agriculture and natural resources*, 1–4. Lima: The Ohio State University Extension Press.

Iho, A., and M. Laukkanen. 2012. Precision phosphorus management and agricultural phosphorus loading. *Ecological Economics* 77: 91–102.

Innes, R. 2000. The economics of livestock waste and its regulation. *American Journal of Agricultural Economics* 82 (1): 97–117.

Jarvie, H.P., L.T. Johnson, A.N. Sharpley, D.R. Smith, D.B. Baker, T.W. Bruulsema, and R. Confesor. 2017. Increased soluble phosphorus loads to Lake Erie: Unintended consequences of conservation practices? *Journal of Environmental Quality* 46 (1): 123–132.

Johansson, R., M. Peters, and R. House. 2007. *Regional environment and agriculture programming model* (No. 1488-2016-124119).

Johansson, R.C., Y. Tsur, T.L. Roe, R. Doukkali, and A. Dinar. 2002. Pricing irrigation water: A review of theory and practice. *Water policy* 4 (2): 173–199.

Kaplan, J.D., R.C. Johansson, and M. Peters. 2004. The manure hits the land: Economic and environmental implications when land application of nutrients is constrained. *American Journal of Agricultural Economics* 86 (3): 688–700.

Khanna, M. 2001. Sequential adoption of site-specific technologies and its implications for nitrogen productivity: A double selectivity model. *American Journal of Agricultural Economics* 83 (1): 35–51.

Kim, C.S., G.D. Schaible, L. Garrett, R.N. Lubowski, and D.J. Lee. 2008. Economic impacts of the US soybean aphid infestation: A multi-regional competitive dynamic analysis. *Agricultural and Resource Economics Review* 37 (1203-2016-95245): 227–242.

Kleinman, P.J., A.N. Sharpley, R.W. McDowell, D.N. Flaten, A.R. Buda, L. Tao, L. Bergstrom, and Q. Zhu. 2011. Managing agricultural phosphorus for water quality protection: Principles for progress. *Plant and Soil* 349 (1–2): 169–182.

Kleinman, P.J., R.M. Fanelli, R.M. Hirsch, A.R. Buda, Z.M. Easton, L.A. Wainger, C. Brosch, M. Lowenfish, A.S. Collick, A. Shirmohammadi, and K. Boomer. 2019. Phosphorus and the Chesapeake Bay: Lingering issues and emerging concerns for agriculture. *Journal of Environmental Quality* 48 (5): 1191–1203.

Kling, C.L., R.W. Arritt, G. Calhoun, and D.A. Keiser. 2017. Integrated assessment models of the food, energy, and water nexus: A review and an outline of research needs. *Annual Review of Resource Economics* 9: 143–163.

Klümper, W., and M. Qaim. 2014. A meta-analysis of the impacts of genetically modified crops. *PLoS One* 9 (11): e111629. https://doi.org/10.1371/journal.pone.0111629.

Knowlton, K.F., M.S. Taylor, S.R. Hill, C. Cobb, and K.F. Wilson. 2007. Manure nutrient excretion by lactating cows fed exogenous phytase and cellulase. *Journal of Dairy Science* 90 (9): 4356–4360.

Kosenius, A.K., and M. Ollikainen. 2019. Drivers of participation in gypsum treatment of fields as an innovation for water protection. *Ecological Economics* 157: 382–393.

Lankoski, J., and M. Ollikainen. 2003. Agri-environmental externalities: A framework for designing targeted policies. *European Review of Agricultural Economics* 30 (1): 51–75.

———. 2013. Counterfactual approach to assessing Agri-environmental policy: The case of the Finnish water protection policy. *Review of Agricultural and Environmental Economics* 94 (2): 165–193.

Lee, H., D.A. Sumner, and A. Champetier. 2019. Pollination markets and the coupled futures of almonds and honey bees: Simulating impacts of shifts in demands and costs. *American Journal of Agricultural Economics* 101 (1): 230–249.

Lewis, D.J., B.L. Barham, and K.S. Zimmerer. 2008. Spatial externalities in agriculture: Empirical analysis, statistical identification, and policy implications. *World Development* 36 (10): 1813–1829.

Lichtenberg, E. 1989. Land quality, irrigation development, and cropping patterns in the northern high plains. *American Journal of Agricultural Economics* 71 (1): 187–194.

———. 2002. Agriculture and the environment. In *Handbook of agricultural economics*, ed. B. Gardner and G. Rausser (Vol. 2A, pp. 1249–1313). Amsterdam: North Holland.

Lichtenberg, E., and D. Zilberman. 1986. The econometrics of damage control: Why specification matters. *American Journal of Agricultural Economics* 68 (2): 261–273.

Lötjönen, S., and M. Ollikainen. 2019. Multiple-pollutant cost-efficiency: Coherent water and climate policy for agriculture. *Ambio* 48 (11): 1304–1313.

Lötjönen, S., E. Temmes, and M. Ollikainen. 2020. Dairy farm management when nutrient runoff and climate emissions count. *American Journal of Agricultural Economics* 102 (3): 960–981.

Marshall, E., M. Aillery, M. Ribaudo, N. Key, S. Sneeringer, L. Hansen, S. Malcolm, and A. Riddle. 2018. *Reducing nutrient losses from cropland in the Mississippi/Atchafalaya River Basin: Cost efficiency and regional distribution* (No. 1477-2018-5724).

McSweeny, W.T., and J.S. Shortle. 1990. Probabilistic cost effectiveness in agricultural nonpoint pollution control. *Journal of Agricultural and Applied Economics* 22 (1): 95–104.

Miller, M.H., E.G. Beauchamp, and J.D. Lauzon. 1994. Leaching of nitrogen and phosphorus from the biomass of three cover crop species. *Journal of Environmental Quality* 23 (2): 267–272.

Mirsky, S.B., M.R. Ryan, W.S. Curran, J.R. Teasdale, J. Maul, J.T. Spargo, J. Moyer, A.M. Grantham, D. Weber, T.R. Way, and G.G. Camargo. 2012. Conservation tillage issues: Cover crop-based organic rotational no-till grain production in the mid-Atlantic region, USA. *Renewable Agriculture and Food Systems* 27 (1): 31–40.

Natural Research Council (NRC). 2010. *Impact of genetically engineered crops on farm sustainability in the United States*, 270. Washington, DC: National Academy Press.

Nichols, D.S. 1983. Capacity of natural wetlands to remove nutrients from wastewater. *Journal (Water Pollution Control Federation)*: 495–505.

Nowak, P., S. Bowen, and P.E. Cabot. 2006. Disproportionality as a framework for linking social and biophysical systems. *Society and Natural Resources* 19 (2): 153–173.

Pannell, D.J. 2008. Public benefits, private benefits, and policy mechanism choice for land-use change for environmental benefits. *Land Economics* 84 (2): 225–240.

Pannell, D.J., and R. Claassen. 2020. The roles of adoption and behavior change in agricultural policy. *Applied Economic Perspectives and Policy*. https://doi.org/10.1002/aepp.13009.

Pannell, D.J., G.R. Marshall, N. Barr, A. Curtis, F. Vanclay, and R. Wilkinson. 2006. Understanding and promoting adoption of conservation practices by rural landholders. *Australian Journal of Experimental Agriculture*. 46 (11): 1407–1424.

Parker, D. 2004. Creating markets for manure: Basin-wide management in the Chesapeake Bay Region. In *Joint annual meeting of the Northwest Agricultural and Resource Economics Association and the Canadian Agricultural Economics Society, Halifax, Nova Scotia, Canada*.

Parker, D.C., and D.K. Munroe. 2007. The geography of market failure: Edge-effect externalities and the location and production patterns of organic farming. *Ecological Economics* 60 (4): 821–833.

Pfeiffer, L., and C.Y.C. Lin. 2014. Does efficient irrigation technology lead to reduced groundwater extraction? Empirical evidence. *Journal of Environmental Economics and Management* 67 (2): 189–208.

Pierpaoli, E., G. Carli, E. Pignatti, and M. Canavari. 2013. Drivers of precision agriculture technologies adoption: A literature review. *Procedia Technology* 8: 61–69.

Pionke, H.B., W.J. Gburek, and A.N. Sharpley. 2000. Critical source area controls on water quality in an agricultural watershed located in the Chesapeake Basin. *Ecological Engineering* 14 (4): 325–335.

Ponnuvel, P., K. Narayanankutty, A. Jalaludeen, and P. Anitha. 2014. Economics of phytase enzyme supplementation in low energy-protein layer chicken diet. *International Journal of Livestock Production* 5 (7): 113–116.

Qiu, Z. 2009. Assessing critical source areas in watersheds for conservation buffer planning and riparian restoration. *Environmental Management* 44 (5): 968–980.

Rao, N.S., Z.M. Easton, E.M. Schneiderman, M.S. Zion, D.R. Lee, and T.S. Steenhuis. 2009. Modeling watershed-scale effectiveness of agricultural best management practices to reduce phosphorus loading. *Journal of Environmental Management* 90 (3): 1385–1395.

Rasmussen, S. 2011. *Production economics. The Basic theory of production optimization*. Berlin: Springer.

Reaney, S.M., E.B. Mackay, P.M. Haygarth, M. Fisher, A. Molineux, M. Potts, and C. Benskin. 2019. Identifying critical source areas using multiple methods for effective diffuse pollution mitigation. *Journal of Environmental Management* 250: 109366.

Reichelderfer, K., and W.G. Boggess. 1988. Government decision making and program performance: The case of the conservation reserve program. *American Journal of Agricultural Economics* 70 (1): 1–11.

Ribaudo, M.O. 1989. Targeting the conservation reserve program to maximize water quality benefits. *Land Economics* 65 (4): 320–332.

Ribaudo, M., and J. Agapoff. 2005. Importance of cost offsets for dairy farms meeting a nutrient application standard. *Agricultural and Resource Economics Review* 34 (2): 173–184.

Ribaudo, M., J.D. Kaplan, L.A. Christensen, N. Gollehon, R. Johansson, V.E. Breneman, M. Aillery, J. Agapoff, and M. Peters. 2003. Manure management for water quality costs to animal feeding operations of applying manure nutrients to land. *USDA-ERS Agricultural Economic Report* 824.

Ribaudo, M., A. Cattaneo, and J. Agapoff. 2004. Cost of meeting manure nutrient application standards in hog production: The roles of EQIP and fertilizer offsets. *Review of Agricultural Economics* 26 (4): 430–444.

Roberts, M.J., and R.N. Lubowski. 2007. Enduring impacts of land retirement policies: Evidence from the Conservation Reserve Program. *Land Economics* 83 (4): 516–538.

Rucker, R.R., W.N. Thurman, and M. Burgett. 2012. Honey bee pollination markets and the internalization of reciprocal benefits. *American Journal of Agricultural Economics* 94 (4): 956–977.

Schnitkey, G.D., and M.J. Miranda. 1993. The impact of pollution controls on livestock-crop producers. *Journal of Agricultural and Resource Economics* 18 (1): 25–26.

Segerson, K. 2013. When is reliance on voluntary approaches in agriculture likely to be effective? *Applied Economic Perspectives and Policy* 35 (4): 565–592.

Selle, P.H., and V. Ravindran. 2007. Microbial phytase in poultry nutrition. *Animal Feed Science and Technology* 135 (1–2): 1–41.

Sexton, S.E., Z. Lei, and D. Zilberman. 2007. The economics of pesticides and pest control. *International Review of Environmental and Resource Economics* 1 (3): 271–326.

Sharpley, A.N., and S.J. Smith. 1994. Wheat tillage and water quality in the Southern Plains. *Soil and Tillage Research* 30 (1): 33–48.

Sharpley, A.N., T. Daniel, G. Gibson, L. Bundy, M. Cabrera, T. Sims, R. Stevens, J. Lemunyon, P. Kleinman, and R. Parry. 2006. *Best management practices to minimize agricultural phosphorus impacts on water quality*. ARS-163. USDA-ARS, Washington, DC.

Shore, M., P. Jordan, P.E. Mellander, M. Kelly-Quinn, D.P. Wall, P.N.C. Murphy, and A.R. Melland. 2014. Evaluating the critical source area concept of phosphorus loss from soils to waterbodies in agricultural catchments. *Science of the Total Environment* 490: 405–415.

Shortle, J.S. 1990. The allocative efficiency implications of water pollution abatement cost comparisons. *Water Resources Research* 26 (5): 793–797.

Shortle, J., and R.D. Horan. 2017. Nutrient pollution: A wicked challenge for economic instruments. *Water Economics and Policy* 3 (02): 1650033.

Smith, D.R., P.A. Moore Jr., C.V. Maxwell, B.E. Haggard, and T.C. Daniel. 2004. Reducing phosphorus runoff from swine manure with dietary phytase and aluminum chloride. *Journal of Environmental Quality* 33 (3): 1048–1054.

Smith, D.R., C. Huang, and R.L. Haney. 2017. Phosphorus fertilization, soil stratification, and potential water quality impacts. *Journal of Soil and Water Conservation* 72 (5): 417–424.

Soane, B.D., B.C. Ball, J. Arvidsson, G. Basch, F. Moreno, and J. Roger-Estrade. 2012. No-till in northern, western and South-Western Europe: A review of problems and opportunities for crop production and the environment. *Soil and Tillage Research* 118: 66–87.

Song, F., and S.M. Swinton. 2009. Returns to integrated pest management research and outreach for soybean aphid. *Journal of Economic Entomology* 102 (6): 2116–2125.

Suter, J.F., G.L. Poe, and N.L. Bills. 2008. Do landowners respond to land retirement incentives? Evidence from the conservation reserve enhancement program. *Land Economics* 84 (1): 17–30.

Sweeten, J.M., L. Erickson, P. Woodford, C.B. Parnell, K. Thu, T. Coleman, C. Reeder, J.R. Master, W. Hambleton, G. Bluhm, and D. Tristao. 2000. *Air quality research and technology transfer white paper and recommendations for concentrated animal feeding operations.* USDA Agricultural Air Quality Task Force, Washington, DC.

Tisseyre, B., C. Leroux, L. Pichon, V. Geraudie, and T. Sari. 2018. How to define the optimal grid size to map high resolution spatial data? *Precision Agriculture* 19 (5): 957–971.

Uusitalo, R., E. Turtola, and J. Grönroos. 2007. Finnish trends in phosphorus balances and soil test phosphorus. *Agricultural and Food Science* 16 (4): 301–316.

van der Salm, C., W.J. Chardon, G.F. Koopmans, J.C. van Middelkoop, and P.A. Ehlert. 2009. Phytoextraction of phosphorus-enriched grassland soils. *Journal of Environmental Quality* 38 (2): 751–761.

van der Valk, A.G., C.B. Davis, J.L. Baker, and C.E. Beer. 1978, November. Natural freshwater wetlands as nitrogen and phosphorus traps for land runoff. In wetland functions and values: The state of our understanding. Proceeding of National Symposium on Wetlands, Lake Buena Vista, FL, 457–467. Minneapolis: American Water Resources Association.

Wagena, M.B., and Z.M. Easton. 2018. Agricultural conservation practices can help mitigate the impact of climate change. *Science of the Total Environment* 635: 132–143.

Waterfield, G., and D. Zilberman. 2012. Pest management in food systems: An economic perspective. *Annual Review of Environment and Resources* 37: 223–245.

Weaver, R.D. 1996. Prosocial behavior: Private contributions to agriculture's impact on the environment. *Land Economics* 72 (2): 231–247.

Wheeler, T., and J. Von Braun. 2013. Climate change impacts on global food security. *Science* 341 (6145): 508–513.

White, M.J., D.E. Storm, P.R. Busteed, S.H. Stoodley, and S.J. Phillips. 2009. Evaluating nonpoint source critical source area contributions at the watershed scale. *Journal of Environmental Quality* 38 (4): 1654–1663.

White, M.J., C. Santhi, N. Kannan, J.G. Arnold, D. Harmel, L. Norfleet, P. Allen, M. DiLuzio, X. Wang, J. Atwood, and E. Haney. 2014. Nutrient delivery from the Mississippi River to the Gulf of Mexico and effects of cropland conservation. *Journal of Soil and Water Conservation* 69 (1): 26–40.

Wilson, G.L., B.J. Dalzell, D.J. Mulla, T. Dogwiler, and P.M. Porter. 2014. Estimating water quality effects of conservation practices and grazing land use scenarios. *Journal of Soil and Water Conservation* 69 (4): 330–342.

Wrenn, D.H., H.A. Klaiber, and D.A. Newburn. 2019. Price based policies for managing residential development: Impacts on water quality. *Resource and Energy Economics* 58: 101115.

Xu, Y., D.J. Bosch, M.B. Wagena, A.S. Collick, and Z.M. Easton. 2019. Meeting water quality goals by spatial targeting of best management practices under climate change. *Environmental Management* 63 (2): 173–184.

Zhang, R.H., and F. Lei. 1998. Chemical treatment of animal manure for solid-liquid separation. *Transactions of the ASAE* 41 (4): 1103.

Zikeli, S., and S. Gruber. 2017. Reduced tillage and no-till in organic farming systems, Germany—Status quo, potentials and challenges. *Agriculture* 7 (4): 35.

Zilberman, D., T.G. Holland, and I. Trilnick. 2018. Agricultural GMOs—What we know and where scientists disagree. *Sustainability* 10 (5): 1514.

CHAPTER 5

Environmental Policy Instruments for Agriculture

5.1 Introduction

Chapter 4 discussed how farming practices selected to maximize farm profits can cause Pareto Inefficient levels of water pollution. The question we turn to now is the choice of policy instruments for pollution control in agriculture. The basic economic theory of pollution policy, outlined in Chap. 2, introduced core economic concepts and findings about pollution policy instruments. These concepts and results provide a starting point for instrument choice for agricultural nonpoint pollution control but are fundamentally limited because the core theory assumes that emissions are measured accurately and controlled deterministically. This is not the case in agriculture.

There are various dimensions of the agricultural nonpoint pollution problem that create challenges for efficient management. First and foremost is the fact that nonpoint emissions cannot be metered routinely and accurately at reasonable cost. The diffuse and complex pathways by which pollutants move from agricultural land to water resources mean that agricultural emissions are for practical policy purposes unobservable. As discussed in Chap. 3, environmental science provides methods to estimate pollution loads, but these are subject to substantial errors and are highly imperfect substitutes for direct measurement. A second challenge is the stochasticity of agricultural nonpoint pollution due to random weather events driving physical processes that determine the timing, volume, and

J. Shortle et al., *Water Quality and Agriculture*, Palgrave Studies in Agricultural Economics and Food Policy,
https://doi.org/10.1007/978-3-030-47087-6_5

types of pollutants moving from barnyards, fields, and pastures to water resources.

Unobservable and stochastic emissions have profound implications for the economics of agricultural nonpoint pollution control. With stochastic pollution, appropriate policy targets must acknowledge that emissions and ambient concentrations have random components and must be selected to manage the resulting risk. Unobservable emissions mean that the go-to emissions-based instruments of standard theory (e.g., emission taxes, emissions trading) are not directly available. Alternative instruments based on what environmental agencies can observe are required.

The first section of this chapter discusses water quality policy goals for agricultural nonpoint pollution control, introduces policy instruments that have been proposed to efficiently manage agricultural nonpoint pollution considering the relevant stochastic processes and monitoring issues, and establishes some essential elements for effective and efficient policies. The chapter then considers the design of a set of instruments for agriculture that are in the Pigouvian tradition in that prices or standards as applicable are set by an environmental regulator. Instruments involving environmental markets in which prices are market determined are considered in subsequent chapters. This discussion essentially is about what "getting prices right" in agriculture would look like and informational challenges that must be met to do so. The discussion also illustrates the use of integrated economic-biophysical modeling systems for policy analysis and for optimizing policy design.

The last part of this chapter presents a review and evaluation of contemporary agricultural nonpoint pollution controls given the criteria and the principles for efficient policy developed in the previous sections. The discussion identifies systemic flaws that limit effectiveness and efficiency. The chapter concludes with a discussion of policy reforms to improve policy performance and sets the stage for subsequent chapters on market-based approaches.

5.2 Uncertainty and Policy Goals

An essential requirement for a pollution policy instrument is that it achieves the desired water quality goal. When emissions can be controlled deterministically, pollution control is appropriately defined in terms of allowable levels of emissions or emissions reductions. When emissions are unobservable and stochastic, actions to control emissions cannot produce specific reductions with certainty. Restrictions on allowable emissions may be violated due to random fluctuations. Accordingly, unobservable and stochastic pollution requires reframing the meaning of and goals for pollution control from the way these concepts were developed in Chap. 2.

5.2.1 *Uncertainty and the Characterization of Nonpoint Pollution Control*

The diffuse nature of nonpoint pollution removes the possibility of observing emissions from specific agricultural locations for regulatory purposes.[1] The aggregate contributions of agricultural sources to pollution loads can be inferred by various techniques but aggregate loads do not provide the site-specific information about the origins of pollutants needed for regulation of individual farms. Loading models can reduce uncertainty about emissions, but as these models are subject to data limitations and model error, they cannot eliminate uncertainty.

There is more to the uncertainty associated with nonpoint pollution than unobservable emissions. An equally important feature is that agricultural nonpoint source pollution and hence most water quality protection practices are subject to stochastic weather conditions (Opalinski et al. 2016). This feature provides another crucial difference to point sources, which work under controlled technologies and experience only minor stochastic changes in their processes.

Heavy rains and floods can dramatically increase runoff and pollution loads, whereas extreme droughts decrease runoff. In a series of stochastic weather events, some will generate a disproportionately large response in runoff, streamflow, and pollution loads resulting in the majority of loads being exported during short periods of time, generally corresponding to high flow events (Gall et al. 2013; Opalinski et al. 2016; Preisendanz et al. 2020). These time periods are sometimes referred to as "hot moments" or "hot events." The sequence of weather events can matter as well as individual events. For example, periods of drought reduce the plant uptake of

[1] Metering pollution for scientific research can be done in small, instrumented plots. Metering for regulatory purposes requires routine accurate measurements for regulated sources.

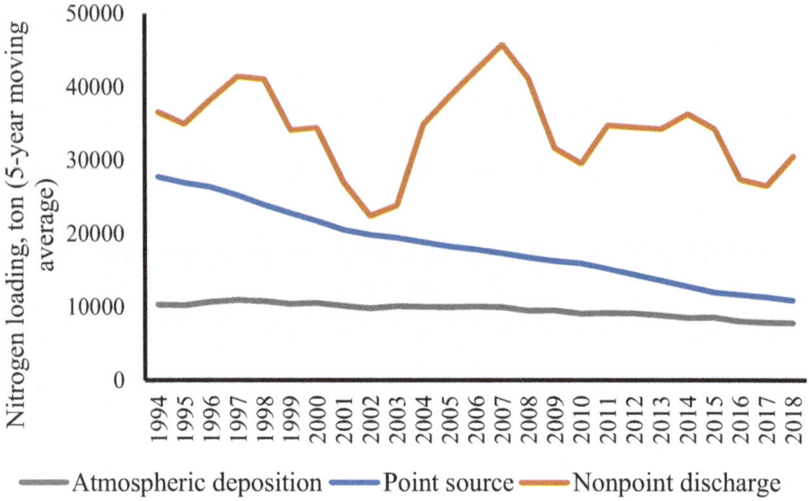

Fig. 5.1 Nitrogen loading to Chesapeake Bay. (Source: Chesapeake Bay Program, Water Quality Standards Attainment and Monitoring [https://www.chesapeakeprogress.com/clean-water/water-quality])

nutrients from soils. The pool of nutrients left in soils available for removal by subsequent rainfall events is greater than it would be otherwise.

Nitrogen loads to Chesapeake Bay illustrate the stochasticity of nonpoint pollution. Figure 5.1 presents five-year moving average values for nitrogen loads at downstream monitoring stations, representing some 40% of the total load from 1990 to 2018. The data upstream from the stations is not divided into pollution sources. The vertical axis denotes the loads in tons per year. The point source loads as well as the atmospheric deposit show a clear downward trend over the twenty-eight years. Furthermore, the five-year moving average of point source loads shows only small fluctuations. The nonpoint loading varies strongly even though averaging over five years evens out the effects of individual years. Phosphorus loads exhibit a similar pattern.

Stochastic weather is challenging for both producers and environmental agencies in their efforts to manage water pollution. Farming is a time-sensitive activity. There are time windows for planting, fertilizing, pesticide applications, and harvesting. Pollution control in agriculture is time sensitive. The type and timing of weather events through the growing season and between growing seasons affect the volume and timing of runoff events and the water quality consequences of pollutants. Addressing weather risk as it affects crop yields is part of farming. Addressing weather risk as it affects water quality adds a new dimension to farm risk management.

Uncertainty about nonpoint pollution loads requires that nonpoint pollution control be framed differently than the control of nonstochastic pollution as described in Chap. 2. With stochastic pollution, pollution control is most appropriately framed as a risk management problem in which pollution controls adjust the frequency and severity of pollution events. We illustrate the concept of managing frequency and severity with Fig. 5.2, which represents the effects of an erosion control practice on the probability distribution (see Box 5.1) of phosphorus runoff from a farm field.[2] The probability distribution is log-normal before and after installation of the practice. The orange curve describes probability distribution with an expected (annual average) runoff over 2 kg/ha prior to the installation of the practice. The blue curve is the probability distribution after installation of the practice. The practice reduces expected runoff to less

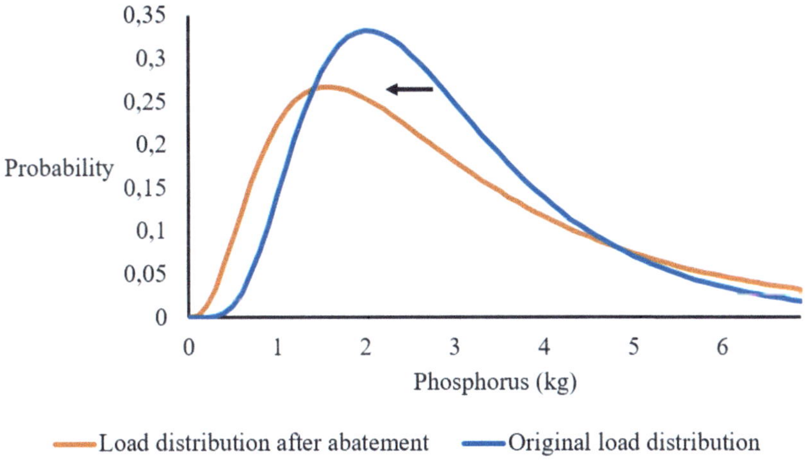

Fig. 5.2 Stochastic phosphorus loads: log-normal distribution of runoff. (Source: Authors' creation)

[2] The figure is a representation of the concept of managing pollution probabilistically and is not based on a specific location BMP.

Box 5.1 Some Probability and Statistics

Because uncertainty is an inherent feature of nonpoint pollution management, statistical and probabilistic concepts are used routinely in research on management methods and policies. We introduce some basic concepts here. Suppose that the nonpoint load (L) at a receptor ranges between zero and some number L_{max}. A probability density function of the load is an equation that gives the relative likelihood of each load between zero and L_{max}. For example, if every value of L, $0 \leq L \leq L_{max}$, is equally likely, then L is ***uniformly distributed***. The relative likelihood of any value in this range is $f(L) = 1/L_{max}$. The only way to "improve" the distribution in this case is to reduce L_{max}. This would eliminate high-load events and increase the probability of low-load events. A common probability distribution used for water pollution management is the ***log-normal*** distribution in Fig. 5.2.

The ***mean*** or ***expected value*** of a stochastic (or random) variable is the average value the variable would take in many observations. For the ubiquitous bell-shaped ***normal distribution***, the mean is the value with the greatest relative likelihood, the peak of the bell. Normal and uniform distributions are symmetric about the mean. Symmetry means that the shape of the distribution to the left of the mean is the same as it is to the right. As illustrated in Fig. 5.2, the log-normal distribution is not symmetric. It is ***skewed*** to the right, meaning that the peak of the distribution is toward the left with a long tail to the right. Because of the long tail to the right, the mean of the log-normal distribution is to the right of the peak.

The ***variance*** of a distribution is a measure of the dispersion of the distribution around the mean – the likelihood that the variable will differ from the expected value. For example, a normal distribution with a smaller variance is narrower about the expected value than one with a higher variance.

The mean, variance, and skew of distributions are referred to as ***statistical moments***.

than 1.8 kg/ha and shifts the distribution toward to the left, implying that most phosphorus runoff events become less likely. These two effects are environmentally desirable. However, the blue curve is slightly above the orange curve in the right tail of the distribution. The thicker right tail implies a slight increase in the probability of higher loads. This feature reflects the fact that conventional erosion control policies, such as riparian buffers (see Chap. 3), reduce the particulate phosphorus runoff but may increase dissolved reactive phosphorus. The same holds true for nitrogen.

5.2.2 *Policy Goals: Cost-Effectiveness and Efficiency for Uncertain Pollution*

Cost-effective pollution management was defined in Chap. 2 as an allocation of emissions (or emissions abatement) that satisfies a given quantitative limit at the lowest possible social cost. With stochastic pollution, cost minimization remains desirable but a question emerges about the appropriate form of the limit. As in Chap. 2, suppose that meeting a water quality goal requires limiting the annual pollution load to some number. Stochastic variation in pollution loads means that the limit may be periodically exceeded thus violating the water quality goal. If this risk matters to environmental agencies, then an appropriate goal must consider in some way the frequency and severity of pollution events (Braden and Segerson 1993; Horan 2001; Shortle 1990; Shortle and Horan 2013). The choice of a goal in turn determines the explicit meaning of a cost-effective allocation (Horan 2001; McSweeny and Shortle 1990; Shortle 1990).

The simplest limit is one that restricts the expected load at the receptor. It is simple in that it does not require explicit management of the variability of pollution, only the expected load, typically expressed as the long term (or steady-state) annual average. The frequency and severity of violations is irrelevant provided that the limit is maintained on average. If the pollutant is uniformly mixed (i.e., the pollution load at the receptor is determined by the sum of emissions from all sources), this limit is implemented as an upper bound on the sum of expected emissions from the contributing sources. This is because the expected value of a sum of random variables is equal to the sum of the expected values of the random variables. A cost-effective allocation would then be one in which (a) individual polluters minimize the cost of controlling their expected emissions and (b) the allocation of expected emissions or emissions abatement across sources minimizes aggregate pollution control costs. Effective policy instruments would need only target expected loads. Ideally, the implementation for stochastic nonpoint sources would use reliable loading models to estimate their expected emissions.

However, environmental agencies are generally concerned with environmental risk and therefore the frequency and severity of pollution events. In general, preferred pollution distributions will generally have smaller means and variances, and greater weight in the left tail (Horan 2001; Segerson 1988). One approach to setting limits for stochastic emissions to achieve water quality goals is the use of *safety rules* that limit the frequency of violations of the water quality goal to an acceptable level (Beavis and Walker 1983; Hanley et al. 1998; Lichtenberg and Zilberman 1988; McSweeny and Shortle 1990; Qiu et al. 2001; Shortle and Horan 2013). For example, if L is the actual pollution load and T is the maximum load water scientists determine to be consistent with the water quality goal set by policy makers, a safety rule would require the probability that L exceeds T be limited to a small number, denoted α: $\text{Prob}(L \geq T) \leq \alpha$. Although an appealing concept, implementing a safety rule would be no simple task as it would require knowing and managing the effects of individual farm practices on the probability of violation. The probability will depend on complex relationships between the expected emissions of individual farms, the variance of their emissions, and possibly other statistical moments like the covariance of emissions between farms. Pollution load modeling and data are not sufficiently evolved to design and implement nonpoint pollution instruments to satisfy safety rules (Box 5.2). Knowledge of the variances of emissions of nutrients and other pollutants and how they respond to changes in farm practices, for example, is very limited. Routine implementation of safety rules is at best an aspiration at this time.

Correspondingly, explicit management of agricultural nonpoint risk in water pollution control is rare and generally simplistic. For example, under the US Clean Water Act, water quality authorities are required to develop total maximum daily loads (TMDLs) for waters that do not meet water quality goals (see Chap. 2). The TMDL determines the maximum amount of a pollutant, usually on an annual basis, that a water body can receive and still meet water quality standards with a *margin of safety* (MOS) (USEPA 2020). To illustrate, suppose again that T is the load required to satisfy the water quality goal and $\bar{L}$ is the limit on the annual average load. Setting $\bar{L}$ equal to T would allow pollution to violate the water quality goal with some frequency depending on the probability distribution of the actual load L. Setting $\bar{L} = T - MOS$ reduces the probability of violations by shifting the expected value of the probability distribution to the left. This policy goal does not explicitly address statistical attributes of the load other than the mean. This makes the MOS a crude approach to risk management.[3]

The MOS approach is also costly because it uses a single dimension, the average annual load, to manage a multi-dimensional problem (Horan and Shortle 2017; McSweeny and Shortle 1990; Shortle 1990; Shortle and Horan 2001). Lower cost outcomes would be obtainable by advances in science that enabled explicit management of variability by converting known unknowns to known knowns (Box 5.2). Consider a producer

Box 5.2 Uncertainty in Nonpoint Pollution Control

In a now popular taxonomy, there are ***known knowns***, ***known unknowns***, and ***unknown unknowns*** in the management of nonpoint pollution from agricultural and urban sources (Lintern et al. 2020). Known knowns include relationships that have been well studied and are well understood. Uncertainty is not absent but is quantified. For example, designing manure storage facilities to manage the risk of breaching during an extreme rain event is a routine engineering problem given data on the severity and frequency of rainfall events.

Known unknowns are relationships that are known to be important to decisions but that are not well quantified and in which new knowledge is recognized to be needed. As we describe in Chap. 4, there is substantial uncertainty about the effectiveness of Best Management Practices (BMPs). Lintern et al. (2020) conclude that current scientific knowledge provides a basic understanding of nonpoint pollution processes and BMP efficiencies at site or individual practice scales but that many known unknowns limit progress in

(*continued*)

[3] Our discussion of managing water quality risk with nonpoint sources is focused on managing the probability distribution of the total pollution load at a targeted receptor (or receptors). At the farm and field scales some kinds of risk can be controlled effectively. For example, designing manure storage facilities to manage the risk of breaching from extreme rain events is a routine engineering problem given data on the severity and frequency of rainfall events is routine engineering design.

Box 5.2 (continued)

reducing nonpoint pollution at large scales. The assessment analyzes 94 studies examining the effectiveness of BMPs in agricultural and urban watersheds. They find that 60% of studies report water quality improvements after BMPs implementations. Importantly, most of the studies reporting improvements do not report actual water quality improvements but simulated improvements based on water quality model simulations. Analyzing studies that are unable to report water quality improvements, the authors find that known unknowns play a large role. Significant known unknowns included lack of knowledge about BMP functioning, lag times in water quality response, the optimal placement and distribution of BMPs within watersheds, and post implementation BMP failures. We have commented previously (Chap. 3) on uncertainty in water quality models for nonpoint pollution management. Lintern et al. (2020) conclude that the differences between simulated water quality improvements and actual water quality improvements may indicate challenges and inaccuracies with existing watershed-scale models related to known unknowns about BMPs and consequent assumptions about their performance.

Unknown unknowns are unexpected or unanticipated risks. Lintern et al. (2020, p. 9167) note that deep uncertainty exists in nonpoint pollution management "...because of unknown or unmeasurable watershed properties, under-recognized or unmeasured processes, and substantial rates of change...that interact in complex ways to generate feedbacks and emergent behaviors we currently are unaware of, and that are not predictable with current watershed models."

Of huge importance for water quality risk management policy is behavioral uncertainty (see Box 5.3). Managing risk requires policy interventions to reduce the use of polluting inputs, increase the adoption of BMPs with social net benefits, shift the location of pollution-intensive production, and convert land to green infrastructure. The known unknowns related to pollutant processes and BMP effectiveness described above apply to the uncertainty of water quality outcomes. Amplifying this uncertainty are known unknowns and unknown unknowns about how producers will respond to alternative policy interventions, and economic unknown unknowns that could affect the structure of demand for agricultural commodities, agricultural production technologies, and competition between agriculture and other land uses.

choosing projects to reduce pollution. If the performance metric is simply the reduction in expected pollution, the producer chooses projects that minimize the cost of reducing the annual average. The effect on variance is not considered. If low-cost practices for reducing expected emissions have greater emissions variability than higher cost practices, then the beneficial effect of reducing the mean at low cost will be offset to some degree by a smaller reduction in variability. Greater reductions in the mean and correspondingly greater control costs will then be required to reduce the probability of high runoff events. If performance is instead based on multi-dimensional measures indicating the effects of control projects on the frequency and severity of pollution events, then projects can be selected to explicitly minimize the costs of risk reductions.

With this caveat, we will generally take a pragmatic view that agricultural nonpoint pollution risk is to be managed analogous to US TMDLs, with policy targets set for long-term annual average loads, as this best conforms to planning and management practice. With this assumption, the implied measure for cost-effectiveness is the cost of reducing annual average load. The measurement of cost-effectiveness when defined in this way is discussed in the Appendix of Chap. 4.

5.2.3 *Co-benefits*

As we have discussed in the previous chapters, agricultural BMPs for water quality protection can provide environmental co-benefits. Riparian buffers, for example, in addition to filtering pollutants from runoff, sequester carbon, provide wildlife habitat, and can moderate downstream flooding. A multi-objective approach to environmental planning that optimizes the multiple benefits less the costs of agricultural practices is appealing in the ory (NRC 2012). In practice, it is difficult to implement for a variety of reasons, including disparate and incomplete measures of outcomes, uncertain cause-and-effect relationships, and undefined objective functions for evaluating trade-offs (NRC 2012). Agricultural conservation policies that serve multiple objectives do not always serve them well (Lichtenberg 2014). Where water quality protection from agriculture is a high priority, policy goals should reflect the priority. However, co-benefits can and should be considered in the analysis of implementation plans and program costs. For example, Wainger et al. (2013) analyze pollution policies affecting the allocation of pollution reductions between high marginal cost point sources and relatively low marginal cost agricultural nonpoint sources to satisfy the Chesapeake Bay TMDL (Chap. 3). They find that policies that discourage trading agricultural nonpoint pollution reductions for point source reductions markedly increase overall control costs and reduce ecosystem service co-benefits from agricultural BMPs.

5.3 Water Pollution Policy Instruments for Agriculture: The Options

Unobservable emissions from individual farms eliminate emissions-based instruments requiring direct observations from the agricultural pollution management tool kit. The question is what kind of instruments should be used to manage stochastic pollution loads? This question entails two domains of choice. One is what compliance base or bases to use. The second is the incentive or regulatory mechanism applied to the compliance base.

5.3.1 Compliance Bases

By compliance base we mean an action or outcome that is under the control of the polluter that is targeted by the policy instrument. This choice has at least two dimensions. One is what the environmental agency can observe for compliance monitoring and enforcement. The other is what is instrumental for management in that there is a direct connection between the compliance base and water pollution. One set of choices for compliance bases is polluting inputs (e.g., fertilizer, pesticides, pharmaceuticals), inputs that reduce pollution (certain feed and soil additives), farming practices or activities that affect water quality outcomes (tillage, nutrient and pest management), and pollution control structures (buffers, grassed water ways, manure storage facilities). Policies that use compliance bases from the set of inputs, practices, and structures are variously referred to as *input-based* policies or *practice-based* policies.

A second set of compliance bases are emissions or performance proxies. This set uses formulas or models to construct farm-specific indicators of pollution control performance that integrate multiple variables affecting pollution loads. Such indicators make use of observations of farm inputs and practices as do practice-based policies but use that information to compute farm performance metrics. Models of various types for estimating emissions were introduced in Chap. 3.

Some performance indicators do not estimate emissions but produce constructs that are indicative of pollution potential. An example is the Phosphorus Index (e.g., Gburek et al. 2000). The index uses data on soil and landscape features, phosphorus fertilization rates, and other farm management practices to indicate the risk of phosphorus movement to waters as very low, low, medium, high, or very high. Policy instruments that use emissions estimates or other indicators that integrate multiple management choices are referred to as *performance-based* policies.

5.3.2 Compliance Mechanisms

By mechanisms we refer to the policy levers used to induce change in the selected compliance base or bases. An environmental agency may choose, for example, to measure performance by an input or indicator, but bringing about desired change requires an intervention that affects choices. In Chap. 2, mechanisms were broadly categorized as command-and-control or economic incentives. Command-and-control involves mandates, rules, or standards restricting choices of inputs or practices, or requiring the use of technologies. Economic incentives operate by affecting polluters' profits related to choices connected to emissions utilizing administratively determined taxes, fines, fees, or subsidies; market-determined price; or liability rules. Another mechanism type that we did not mention in Chap. 2 but that has a history of use in agriculture is normative messaging and behavioral nudges.

It is useful to distinguish economic incentives by the implied property rights in pollution and methods by which incentives are set. A widely accepted principle of pollution management is that polluters do not have a right to degrade the environment. A corollary is the polluter-pays-principle, assigning the cost of pollution prevention and control to the polluter. The predominant approaches to water pollution control in agriculture have eschewed this approach by relying on voluntary approaches. Producers often operate under a presumptive right or entitlement to pollute (Shortle et al. 2012; Bromley and Hodge 1990). A corollary of this is the pay-the-polluter approach in which producers are compensated for the costs they incur to control pollution.[4] This distinction between

[4] What we refer to as pay-the-polluter is sometimes referred to as the "beneficiary-pays" principle. Those who benefit from pollution reductions compensate polluters for their costs.

polluter-pays and pay-the-polluter in agriculture is a generalization as there are mixed approaches and variations of degree, but it is a useful one for characterizing instruments.

Methods for setting incentives include administrative (or legislative) procedures, markets, and liability rules. Administrative methods set incentives by fiat. Examples are subsidy rates for pollution practices, tax rates on polluting inputs or emissions, or tax preferences for BMP investments. Market-based methods use voluntary exchange of property rights in markets to set prices. The primary examples in the context of water quality are water emissions trading, usually described as water quality trading, and water quality auctions. Water quality auctions are procurement auctions in which an environmental agency or other entity utilizes an auction to purchase pollution reductions. Liability rules in the water quality domain allow entities that are damaged by water pollution to seek compensation from the source (Bromley 1978). An example would be liability for a fish kill caused by a manure lagoon failure that results in a massive manure load entering a stream. A polluter's liability exposure provides the incentive to take pollution control measures.

5.3.3 *Policy Instruments*

Our primary interest is in command-and-control mechanisms, incentives set by administrative fiat, and market mechanisms. Categories of instruments considering these methods, property rights, and compliances bases are: (1) *regulatory standards* applied to compliance bases, (2) *pay-for-practices* (e.g., subsidies for control practices, land retirement auctions), (3) *pay-for-performance* (e.g., fiat subsidies for performance-based instruments, auctions for performance-based instruments, water quality trading markets, water quality auctions), (4) *practice penalties* (e.g., fiat charges of fines for non-preferred practices), (5) *emissions proxy penalties* (e.g., fiat charges on estimated emissions, water quality trading markets). The two *pay-for* categories in this taxonomy apply to mechanisms in which producers have implicit rights to pollute and receive compensation for their actions. Categories (1), (4), and (5) are *polluter-pays* approaches. Water quality trading is placed in categories (3) and (5) because the design of trading programs affects whether producers are paid for pollution reductions or bear the cost of pollution reductions. A sixth category that does not fall neatly in our taxonomy involves information programs. These are "soft approaches" that can be used as the main mechanism or as supplement to any of the other approaches. Table 5.1 lists the categories and examples within each.

Table 5.1 Agricultural policy instruments

Polluter-Pays – Regulatory Instruments

Input Standards/Mandates:

- Restrictions on the type, quantity, and timing of allowable use of inputs positively related to pollution (e.g., pesticides, fertilizers, pharmaceuticals, water in irrigated agriculture)
- Mandated use of inputs negatively related to emissions (e.g., feed additives reducing animal excretion of nutrients, required application of gypsum to reduce phosphorus losses from cropland)
- Restrictions on the allowable uses of environmentally sensitive lands (bans on cultivation of wetlands, highly erosive lands, riparian zones)
- Required land uses (grassed waterways, riparian buffers differentiated by width, vegetation types, and allowable uses)

Practice Standards/Mandates:

- Restrictions on farming practices (e.g., fall-tillage bans, required setbacks from water bodies for tillage operations or manure applications, prohibitions on manure spreading on frozen ground)
- Required use of farming practices (e.g., soil testing, low-tillage practices, nutrient management plans, manure storage, incorporation of land-applied manure, livestock exclusions from streams)

Polluter-Pays – Economic Incentives

Input/Practice-Based Incentives:

- Taxes on polluting inputs (e.g., fertilizer taxes) or non-preferred practices (e.g., intensive tillage)

Performance-Based Incentives:

- Taxes on estimated emissions
- Water quality trading (farmers capped, initial endowments less than baseline emissions)
- Liability rules

Pay-the-Polluter

Pay-for-Practices:

- Cost-sharing financial assistance for voluntary implementation of structural and nonstructural Best Management Practices (BMPs)
- Land-area-based payments for installation of BMP packages
- Profits tax reductions based on BMP investments

Pay-for-Performance:

- Flat-rate payments for reductions in pollution relative to a baseline
- Water quality procurement auctions
- Water quality trading (farmers endowed with baseline emissions)

Information programs

- Normative messaging
- Educational programs
- Technical assistance (advising)

Source: Authors' creation

"Nudge" is a behavioral science concept that refers to a small non-disruptive intervention intended to improve decision making in situations in which cognitive biases tend to lead people to make economic or other mistakes (Thaler and Sunstein 2008). Nudges are often implemented by providing information or improved frameworks for decision making. They are suited to programs that encourage voluntary actions, like adopting new technology or practices, when the action is beneficial to the adopter (Ferraro et al. 2017). In cases where producers are making behavioral mistakes, for example, using fertilizers in excess of profit-maximizing levels or failing to participate in subsidized conservation programs that would be economically beneficial to them, nudges offer an approach to induce behavioral change (Ferraro et al. 2017).

Social comparison is a form of nudging that seeks to affect behavioral change by providing decision makers with information about how their choices compare to those of others. Social comparisons of residential water use for example have been demonstrated to increase water conservation (e.g., Ferraro and Price 2013). The same is true for home energy use (e.g., Allcott and Rogers 2014). The method relies on the tendency of people to follow norms. Messages that communicate norms can affect behavior toward the norm. Experimental behavioral research provides support for using programs to communicate social norms and provide comparisons to promote pollution control practices in agriculture and participation in subsidy programs (Ferraro et al. 2017).

Nudges have a definite role in the mix of instruments, especially those that rely on voluntary adoption, but their potential to address the suite of problems associated with current water quality policies is limited. As is evident from Chap. 4, profit-maximizing producers who make no behavioral mistakes will be sources of externalities without mechanisms to restrict their choices or incentives to implement pollution control practices. Furthermore, the conventional voluntary programs for agricultural nonpoint pollution control that behavioral nudges could enhance have fundamental limitations as effective and efficient pollution control mechanisms that prompt the search for significant policy reforms (OECD 2012; Segerson 2013; Ribaudo and Shortle 2019; Shortle et al. 2012).

Liability rules are high-transaction-cost mechanisms as they require lawsuits to establish responsibility for pollution damages. The high transaction cost and event orientation (e.g., manure lagoon failures) limit the appeal of liability rules as a mechanism for managing pollution from routine economic activities like farming. The much-discussed physical characteristics of nonpoint source pollution make this doubly true because they make it difficult or impossible to trace the specific sources of a pollution problem (Ribaudo et al. 1999). Furthermore, efficient farm decision making under liability rules requires the farm manager to predict the likely water quality consequences of their actions. Given that this task is a large challenge for experts, it is clearly beyond the capacities of typical farm managers.

Another policy instrument we have not included in this discussion that warrants mention is research and development (R&D) policy. R&D is subject to market failures that lead to a suboptimal supply without public support for basic and applied research (Goulder and Parry 2008). We set R&D policy for agricultural nonpoint pollution control aside as a complex topic beyond the scope of this book but underscore the importance of R&D to improving the environmental performance and overall social efficiency of agriculture. Examples of relevant R&D domains include the development of substitutes for pollution-intensive agricultural commodities (e.g., industrial proteins Box 3.4); advances in genetically modified crop and livestock commodities to reduce use of polluting inputs (e.g., Box 4.8); development of new farming, soil, and manure management methods that reduce the use of polluting inputs or reduce the movement of pollutants from fields and barnyards to water resources (e.g., precision agricultural production practices, manure, and soil amendments); and advances in watershed modeling and geospatial information technologies needed for cost-effective pollution management.

5.4 Some Useful Concepts: Adverse Selection, Moral Hazard, Spatial Targeting, Spatial Differentiation

The list of compliance bases and mechanisms provides an *a la carte* menu of choices. Some items go together naturally, some do not. For those that do, how well they work depends on how they are implemented. Differences in the economic and water quality outcomes of instruments will depend on inherent features of the instrument and the specifics of the implementation. For example, a pay-for-practices subsidy instrument would provide eligible producers with a menu of subsidy-eligible practices, criteria for practice selection, payment rates for the practices, maintenance requirements, and a monitoring plan. These specifications will affect the potential water quality outcomes, producers' participation decisions, and budgetary and social costs. Much of this chapter and Chaps. 6, 7, and 8 are devoted to examining the design and performance of various instruments, some that are used in practice, and some that are promising but which have not yet found significant use if any. Before moving on to that discussion, we present some additional concepts that are used extensively in economic research on policy design and policy evaluation.

5.4.1 Asymmetric Information

Unobservable emissions, stochastic weather, and imperfect understanding of the biophysical processes connecting farms to water quality are important sources of uncertainty that agricultural nonpoint policies must contend with. Additional sources of uncertainty relevant to policy design stem from the fact that producers have specialized information about their lands, farming operations, and how alternative technologies will work on their farms that is not known by environmental agencies. Producers also know details of their farming activities that affect pollution loads that are difficult and costly to monitor and that are therefore unknown to regulatory agencies. This differential knowledge structure between producers and regulatory agencies is referred to as *asymmetric information*. It results in agencies having uncertainty about how polluters will respond to policy interventions, which adds to uncertainty about environmental outcomes of policy interventions and poses significant challenges for designing instruments to minimize social costs.

Box 5.3 Behavioral Uncertainty

There is much emphasis in nonpoint pollution management on uncertainty about the physical environment – unobserved emissions, stochastic processes, and uncertainty about cause-and-effect relationships. Another important source of uncertainty for policy design is *behavioral uncertainty*. Depending on the water quality problem, thousands or tens of thousands of farm decision makers make scores of choices that determine pollution control costs and outcomes in any year. Economics assumes profit maximization as a central tendency in farm decision making but there can be substantial unknowable heterogeneity in individual preferences, human capital, ingenuity, technical, environmental, and economic conditions that influence outcomes. The responses of individual producers and agricultural systems to policy interventions can be predicted but cannot be known beforehand. This uncertainty cannot be eliminated by economic analysis of possible responses, but the likelihood of policy failures and unintended adverse consequences can be diminished.

Asymmetric information about polluters' costs contributes to *behavioral uncertainty* (Box 5.2) and the problem of *adverse selection*. Cost-effective policy instruments allocate pollution reductions to sources according to their relative control costs. If a regulatory agency knows costs, it can then assign the cost-effective allocation. But if costs are not known, due to asymmetric information, then the agency is unable to know the least-cost allocation. In Chap. 2 we concluded that emissions standards cannot be designed to be cost-effective because of asymmetric cost information. The result is an adverse selection problem in which high-abatement cost polluters do too much pollution control and low-abatement cost polluters do too little. This adverse selection problem is solved in Chap. 2 by pricing observable emissions or contributions to ambient pollution, which induces polluters to select the least-cost allocation. Solving the adverse selection problem becomes more difficult with greater complexity in the physical and economic environments. The economic mantra of pricing pollution is partly about internalizing external environmental costs, but it is also about addressing adverse selection in the allocation of pollution control among alternative sources. The complexity of agricultural nonpoint pollution makes for a complex pricing problem (Shortle and Dunn 1986; Shortle and Horan 2001, 2013, 2017).

Moral hazard occurs when producers take actions affecting environmental outcomes that the environmental agency cannot detect without invasive and costly monitoring. Optimal regulation of emissions requires measuring emissions. This is infeasible for agricultural nonpoint pollution. The next best choice is to observe and regulate or price all the decisions a farm manager makes that are known to affect emissions, but this too is infeasible for some actions may be readily monitored but others may not. Choices that the agency would like the polluter to make for effective and efficient pollution control may therefore be made in contrary ways if it is to the polluter's economic advantage. The resulting efficiency loss is the problem of moral hazard. The opportunities for moral hazard in agricultural nonpoint pollution are plentiful because the specific timing, level, and form of some pollution-relevant actions are costly to observe.

The degree to which moral hazard is a problem will be influenced by the regulatory framework, including incentives for verifiable measurements and noncompliance penalties. Accordingly, the inducements for moral hazard are a consideration in instrument design. Self-reporting is an invitation to moral hazard. We emphasize that moral hazard does not result from unethical behaviors. In the economic sphere, people make choices based on costs and returns as allowed within the regulatory framework and its penalty systems (Fraser 2002; Ozanne et al. 2001).

5.4.2 *Spatial Targeting*

A topic of enormous importance to agri-environmental policy is spatial targeting. In the context of water pollution control, spatial targeting is about where to focus water quality protection activities and resources. There are two aspects of this question. One is the selection of which water quality problems to address and to what extent. Research on cost-effective policy design typically takes the environmental problem as given. But with scarce resources for policy development, implementation, and administration, choices must be made about which problems to prioritize (Babcock et al. 1997; Wu et al. 2001). The desirability of spatial targeting for effective and efficient environmental management is underscored by the common use of policies that provide producers with implicit rights to pollute and achieve pollution reductions through pay-for-practice or pay-for-performance policy instruments. Allocating scarce budgetary resources to problems with relatively high costs and low benefits limits environmental and economic efficiency gains (Lichtenberg 2014; Shortle et al. 2012).

For economic efficiency, economics recommends prioritizing water quality problems in order of their benefit-cost ratios (benefits of water quality improvements divided by the costs) until the budget limit is reached. To maximize the net benefits from public spending from a given budget requires maximizing benefits. To do so, resources should be allocated to places that provide the greatest benefits per dollar spent. An illustration of spatial variations in benefit-cost ratios for water quality protection projects is presented in the Appendix to this chapter. Targeting based on benefit-cost ratios is, however, a challenge as it requires information on benefits and costs that may be highly uncertain. Alternative targeting approaches based, for example, on project costs or environmental impacts, will limit economic efficiency gains and the social returns from environmental investments. When targeting is based on costs or environmental impacts, the amount of spatial variability in costs relative to benefits and how benefits and costs are correlated are the most important factors influencing the magnitude of efficiency losses (Babcock et al. 1997; Wu et al. 2001).

Benefits and costs of water quality policies are substantially affected by the policy instruments used to achieve pollution reductions. Accordingly, spatial targeting of water quality problems based on economic criteria would be influenced by the policy instruments used to pursue water pollution control.

A second dimension of spatial targeting is prioritizing pollution source locations within watersheds. Like the spatial targeting of water quality problems, spatial targeting of source locations serves effectiveness and efficiency by allocating resource to fields or farms that provide the greatest return for the money spent. For example, where in the Baltic Sea, Chesapeake Bay, Lake Erie, or Mississippi basins, all high-priority problems, should resources for policy development, implementation, and administration be devoted? This question can be addressed at multiple scales. Pollution controls in some tributaries will be more cost-effective or have higher benefit-cost ratios than others. Pollution controls in some sub-watersheds of major tributaries will be more cost-effective or have higher benefit-cost ratios than others. At the smallest scale, pollution control in some fields or portions of fields may be more cost-effective than others.

As discussed in Chap. 4, agricultural pollution loads come from land areas, but not all land areas in a watershed contribute equally to downstream water quality problems. Disproportionality is often the case with a small fraction of the land area contributing a large fraction of the pollution loads. There is ample research evidence that targeting resources spatially to source locations that contribute significantly to water quality problems is essential for budgetary efficiency and economic cost-effectiveness (Amin et al. 2019; Braden et al. 1989; Diebel et al. 2008; Hansen and Hellerstein 2006; Kaufman et l. 2014; Marshall et al. 2018; Nowak et al. 2006; Ribaudo 1989; Ribaudo et al. 2014). Different pollution control instruments are more effective in spatial targeting of source areas than others (e.g., Talberth et al. 2015). Accordingly, the capacity for spatial targeting is considered in instrument choice.

The Appendix to Chap. 4 illustrated the effect of location on the cost-effectiveness of a suite of BMPs in Mississippi/Atchafalaya River Basin (MARB). The case study of the potential social costs of reducing nutrient loads to the Gulf of Mexico from agriculture in the MARB by 45% presented in Box 4.2 estimates costs-savings to consumers in the spatially targeted scenario of 1.4 billion USD annually. Case studies illustrating significant benefits of spatial targeting in tandem with selection of cost-effective practices at multiple scales in the Chesapeake Bay watershed are presented in the Appendix to this chapter.

5.4.3 Spatial Differentiation

The choices of which problems to manage and where to devote regulatory actions and resources to manage them are partly planning problems, but as we said above, policy instrument design is also important. Consider again voluntary compliance, pay-for-practice policy instruments. Policy makers can make payment rates independent of producers' locations or they can differentiate payments spatially in ways that encourage participation by producers in critical source zones. The latter serves economic efficiency, the former does not. Differentiation of prices or quantity controls spatially is an attribute of policy instruments that serve to induce cost-effective responses within heterogenous agricultural landscapes.

5.5 Polluter-Pays Instruments in the Pigouvian Tradition

In this section we take up the design of the polluter-pays instruments. From Table 5.1 the principal types are (1) prices on inputs or practices connected to water pollution, (2) quantity controls (standards/mandates) applied to inputs or practices connected to water pollution, (3) prices applied to estimated emissions (or emission proxies), and (4) quantity controls applied to estimated emissions. Table 5.1 indicates various ways in which prices can be set. In this section we assume, in the Pigouvian tradition, that prices or standards are administratively determined by an environmental agency. Environmental markets are considered in subsequent chapters. To develop an appreciation of how the environmental agency's knowledge of polluters' choices and cause-effect relationships matters to policy design, we begin the discussion with an ideal information environment in which the environmental agency has no uncertainty.

5.5.1 Policy Design with Perfect Information

In an ideal information environment, the agency:

- Knows the biophysical relationships between producers' choices and water quality outcomes. We refer to this relationship as the loading function.
- Knows agricultural production technology.

- Can observe without error all physical characteristics of farms' land resources, the prices paid and received by producers, and all choices that producers make that affect their profits and water quality outcomes.
- Can predict the weather without error.

Under these assumptions the environmental agency has no uncertainty about emissions even though it cannot observe them. Knowing the loading function in an ideal information environment is equivalent to knowing emissions as they can be predicted without error. The agency faces no problems from asymmetric information or adverse selection. There is no behavioral uncertainty.

In the ideal information environment, the agency has sufficient information to design any of the four polluter-pays instrument types to meet a pollution load target at least cost (Griffin and Bromley 1982; Shortle and Dunn 1986; Shortle et al. 1998). The agency can determine the least-cost input choices for each farm if it chooses to use input-based quantity controls. It can determine the least-cost estimated emissions for each farm if it chooses to use an estimated emissions standard. It can determine the input tax rates (or subsidy rates for inputs that are pollution reducing) for each input on each farm that would be required to induce the producer to choose the least-cost input choices. And it can determine the tax rate on estimated emissions that would be required to induce producers to choose the allocation of estimated emissions that achieve the desired limit at least cost.

If the agency sought to maximize the social net benefits from pollution and, in the ideal information environment, knows the social damage cost function, the following rules would be followed:

- In an input-based Pigouvian tax approach, every input affecting pollution would be taxed at a rate equal to the social marginal damage cost resulting from the use of the input. The tax rate would be negative, implying a subsidy, for pollution-reducing inputs.
- In an input-based quantity control approach, upper bounds would be placed on the use of polluting-increasing inputs and lower bounds on the use of pollution-reducing inputs. The bounds would be the same quantities that result from imposing the Pigouvian input taxes.

- If the instrument is a tax on estimated emissions, the tax would be set at the Pigouvian rate, and thus equal the social marginal damage cost.
- If the instrument is a standard on estimated emissions, the standard would limit estimated emissions to the Pareto Efficient level.

Whether the instruments are selected to achieve the Pareto Efficient outcome or designed to achieve some other policy-determined limit at least cost, the following would be true:

- The tax rates or quantity controls in optimized input-based instruments would be spatially differentiated across fields and farms due to heterogenous responses of emissions and ambient concentrations at receptors to inputs at different locations. The instruments would also be differentiated by crop and rotation types as these influence the marginal responses of emissions to inputs.
- The input tax rates or quantity controls are set so that input use of all types across all regulated farms result in an aggregate emissions level equal to the policy limit.

The estimated emission standards vary from farm to farm due to variations in abatement costs. Tax rates on estimated emissions would equal the marginal damage costs. Marginal damage costs would generally be farm specific due to imperfect (nonuniform) mixing.

Box 5.4 Ambient Incentives

A radical alternative to policy instruments based on farm practices or farm-level performance indicators is the ambient tax-subsidy instrument proposed by Segerson (1988). The instrument is in the Pigouvian tradition of pricing negative externalities, but it moves compliance monitoring from farms to the affected waters. The instrument taxes excess ambient pollution when it is greater than a set target level and subsidizes the ambient pollution deficit when ambient pollution falls below the target. All polluters pay a tax or receive a subsidy, depending on the case, with the amount they pay or receive completely independent of their own actions. Essentially, it rewards or punishes polluters as group. Accordingly, there is no

(*continued*)

Box 5.4 (continued)

monitoring of individual behavior. A producer who free rides on the effort of others will enjoy the payment if ambient pollution falls below the target. A producer who makes every effort to reduce pollution will pay the tax if ambient pollution is above the target. As a group, polluters have an incentive to reduce ambient pollution, but individuals have an incentive to free ride. The specific structure of Segerson's incentive design eliminates incentives for free riding, resulting in an efficient allocation of emissions.

The premise of the instrument is that the environmental agency can accurately and routinely measure ambient pollution at the water receptor but not the abatement of individual farms. The setup is one of moral hazard with asymmetric information about emissions. The functioning of the instrument is based on certain assumptions. One is that the environmental agency knows polluters' abatement cost functions. This information is essential for setting incentives to produce a cost-effective allocation. Also, polluters are assumed to be able to predict the water quality impacts of their own emissions. Essentially, they are assumed to know their loading functions and the loading functions they know predict without error. Finally, polluters can predict the emissions of others, which seems an impossible task (see Box 5.3). Subsequent research has explored alternative designs and applications explicitly concerned with the agricultural context (e.g., Horan et al. 1998, 2004). Another extension of interest given the reliance on voluntary compliance approaches to agricultural nonpoint pollution control is Segerson and Wu (2006) in which an ambient tax is proposed as a threat to increase the effectiveness of voluntary initiatives.

The information requirements of ambient taxes suggest that they are most suited to water quality problems in small watersheds with a small number of relatively homogeneous producers, well-understood relationships between farming practices and pollution loads, easily monitored ambient water quality conditions, and short lag times between management activities and water quality outcomes (Weersink et al. 1998).

Framing farm decisions as we did in Chap. 4, each of the instrument types when optimized results in the (a) optimal intensity of variable input use (both productive and damage control inputs), (b) optimal use of BMPs with positive net social benefits, and (c) optimal choice of crops for each field and farm. The instruments also optimize the amount and location of land in production.

5.5.2 *A Conceptual Example of Policy Design with Perfect Information*

To develop a concrete understanding of what we have described above, we present a conceptual example of nitrogen pollution control by an agency with perfect information. Nitrogen pollution in the example is a function of the nitrogen fertilization rate as in the farm model presented in Chap. 4, but we add a pollution control input in the form of a buffer strip.[5] Fertilizer is an intensity decision. The buffer is a green infrastructure BMP decision.

In Chap. 4 the social net benefits from crop production were producers' net returns (Ricardian rents) less the social damage costs from pollution. With the addition of a riparian buffer to the producers' choices, there are additional costs. One is the cost of establishing and maintaining vegetative cover on the riparian land removed from crop production for the buffer. The second cost is the opportunity cost of taking land out of crop production. With these costs, a producer's net benefits are the net returns from crop production from land that remains in production (crop revenues less production costs), less the opportunity costs (i.e., forgone profits from land removed from production) and the buffer establishment and maintenance costs. We assume there are no private net benefits to producers from buffer installation. Accordingly, producers have no economic incentive to convert land to buffer in the absence of policy.

To set the quantity or price-based instruments introduced in Sect. 5.5.1 optimally, the environmental agency must first determine the fertilization rate and buffer area that maximize the social net benefit. Once this is determined, the question is how to set the quantity controls or tax/subsidy rates to implement the solutions. Following the marginal analysis logic presented

[5] The model is based on Lankoski and Ollikainen (2003).

in the previous chapters, and assuming for simplicity that the fields have a shape of a square, the optimal fertilizer and buffer choices are characterized by two interdependent conditions:

$$pMP = c + MDC\left(\text{optimal fertilizer rate}\right) \tag{5.1a}$$

$$MSB = \pi + k \ \left(\text{optimal buffer area}\right) \tag{5.1b}$$

The first condition, (5.1a), is for the optimal fertilization rate. As in Chap. 4, this condition requires the marginal revenues from fertilizer use (pMP), to equal the price of fertilizer (c) plus the marginal damage cost (MDC) of the resulting pollution. As the pollution load now depends on the buffer as well at the fertilizer application, the marginal damage cost depends on both choice variables. The effect of the buffer is to reduce the nitrogen reaching the waterbody, and therefore the pollution damage cost. This reduces the marginal damage cost of fertilizer so that the optimal fertilization level is higher than it would be without the buffer. The second condition, (5.1b), is for the optimal buffer area. The term *MSB* in this condition denotes the social marginal benefit of the buffer. As a pollution-reducing input, the buffer provides benefits by reducing the nutrient load. The *MSB* of the buffer is the forgone marginal damage cost. The optimal buffer sets equal the marginal social benefit of the buffer with the marginal cost of the buffer. The marginal cost of the buffer is the forgone net returns (π) from crop production on the land allocated to the buffer plus the marginal establishment cost (k). Note also that only a part of the field will be allocated to buffer, as the MSB is decreasing in the size of the buffer due to diminishing marginal effectiveness of buffer width.

Figures 5.3 and 5.4 depict the optimal solution. The optimal fertilizer use is denoted n^* and the size of the buffer strip (as a share of land) is b^*. The figures illustrate the simultaneous choice of the buffer strip and fertilizer application rate.

An optimal input-based price instrument sets prices for fertilizer and buffers. The fertilizer tax provides an incentive to reduce the fertilizer application rate. The buffer subsidy provides the needed incentive to convert cropland to buffer. If the fertilizer tax is placed at the rate τ per unit of fertilizer use, increasing the per unit cost of fertilizer applications to $c + \tau$, the profit-maximizing producer will choose the fertilizer rate at which marginal revenues equal the augmented fertilizer cost, i.e., pMP, $= c + \tau$. If the tax rate is set equal to MDC, i.e., $\tau = MDC$, then the

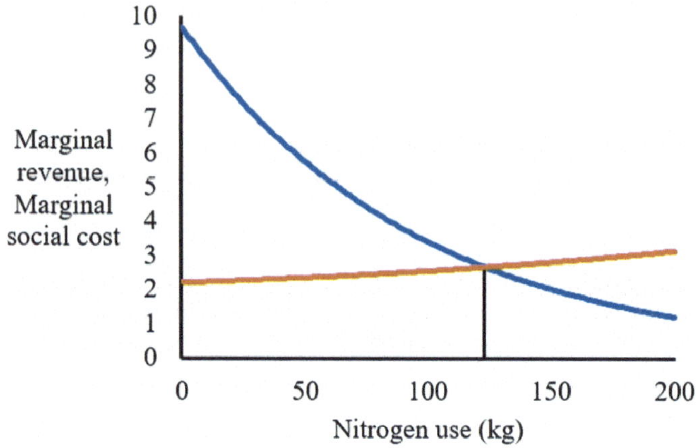

Fig. 5.3 Nitrogen fertilizer intensity with optimal buffer strip. (Source: Authors' creation)

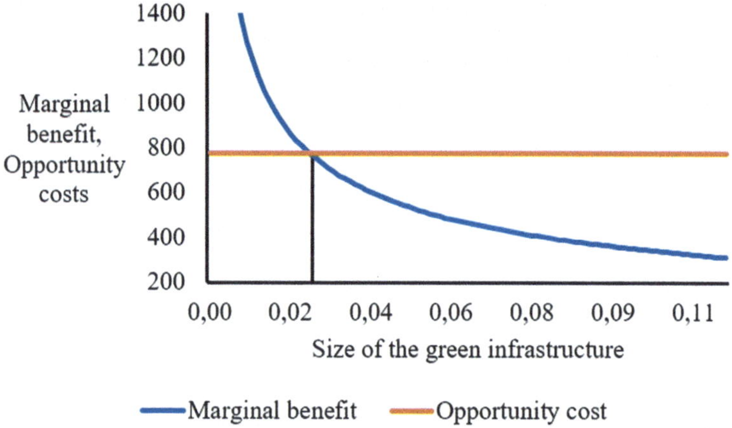

Fig. 5.4 Buffer strip with optimal nitrogen fertilization rate. (Source: Authors' creation)

social optimality condition (5.1a) is met. Similarly, suppose land placed in buffer is subsidized at the rate *s* per hectare. The profit-maximizing producer will choose the buffer so that the producer's marginal benefit of conversion, which is the subsidy payment per hectare, is equal to marginal establishment cost plus the marginal opportunity cost. In this case, we have $s = \pi + k$. If the subsidy rate is set to *MSB*, i.e., $s = MSB$, then the profit-maximizing choice will satisfy the socially optimality condition (5.1b). The Pigouvian input tax policy therefore consists of a tax on fertilizer that is equal to the marginal damage cost and a subsidy on buffers equal to the marginal damage costs avoided: $\tau = MDC(n^*, b^*)$, and $s = MSB(n^*, b^*)$.

The optimal input standards in an input-based quantity approach would restrict the producer's choices to the optimal levels: $n \leq n^*$ and $b \geq b^*$. A Pigouvian tax on the estimated pollution load would tax the estimated load at the *MDC* that results when fertilizer and buffer are optimized. The optimal estimated emissions standard would limit the estimated emissions to the level that results when the fertilizer and buffer are optimized.

5.5.3 A Numerical Example with Multiple Sources and Nonuniform Mixing

A key feature missing from the single-farm analytical model is spatial heterogeneity and the resulting spatial differentiation of instruments. Heterogeneity can be most easily captured by a numerical illustration with multiple land resources. Instead of one farm, consider three farms producing the same crop but with different land qualities located at different distances from the receiving water body. For simplicity, the land within the farms is homogenous with respect to productivity and pollution propensity.

As in Chap. 4, the crop yield function is of the Mitscherlich form but differs from farm to farm due to differences in soils. The edge-of-field loading function is the Simmelsgaard (1991) nitrogen loading function, where the nitrogen loads increase with the nitrogen application. Loads from farms are not uniformly mixed. Delivered loads from a farm are the edge of farm emissions multiplied by the applicable delivery ratio. To illustrate the crucial role of location and land quality in policy design, consider two scenarios, one in which the farm with the best land quality is located closest to the receiving water body and a second in which it is located the farthest. The location assumption is reflected in the delivery ratio. In the first scenario it is comparatively high for high-quality land and in the second it is comparatively low for it.

Absent policy intervention, fertilizer and buffer are chosen on the three farms to maximize net returns. Given no net returns from buffers, the private buffer allocations are zero. The fertilization rates (n^0) and corresponding edge of farm emissions $\left(r_N^0\right)$ are as follows: on the farm with the most productive land, n^0 = 169 kg/ha and the load at the farm edge is $r_N^0 = 24.4$ kg; for the medium productivity farm the corresponding numbers are n^0 = 140 kg/ha and $r_N^0 = 19.8$ kg at the farm edge; and for the farm with the lowest productivity land the numbers are n^0 = 127.8 kg/ha and loads $r_N^0 = 18.2$ kg at the farm edge.

Now consider two instrument designs: input-based standards and input prices. To compute the policies, the first step is to solve for the socially optimal nitrogen application rates and buffer strips. This solution provides the standards for the inputs. We then determine the prices that result in fertilizer and buffer choices that lead to the optimal expected loads at the field edge and the receptor.

Results for the standards and prices, and the resulting edge of farm and delivered pollution levels are shown in Table 5.2. The optimal fertilizer and buffer levels vary substantially across farms reflecting differences in land productivities and the spatial location of the farms relative to the receptor point. The fertilizer reduction relative to unregulated private optimum is 30 kg for the most productive farm when located near the receptor point. Similarly, the nitrogen tax (€/kg) and buffer strip subsidy (€/ha) are strongly differentiated across the parcels. The location of farms matters to both the nitrogen intensity and the buffer sizes.

Table 5.2 Input-based standards versus prices: an illustration

Land quality	*Max yield kg/ha*	*Delivery ratio (β)*	*Nitrogen fertilizer kg/ha*	*Buffer strip % land area*	*Nitrogen tax €/kg*	*Buffer subsidy €/ha*	*Edge-of-field runoff kg*	*Delivered runoff to receptor kg*
Scenario 1: High-quality land has high delivery ratio								
High	5682	1.0	126.5	0.024	0.96	691.7	13.7	13.7
Medium	4182	0.7	110.6	0.021	0.61	462.6	12.4	8.7
Low	3682	0.5	105.7	0.015	0.44	390.3	12.5	6.2
Scenario 2: High-quality land has low delivery ratio								
Low	3682	1.0	94.9	0.047	0.69	389.1	9.9	9.9
Medium	4182	0.7	110.6	0.021	0.61	462.6	12.4	8.7
High	5682	0.5	140.8	0.008	0.59	718.8	16.9	8.4

Source: Authors' creation

5.5.4 *Defining and Coping with Complexity: Part I*

The numerical example above had one crop, two inputs, and three farms. Real-world problems will entail multiple crops, inputs and practices, farms, and heterogeneous land resources. The combinations of potential crops, crop rotations, productive inputs, pollution control practices, and land resources imply that optimal policies would be massively information intensive and complex to develop, implement, and enforce. Imagine, for example, a thick manual specifying the prices or quantity controls for many inputs and practices conditional on crops and field locations for possibly thousands or tens of thousands of farms that must be routinely updated to account for changes in crop and input prices and technology. Optimized policies of the kind we have described are simply implausible, and this is before we consider asymmetric information, stochastic weather, and other sources of uncertainty. Plausible policies will limit the number of variables that are monitored and regulated by relevant agencies and reduce the spatial differentiation of prices or quantity controls across farms (Shortle et al. 1998; Shortle and Horan 2017). Both steps will reduce public transaction costs from policy development and administration, but with a loss of cost-efficiency in pollution control. Policies that achieve Pareto Efficient solutions are referred to in economics as *first-best*. Policies that, due to restrictions of the type we are describing, are optimized subject to those restrictions are referred to as *second-best*. Plausible policies are second-best.

To illustrate the costs from reducing the number of regulated variables or reducing farm to farm variation in policy parameters (i.e., tax or subsidy rates or specific quantity controls), consider some simpler policy designs than those in the previous section. To reflect the effects of reducing the number of variables regulated, we consider a policy that subsidizes buffer installation only, allowing the fertilizer rate to be selected without regard for the social cost. Of the two variables, buffer installation is the easier to monitor at the farm level. To illustrate the effects of reducing spatial differentiation in policy application (i.e., farm to farm or field to field variations), consider the buffer subsidies in one case to be spatially differentiated and based on the marginal pollution damage costs and in a second case as uniform across parcels, defined as the mean of the spatially differentiated subsidies. These scenarios can be labeled, respectively, as second-best.

For the scenarios, we consider as a benchmark the first-best Pigouvian tax levied on estimated loads. The Pigouvian tax is spatially differentiated, reflecting the marginal damage (10€/kg in Table 5.2) and the location of farms, as the latter affects the delivery ratio to the receptor. The Pareto Optimum leads to a reduction of nitrogen loads by 18.7 kg at the receiving water body. The abatement costs are 69.2 €, resulting in the cost of 3.7€/kg of reduced nitrogen.

Table 5.3 provides the results for the differentiated and uniform second-best buffer subsidies. It reports the land share of the buffer, the subsidy paid, the abatement cost, and the reduction of nutrient loads at the receiving water body. In the case of the optimal buffer subsidy, the maximum subsidies per ha vary from 875.7€/ha to 447.9 €/ha, while in the case of uniform they are the average of the differentiated payments, 626.1€/ha. Under the second-best, differentiated buffer subsidy, the share of the buffer decreases with distance and is always greater than at the social optimum. The overall reduction of loads is 13.8 kg, clearly less than under the Pigouvian tax, and total abatement costs 61€.

The outcome under the uniform subsidy policy is much worse. This comparison shows that the performance of alternative policies within the second-best category can vary substantially. The costs of the uniform subsidy are nearly double those of the differentiated subsidy. This difference reflects in part adverse selection, as information about land productivity and its effects on costs is not utilized in the design. Compared to the differentiated subsidy policy, the increased cost also reflects failure to address imperfect substitution across source areas due to differences in delivery ratios. The uniform subsidy for the medium- and low-quality lands is higher than the optimal subsidy, resulting in overinvestment in buffers. But the delivery ratios for these fields are comparatively small so that total abatement is quite modest (10.8 kg).

Economics research generally recommends the use of performance-based policies for efficient pollution control. Key reasons include a focus on ends rather than means in monitoring and enforcement and providing polluters with incentives in their abatement choices to minimize pollution control costs. This discussion suggests another reason in the context of agricultural nonpoint pollution control. Practice-based policies require not only monitoring choices but putting quantitative limits or prices on specific choices, ideally at the farm and field level. Performance-based policies do not eliminate the need for practice monitoring as producers'

Table 5.3 Second-best buffer strip subsidies: a comparison of costs and nutrient loads reduction at the receiving water body

Land quality	*Delivery ratio*	*Differentiated buffer strip subsidy*				*Uniform buffer strip subsidy*			
		Buffer strip		*Abatement*	*Reduction*	*Buffer strip*		*Abatement*	*Load reduction kg*
		Share	*Subsidy €*	*Cost €/ha*	*of loads kg*	*Share*	*Subsidy €*	*Cost €/ha*	
High	1	0.04	36.8	35.2	7.9	0.0	0.0	0.0	0.0
Medium	0.7	0.03	17.7	16.9	3.8	0.1	49.2	44.5	5.9
Low	0.5	0.02	9.2	8.9	2.1	0.2	79.8	67.4	5.0
Totals	Totals			61.0	13.8			111.9	10.8

Source: Authors' creation

choices must be measured to assess performance. But they can simplify policy design by eliminating the need for complex input- or practice-specific regulations or prices.

5.5.5 *Defining and Coping with Complexity: Part II*

The first- and second-best instruments above have been designed with the knowledge of producers' costs and choices, and a loading function that eliminates uncertainty about pollution loads given observations of producers' choices. The first-best policies fully utilize the information, the second-best do not. If the agency does not know the producer's costs, it is unable to calculate any of the policies we have described. This also is true if there is uncertainty about the loading function and stochastic drivers. Uncertainty does not preclude rational policy design, but it complicates choices significantly and reduces potential efficiency gains.

A key challenge posed by nonpoint pollution is the large number of candidate policy options. As evident from our taxonomy of policy instruments, there are multiple combinations and mixes of possible compliance bases, regulatory mechanisms, and specific design details. Our discussion of first- and second-best instruments indicates that instruments that have potential economic and ecological merit and are administratively plausible cannot be fully ranked analytically. Empirical assessments are required. One consequence of uncertainty is to greatly increase the size of the policy evaluation challenge by exploding the number of possible outcomes from alternative policy choices. With asymmetric information and uncertainty about polluters costs, loading functions, and other relevant relationships there will be for any combination of compliance base, regulatory mechanism type, and corresponding design details, a range of possible economic and ecological outcomes that occur depending on what values or specifications uncertain variables or relationships take.

To illustrate, consider a very simple analysis of a fertilizer tax levied on a single producer. If the producer's marginal revenue curve and fertilizer costs are known, then the producer's response to any setting of the tax can be determined. If the loading function is known, the corresponding impacts on water quality can be determined. But if the marginal revenue curve is unknown, the producer's response to the tax is unknown. A specific tax rate would have a range of outcomes. Correspondingly, if the loading function is unknown, any fertilizer response to the specific tax rate could produce in turn a range of water quality outcomes (see Box 5.5).

Box 5.5 Stochastic loads: Monte Carlo simulations

The loads in Table 5.3 are calculated using the Simmelsgaard's nitrogen loading function. This function, which predicts annual average loads, was estimated using data from field observations for a limited range of weather conditions realized during the field observations. As the limited weather history captured by the observations does not represent the full range of the distribution of possible weather, such observations cannot capture the full variation in stochastic weather conditions. Monte Carlo simulations can be used to show actual, realized loads over a broader range of weather conditions. Table 5.4 presents statistics on the distribution of the realized edge of farm loads in 5000 Monte Carlo simulations.

Realized loads are reported using min-mean-max. In addition, Table 5.4 shows the proportion of the realized loads in the tails of the distribution, equal or below 5% and equal or above 95%. The statistical means of the realizations for all qualities and locations are greater than the expected loads predicted by Simmelsgaard's function. Looking at the 5 % tail proportion, only about 5% of the lowest realizations are close to the expected loads. The same will be true of loads at the receptor as they are proportional according to the delivery ratios. What explains the difference? The measured minimum and maximum values used in Monte Carlo simulations are 12.1 kg/ha minimum, 15 kg/ha mean and 25 kg/ha maximum. These values

Table 5.4 Stochastic nutrient loads: expected vs. realized edge-of-field loads

Land quality	*Max yield*	*Delivery ratio* (β)	*Expected loads*	*Realizations of stochastic nutrient loads*				
				Min	*Mean*	*Max*	*5%*	*95%*
Scenario 1: High-quality land has high delivery ratio								
High	5682	1.0	13.7	11.07	15.84	22.75	12.28	20.49
Medium	4182	0.7	12.4	10.07	14.41	20.70	11.17	18.64
Low	3682	0.5	12.5	10.14	14.51	20.84	11.25	18.76
Scenario 2: High-quality land has low delivery ratio								
Low	3682	1.0	9.9	8.02	11.47	16.48	8.90	14.84
Medium	4182	0.7	10.86	10.07	14.41	20.70	11.17	18.64
High	5682	0.5	16.9	13.68	19.57	28.10	15.17	25.30

Source: Authors' creation

(*continued*)

Box 5.5 (continued)

indicate that the tails of the distribution are heavily tilted towards high loads. Together Tables 5.3 and 5.4 demonstrate important differences in *ex ante* planning and *ex post* measurement providing a challenge to policies drawing on expected values, as pointed out initially by Shortle and Dunn (1986).

Suppose for simplicity that there are ten tax rates under consideration. With no uncertainty, there are ten possible tax and water quality outcomes. Suppose that with uncertainty there are ten possible realizations of the tax response (i.e., fertilizer application rates) for each tax rate and another ten possible realized water quality outcomes for every fertilizer response. There are then 100 possible fertilizer response outcomes and 1000 water quality outcomes. Multiply this case by multiple sources of uncertainty in farm behavioral responses for multiple water quality relevant choice variables, multiple sources of uncertainty in the response of water quality conditions to farm management choices, and many farms and it becomes apparent that the complexity of empirical evaluation explodes with uncertainty in economic behavior, ecological outcomes, and damage costs.

Empirical methods for addressing uncertainty are well developed, though not necessarily easy to implement. The topic is beyond the scope of this book, but we mention some approaches to illustrate. When using deterministic models like REAP (see Boxes 4.2, 4.5, 4.6) for policy evaluation, uncertainty about model parameters can be analyzed using sensitivity analyses of various types to generate a range of possible outcomes. Uncertainty about economic factors that drive the model but are external to the model, like income driving the demand for agricultural commodities, can be analyzed using scenario analyses. Alternatively, optimization models may be explicitly probabilistic rather than deterministic, incorporating probability distributions of unknown parameters, and solved to maximize expected social net benefits or minimize expected social costs (e.g., Hanley et al. 1998; Horan and Shortle 2017; McSweeny and Shortle 1990; Nguyen et al. 2013; Qiu et al. 2001; Shortle and Horan 2001, 2013).

Policy design and choice can be done in various ways to cope with uncertainty related to asymmetric information, moral hazard, weather uncertainty, and imperfect information about biophysical relationships.

We described some methods earlier in the chapter in our discussion of pollution load targets. Another way is to select policies that require less information to achieve environmental targets. Setting administered prices to achieve environmental goals, like the fertilizer tax above, requires information on the response of producers to the tax rate. In Chap. 2 we described a key appeal of the use of markets to set prices on pollutions is their capacity, at least in theory, to achieve environmental goals efficiently without regulators knowing abatement costs by creating incentives for polluters to use their private information in ways that facilitate cost minimization. We take up market-based policies in Chaps. 6, 7, and 8. Another way to cope with uncertainty in policy design is to explicitly address uncertainty in setting policy parameters. An example of administrative pricing to address uncertainty is illustrated in Box 5.6.

Box 5.6 A Pigouvian Tax for Agriculture?

The Pigouvian tax introduced in Chap. 2 imposes a tax on emissions equal to the marginal social damage cost of emissions in the Pareto Efficient solution. Such a tax cannot be applied in agriculture because emissions are unobservable. Furthermore, the Pigouvian tax assumes that emissions are not stochastic but instead completely under the control of the polluter. This too is not the case in agriculture. We have introduced the concept of pollution instruments for agriculture that tax estimated emissions (Sects. 5.3 and 5.5). Would imposing tax equal to the marginal cost of estimated emissions have the efficiency properties of a Pigouvian tax?

The answer is no. One reason is that a tax on estimated or mean emissions is inherently second best at best as it manages only expected pollution directly (Shortle and Dunn 1986; Shortle and Horan 2001). A tax on estimated pollution provides direct incentives for polluters to adjust their pollution control choices to manage mean emissions but not the variability of pollution. When the marginal damage cost of pollution is increasing, the expected cost of pollution increases with the variability of emissions, and so managing variability can have important damage cost implications. A tax that fails to manage emissions variability cannot fully internalize the expected marginal social costs of pollution. A tax on estimated emissions therefore cannot be first-best in the way the Pigouvian tax is.

(*continued*)

Box 5.6 (continued)

This limitation does not necessarily mean that incentives applied to estimated emissions are a bad idea. It may be the best option, provided the costs of ignoring variability are not too great and given that most plausible agricultural nonpoint pollution control incentives will be second-best or worse. A theme of this book is that performance-based instruments, such as a tax on estimated emissions that addresses mean performance, are preferred when feasible for a variety of reasons.

What would be the optimal rate for a tax on estimated emissions? Shortle and Horan (2001) demonstrate that given uniform mixing of pollutants, the optimal tax rate equals the expected marginal social cost of the tax-induced change in estimated emissions plus an adjustment that partially accounts for the damage cost impacts of emissions variability. This adjustment may be positive or negative. It is a positive risk premium (negative risk reward) if the variability of agricultural emissions increases (decreases) with estimated emissions, as reducing mean emissions will have the side benefit of also reducing emissions variability. The tax is therefore greater (smaller) than the expected marginal damage cost when the variability of emission increases (decreases) as expected emissions increase. This tax is economically better than one that does not consider the effect on emissions variability, but not as efficient as one that is combined with a separate incentive to explicitly manage variability.

5.5.6 *Polluter-Pays Policies in Practice*

Examples of mandatory policies of the types we have described exist (OECD 2012, 2017; Shortle and Horan 2013). Most notable is the significant regulation of pesticides in the form of restrictions on available types and regulations on application rates and practices. Another notable example is the regulation of large, confined animal operations in the US. Several European countries levy environmentally based taxes on pesticides, fertilizers, and environmental performance proxies. However, as discussed in Chaps. 1 and 2, mandatory polluter-pays policies for agricultural nonpoint pollution control are rare, and when they exist, they are generally mild in comparison to the types of regulatory regimes applied to point sources.

5.6 Pay-the-Polluter and Other Voluntary Compliance Policies

Voluntary compliance encouraged by subsidies is the dominant policy approach to agricultural nonpoint pollution control. Further, while economic incentives are used extensively to encourage voluntary adoption of pollution control practices, there is generally very little use of watershed management science (i.e., loading functions) in the design of these programs. Current policies are not achieving established water quality objectives, are not economically efficient, and impose significant budgetary costs.

Guided by the Pigouvian tradition, paying polluters and other voluntary compliance policies are not the first instruments economists think about when addressing environmental externalities. The large role of voluntary compliance mechanisms in agriculture has various origins and reflects a variety of considerations. In addition to the inherent challenges of controlling nonpoint pollution, there are traditions of agricultural policy that have been carried forward. These include presumptive rights and traditions of using incentive payments to subsidize the use of socially preferred farming practices. This is definitely the case in the US where the major federal policies for agricultural nonpoint pollution control have their origins in soil and water conservation policies developed in the 1930s (Classen et al. 2008; Shortle and Uetake 2015).

5.6.1 Voluntary Compliance Programs: Approaches

The obvious purpose of voluntary compliance programs for water quality protection is to get producers to adopt practices that reduce pollution. There are various approaches to promote voluntary compliance in agriculture (OECD 2017; Pannell and Claassen 2020; Segerson 2013).

Information Programs One approach is to encourage and facilitate unilateral actions by producers or landowners using information programs to make producers aware of problems, inform them of things that can be done and how to do them, and to provide nudges to encourage action. An additional step may be to provide producers with technical (advising) services to help them develop and implement plans for their operations. These too can be sponsored by agricultural and environmental agencies or by environmental and farm organizations. The benefits to producers are generally nonmonetary but there can be monetary benefits when information programs lead to adoption of new farming practices that are both profitable and environmentally beneficial. Conservation tillage and nutrient management are examples (Segerson 2013).

Financial Incentives Voluntary approaches based on providing behavioral nudges and technical information require no enforceable performance compliance. In contrast are voluntary compliance approaches in which producers enter enforceable contracts with government agencies in return for financial payments. Incentive payment types include area-based payments for adopting BMP packages, cost-sharing subsidies for adoption of BMPs, both pay-for-practice mechanisms, and conservation and water quality auctions, which are pay-for-performance mechanisms. Participation is motivated to varying degrees by private benefits when practices are beneficial to producers and the incentive payment.

Threats The inverse to compensation for adoption of pro-environmental behaviors is fear of costs from failure to adopt pro-environmental behaviors (Segerson 2013). An expectation that a collective failure to reduce water pollution among a well-defined group will lead to the imposition of regulations or increased taxes is one kind of motivation. An example is the Florida Everglades Agricultural Privilege under which producers face higher tax payments when water quality standards for phosphorus are not met. Cross-compliance requirements also create a threat structure (Segerson 2013). Cross-compliance makes eligibility for income support programs contingent on adoption of conservation or water quality protection practices. The threat of non-adoption is the loss of eligibility for income support policies. Cross-compliance provisions in US agricultural programs that make eligibility for participating in programs that provide benefits contingent on conservation behaviors have been successful in protecting wetlands from conversion to agriculture production and reducing soil erosion from highly erodible lands (Claassen et al. 2004; Claassen 2012). In the EU Common Agricultural Policy, producers must fulfill environmental cross-compliance conditions to be eligible for the decoupled area payments providing basic income support to producers. These are common EU standards on good agricultural and environmental practices that provide a minimum environmental protection of water and biodiversity.

Information programs and financial incentive programs are the principal tools for voluntary approaches. As noted in Sect. 5.3.3, nudges and information programs have a definite role in the mix of instruments for agriculture, especially one that relies on voluntary adoption, but their potential to address the suite of problems associated with current water quality policies is inherently limited given that pollution abatement is costly.

The major incentive-based voluntary programs in the US are of two broad types: programs for working lands and land retirement programs. Working-lands programs are intended to increase the utilization of resource conservation and environmental protection practices on working farms. Land retirement programs are intended to convert land used for crop and livestock production activities to conservation uses (see Box 5.7). Programs for working lands typically provide subsidies for implementation of farming practice that serve resource conservation or environmental goals. The main working-lands program is the Environmental Quality Incentives Program (EQIP). State governments also offer various types of voluntary programs, including incentive payments. In the European Union, member states are responsible for providing voluntary agri-environmental programs. These generally provide area-based subsidies and investment subsidies as cost-sharing for nonproductive environmental investment subsidies for taking actions or implementing farming practices to promote biodiversity, water quality, and climate mitigation.

Box 5.7 US Conservation Programs

Voluntary conservation programs to promote and subsidize agricultural nonpoint pollution control practices are provided by the federal government through the US Department of Agriculture's (USDA) Natural Resource Conservation Service (NRCS) and by the states. The federal programs are by far the most important. The major federal working-lands programs are the Environmental Quality Incentives Program (EQIP) and the Conservation Stewardship Program (CSP). EQIP provides technical and financial assistance for

(*continued*)

Box 5.7 (continued)

the adoption of conservation practices serving multiple objectives, including water quality protection. Common practices include nutrient management, conservation tillage, cover crops, field-edge filter strips, and fences to exclude livestock from streams. Program priorities are set nationally, but state administrators (NRCS State Conservationists) have some discretion to address state-level priorities. Agricultural nonpoint pollution control is a national priority and a common state priority. EQIP is a cost-share program in which federal support is a share of the cost of implementing a practice, including opportunity costs. Percentages vary by practice but can be as much as 75%. Contract lengths vary up to 10 years. Practices are subject to national NRCS technical standards adapted for local conditions. CSP provides technical and financial support for ongoing and new conservation efforts on farms that meet farm-wide stewardship requirements. CSP provides producers with an annual payment for conservation activities.

The major land retirement program is the Conservation Reserve Program (CRP). CRP provides multi-year contracts to convert cropland to permanent vegetative cover (grasses, trees). The CRP program has two components. One is an annual competitive enrollment in which USDA solicits bids from landowners seeking to participate. These are ranked using an Environmental Benefits Index (EBI). EBI factors for the CRP vary over time but include consideration of a variety of conservation and environmental factors. CRP participants receive annual rental payments, cost-share assistance for establishment, and in some cases additional incentive payments. The second component of the CRP program provides noncompetitive funding for installation of specific practices (e.g., buffers) on environmentally sensitive lands. State priorities can receive consideration in the CRP program through the CRP Enhancement Program (CREP). In participating states, USDA enters agreements with states to provide financial support to producers to address targeted state conservation issues. A second land-oriented program, the Agricultural Conservation Easement Program, pays producers to maintain or restore wetlands or grasslands.

(*continued*)

Box 5.7 (continued)

USDA's Regional Conservation Partnership Program (RCPP) provides funds for projects implemented at regional or watersheds scales by eligible nonprofit organizations or government agencies. Funds can be used for land conversion, subsidies for conservation practices on working lands and technical assistance to producers. Funds are allocated partly by competitive grants, by allocation formulas to states, and to large regional priorities.

Between 1996 and 2011, inflation-adjusted USDA conservation spending grew by roughly 50% (USDA-ERS 2020). Spending on the six largest conservation programs has been on the order of \$6.0 and \$6.5 billion annually since 2011 and is projected to continue within that range during 2019–2023. Almost all of this funding is for EQIP, CSP, and CRP. The share going to the working-lands program has increased substantially since the late 1990s largely at the expense of land retirement programs. Water quality protection has increased its share of USDA conservation spending since the 1990s (USDA-ERS 2020).

5.6.2 *Voluntary Program Design Challenges*

Designing voluntary programs to achieve water quality goals cost-effectively entails all the challenges that we have described for mandatory polluter-pays programs. These include spatial targeting for program resources, spatial differentiation of incentives in heterogenous landscapes, and the choice between practice and performance measures. We comment on these features and others that are unique for incentive-based voluntary programs.

Practice Selection and Payment Schedules Pay-for-practice programs commonly give participating producers a menu of conservation practices to choose from (e.g., Kaufman et al. 2014; Zimmermann and Britz 2016). Depending on the rules, some practices may be obligatory, while others are at the producers discretion. Producers' participation in programs and their choice of practices are influenced by the private economic benefits they receive, their attitudes toward the public goods provided by programs, and the payments they receive relative to the direct out-of-pocket and opportunity costs of adopting practices (Pannell et al. 2006; Pannell and Claassen 2020; Segerson 2013). The presence of private benefits is generally essential for producer participation in cost-sharing programs.

An advantage of an *a la carte* menu is to provide for differentiated actions, enabling producers to choose practices that are best suited to their lands. In cost-share programs, payment schedules typically indicate the share of cost that the sponsoring agency will cover. This share may vary by practice type. The theory of cost-sharing subsidies is that producers receive private benefits from practices but that the costs exceed them. Adoption requires a subsidy, warranted by the jointly produced public benefits. Cost-efficiency problems occur if the menu does not include the socially efficient practices for the location and if the payment schedule does not at least compensate for differences between private benefits and the total practice cost.

The challenges for environmental area-based subsidies as found in the EU are like those of cost-sharing: in most cases adoption of BMPs entail costs that are not covered unless society subsidizes it. Using uniform area payments is insensitive to cost differences between farms and determining the amount of subsidy is difficult with an obvious risk that low-cost farms are not necessarily those who cause most of the pollution. Given asymmetric information, adverse selection problems will occur. Problems can include nonparticipation by farms that ought to participate and/or low cost-efficiency by farms that do.

Information Rents Budgetary efficiency in pay-for-performance or pay-for-practice programs is maximized by paying producers no more than their out-of-pocket and opportunity costs. Programs that pay at fixed rates will invariably pay some producers more than their costs. The surplus is called an *information rent*. Information rents subtract from the budget available to purchase environmental improvements and reduce cost-efficiency. Auctions are one method for reducing information rents. We take these up in Chap. 7.[6]

[6]Though information rents are an unproductive use of resources, the additional farm income that comes when payment exceed costs is considered by some to be a positive feature of voluntary incentive programs (Segerson 2013).

Multiple Objectives With a few exceptions, agri-environmental programs are intended to serve multiple public policy objectives and the practices included within them are selected accordingly. If practices are perfect complements in achieving objectives, then any one practice serves all goals (Lichtenberg 2014). Variations in cost and effectiveness with respect to specific goals in the set implies that rankings of the cost-effectiveness of practices across goals will not be uniform. A selection of practices to achieve a non-water quality objective would serve a water quality objective but not at least cost.

The *Tinbergen Rule* of public policy design recommends that unless policy objectives are perfect complements, policy makers seeking to achieve multiple objectives effectively and efficiently should have at least one instrument for every objective. But perfect complementarity across public policy goals is not the case. In the standard domain of agricultural conservation practices, there are practices that have some degree of complementarity for some goals, some that are independent, and some that are in conflict. An example of the latter is that practices that reduce particulate phosphorus losses from farmland can increase dissolved phosphorus. A similar problem can occur with land retirement programs. Land retirement programs may similarly serve multiple criteria. Land selected to meet one criterion may be low-performing or inefficient for water quality protection (Lichtenberg 2014). Land selection criteria used in the US CRP program favors geographic distributions that bias spending to regions that are not high national priorities for water quality protection, resulting in a low cost-effectiveness of the CRP for water quality protection (Lichtenberg 2018).[7] Claassen, Cattaneo, and Johansson (2008) suggest that improvements in environmental cost-effectiveness of the CRP could be improved by a greater emphasis on wildlife habitat and water quality benefits when accepting bids for enrollment in the program. Given less than perfect complementarity between program goals, targeted programs for

[7]The overall efficiency of multi-objective programs like the CRP and EQIP would consider the overall benefits provided versus their costs. Biases in favor of non-water quality objectives might compensate to some degree for inefficiencies in water quality protection if large benefits were obtained from the non-water quality objectives. Water quality is, however, likely the most significant environmental market failure associated with agriculture. Underweighting water quality protection in multi-objective programs likely decreases the overall efficiency of the spending.

individual objectives, including water quality, can generally be expected to be more efficient than multi-objective programs.

Additionality As noted above, practices included in voluntary conservation programs generally provide a mix of private and public benefits. A necessary condition for water quality improvements to be attributed to a producer's BMP adoption is that the producer would not have selected the package without the public subsidy. Practices that meet this condition are said to be *additional* because their adoption reduces pollution relative to what it would be otherwise. Practices with high adoption costs relative to the private benefits they provide a producer are less likely to be adopted without subsidy and therefore more likely to be additional compared to practices with high private benefits relative to adoption costs (Claassen et al. 2018). Like information rents, allocating scarce public funds to nonadditional practices reduces the effectiveness and efficiency of voluntary programs. In the US, conservation tillage, a routinely subsidized practice, is an example of a practice with generally high private benefits that many producers have adopted without any financial assistance from government programs. Structural practices such as fencing to exclude livestock from streams and riparian buffers are examples of effective practices that have high implementation costs and low private benefits. Such practices have been found to be mostly additional in US programs (Claassen et al. 2018).

On-Farm Spillover Effects Programs for BMP adoption on working lands and for land retirement can have impacts on water quality and other conservation outcomes beyond the direct effect of adoption or land conversion. One reason for this on working lands is the existence of complementary or conflicting interactions between practices at the farm level (Cooper 2003; Dorfman 1996; Fleming 2017; Fleming et al. 2018; Khanna 2001; Lichtenberg 2004a, b; Wu and Babcock 1998). With spillover effects, incentive payments for one practice may "crowd in" the adoption of complementary practices or "crowd out" substitute practices (Fleming et al. 2018). An often-cited example of complementary practices is increased use of herbicides with the adoption of no-till. Another type of spillover in slippage. Slippage occurs when subsidies or payments for practices on working lands create incentives for producers to convert their land in less intensive uses, like hay, to more intensive cultivation (Fleming et al. 2018; Khanna et al. 2002; Lichtenberg 2004b; Lichtenberg and Smith-Ramirez 2011).

Research on on-farm spillovers demonstrate the phenomenon and its relevance to outcomes. Some spillovers may be beneficial, some may be adverse. When on balance they are adverse, the overall additionality of voluntary programs is diminished. An illustration is provided by a study of the impacts of a cover crop cost-share subsidy program in the state of Maryland (US) to reduce nitrogen loads to the Chesapeake Bay. Fleming et al. (2018) use survey data to analyze producers' participation in the cost-share program and the effects on production choices. The study found significant direct additionality from incentive payments for crop cover practices to reduce nitrogen emissions, a complementary effect on land in conservation tillage but a large slippage effect due to conversion of land from vegetative cover, such as hay or pasture, into cultivated cropland that would substantially offset the nitrogen reductions due to the cover crop subsidy payment. Slippage not only occurs through the active conversion of land from vegetative cover, but also through encouraging the continuation of land in intensive crop production for longer periods than otherwise would be. The slippage effect can therefore operate also through maintaining land in crop production that otherwise might be rotated out of it (to alfalfa or other types of hay) or converted to perennial hay or pasture.

Participation in voluntary programs is a commonly used as an indicator of program accomplishment. Because of slippage and other induced economics responses, participation can be misleading as environmental outcomes are offset to some degree. To illustrate, Fig. 5.5 shows farm participation rates in European agri-environmental schemes in 2013. The participation rates vary enormously. In this group Finland is a standout with the highest participation rate in 2013. The program provided water quality benefits, but research suggests the net benefits in Finland were close to zero (Laukkanen and Nauges 2014) or even negative (Lankoski and Ollikainen 2013). According to Laukkanen and Nauges (2014), the program provided water quality benefits by reducing both nitrogen and phosphorus loading. However, the costs were so high that they exactly offset the estimated monetary benefits from water quality improvements. Lankoski and Ollikainen (2013), on the other hand, found that the program increased nitrogen loading. They estimated that the generous

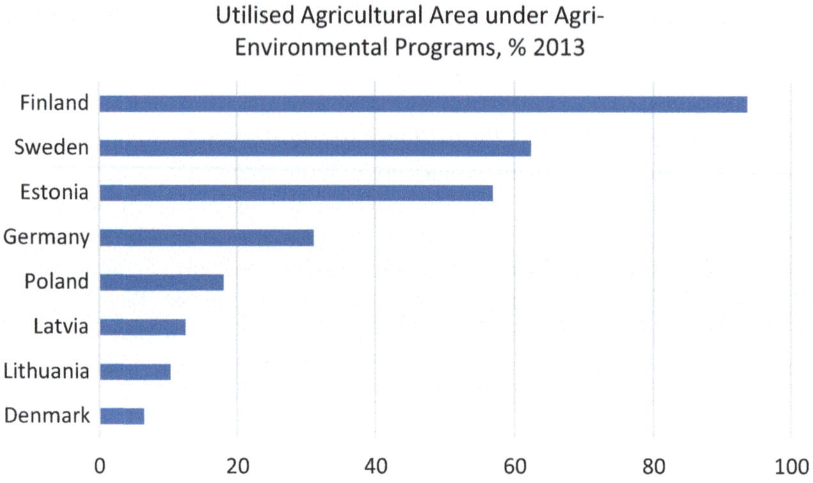

Fig. 5.5 Participation rates in voluntary agri-environmental programs in percent of utilized agricultural area. (Source: Ollikainen, M., Hasler, B., Elofsson, K., Iho, A, Andersen, H-A, Czajkovski M., and Peterson K. 2019. Toward a Baltic Sea Socioeconomic Action Plan. *Ambio*, 48, p.1380)

payments' effect of keeping marginal land in production and favoring nitrogen-intensive crops dominated the effect of fertilizer reductions.

Leakage Economic spillovers also result from aggregate responses of producers that affect input or output prices that in turn induce changes in land use or farming practices. Leakage generally refers to an increase in pollution-producing output in economically linked regions due to increased pollution controls in one or a subset of the regions. One form of leakage occurs when programs to convert environmentally sensitive cultivated land to environmentally preferred uses in some locations create incentives to convert land to more intensive cultivation in others. For example, if programs to remove land from cultivation reduce crop supplies leading to increased crop prices, then land that was not profitable in cultivation may become profitable (Roberts and Bucholz 2006; Wu 2000, 2005).[8]

[8] The terms leakage and slippage are sometimes used synonymously. For example, Wu (2000, 2005) describes as slippage what we are describing as leakage. Following Fleming, Lichtenberg, and Newburn (2018), we find it useful to distinguish between spillovers that are internal to the farm as slippage and the analogous external spillovers from price changes as leakage.

Price-induced changes affecting pollution in other regions may result from changes in crops mix and production practices as well as conversion of environmentally sensitive lands to crop production. The potential for such leakage is illustrated by the USDA Economic Research Service study reducing nutrient losses to the Gulf of Mexico from the Mississippi-Atchafalaya River Basin (MARB) (Marshall et al. 2018). Increased pollution controls in the MARB increase production costs and reduce commodity supplies from the region. Reduced supplies increase commodity prices, leading to an intensification of production in regions outside the MARB. Without additional constraints on nutrient loadings to water bodies outside the MARB, non-MARB cropland in production expands by 2.1% to 4.0%. As a result, nutrient and sediment losses increase 1.2% to 5.0% across non-MARB regions for the two scenarios considered in the study.

Research on additionality, slippage, leakage, and other indirect effects highlight the importance of understanding individual and aggregate behavioral responses to policy interventions in assessments of program performance and design. Research estimating the degree to which interactions offset policy goals is incomplete in its coverage of policies and countries, and difficult because of the complexity of the relationships. However, the existing research suggest that the effects can be substantial and help to explain the disappointing performance of water quality initiatives. These economic responses also highlight the importance of assessing water quality programs based on water quality outcomes rather than by expenditures on practices, the number of participating producers, or the number of hectares enrolled in programs.

Spatial Targeting at Multiple Scales As we have discussed, spatial het erogeneity in the propensities of land areas to contribute pollution, the sensitivity of water resource to pollution, and the cost of pollution reduction make spatial targeting crucial to cost-effective control and maximizing the benefits of public spending for water quality protection.

Targeting is not an inherent defect of voluntary compliance programs except in one respect. Polluters can choose whether to participate or not. Participation of lands in high-priority locations for water quality protection is therefore at the discretion of the landowners. It is an actual defect when program designs fail to target program resources to priority locations defined at the fine scales relevant to effective and efficient water quality management, which is often the case. As noted above, the US CRP parcel selection method does not allocate CRP expenditures according to water quality priorities. Allocations of US working-lands funding to states are not based on water quality criteria but farming activity criteria (Lichtenberg 2014, 2018). The US RCP offers greater opportunities for funding at multiple scales, by prioritizing national problems and facilitating watershed-based approaches (Lichtenberg 2018). The US EQIP program offers some potential for targeting water quality priorities within states in principle but EQIP programs in practice are often operated as first-come first-serve (USGAO 2017).

Pay-for-Practice Versus Pay-for-Performance Practice-based mechanisms are generally not expected to be as efficient in their use of public resources as pay-for-performance mechanisms (Ribaudo et al. 1999; Ferraro and Simpson 2002; Savage and Ribaudo 2016). Producers who can provide the most abatement at the lowest cost have the largest economic incentive to act. This implies that self-selection by producers into a voluntary pay-for-performance program favors those who can provide the most efficient outcomes.

Noncompliance Both mandatory and voluntary programs entail incentives for noncompliance because they entail taking costly actions on farms. Moral hazard is strongly present in agriculture, as many BMPs are costly and their adoption is unverifiable. In an ideal case, producers and agencies would make environmental contracts that are self-enforcing (see Wu and Babcock 1998; Moxey et al. 1999). This kind of agreement would eliminate noncompliance. Unfortunately, the compliance requirements of agri-environmental policies are rarely self-enforcing. Therefore, noncompliance risk is always present and enforcement policies are needed.

Agencies have two main tools to promote compliance: the inspection probability and the size of the fine if violations are discovered (Fraser 2002, 2004; Lankoski et al. 2010). An additional possibility is to use compliance rewards (Yano and Blandford 2009). These studies are in line with the general message of economics: random inspection is essential for policies that rely on self-reporting, and the likelihood of compliance increases with the detection probability and the expected penalty for noncompliance. Monitoring and enforcement issues vary with the instrument type.

For example, farm-based practice regulations require record keeping and on-site inspections. Even then, some management activities can be difficult to document (Jackson-Smith et al. 2010). In contrast, taxes or subsidies on inputs can be levied at points in the supply chain that reduce monitoring costs (Lankoski et al. 2010).

Box 5.8 Water Pollution and Climate Change

Climate is the key determinant of where crops and livestock are produced, how they are produced, and the productivity of agriculture. The effects are direct, for example, the availability of water and warmth needed for plant growth, and indirect, for example, the effects on pest risks. Climate is also a key determinant of water supplies and a range of factors affecting water quality. Climate change is therefore of fundamental importance to the topics of this book. Projected climate change will have profound impacts on agriculture, water resources, and the impacts of agriculture on water quality.

The impact of global climate change on agriculture's impacts on water quality in any location will depend on several interrelated responses. One is the change in the regional climate affecting agriculture and the water resources. A second is the response of the water resource to regional climate change separate from the effects on pollution loads from agriculture. Climate change alone will affect precipitation amounts, forms, and timing; evapotranspiration; water temperatures; and other variables that control water resources, water quality, and the response of water quality to pollution from human activities (e.g., Meier et al. 2011; Najjar et al. 2010). A third response is the effect of the regional climate on the volumes and timing of runoff events and other variables that influence the composition, timing, and volume of pollution loads from agriculture within the contributing area to the water resource. Climate change is driven by an increase in solar energy retained in earth's systems. This increase is exhibited in rising temperatures, but also in an acceleration of hydrologic processes. Other things equal, regions that experience increases in precipitation and the intensity of rainfall events will have

(*continued*)

Box 5.8 (continued)

increased volumes of runoff and pollution loads from agricultural lands. There are, however, many factors that will influence pollution loads and their impacts on water quality. Elevated soil temperatures, for example, accelerate the decay of organic matter, which can result in an increase in nitrates and dissolved phosphorus loads. Yet, increasing temperatures also increase evapotranspiration, crop growth, and nutrient uptake, which can have the opposite effect (Li et al. 2011; Bouraoui et al. 2002). More frequent freeze and thaw cycles in colder climate can increase the loading of dissolved nutrients and the emissions of nitrous oxide, a potent greenhouse gas (Bechmann et al. 2005; Koponen and Martikainen 2004). Changes in the seasonality of runoff can change the timing of delivery of nitrogen loads to waters with seasonal eutrophication problems driven by the combined effects of nitrogen and temperatures (Shrestha et al. 2012).

A fourth response is the effect of regional climate change on agricultural production. Climate-driven changes in the profitability of crop and livestock production lead to changes in agricultural land use as producers adapt, and, accordingly, to changes in agricultural landscapes (Hertel and de Lima 2020). Adaptations may involve changes in production systems using existing technologies – e.g., choices of crops, rotations, fertilization practices, pest management practices, and irrigation (Hertel and Lobell 2014). Adaptation may also involve the development and use of new technologies, e.g., new heat and drought-resistant varieties. Agricultural adaptations will depend not only on changes in growing conditions but also on changes in prices in agricultural and other markets. Beyond the region, global climate change matters to regional outcomes as it will affect market demands and supplies of globally traded agricultural commodities which will in turn influence producers' choices. The overall effects of adaptation and market responses may be to amplify or mitigate the direct effects (e.g., Abler et al. 2002).

There is substantial uncertainty about the impacts of climate change on water pollution loads from agriculture due to uncertainty about climate change at the small spatial scales needed to predict both agricultural and water quality impacts (e.g., Arheimer et al. 2012), uncertainty about the impacts of climate change on agricul-

(*continued*)

Box 5.8 (continued)

tural prices and production (e.g., Hertel and Lobell 2014), and uncertainty about the effects of climate change on the performance of agricultural BMPs (e.g., Lintern et al. 2020). These include agricultural policy responses that influence agricultural adaptation for better or worse (e.g., Müller et al. 2017), and water quality policy responses. In general, the expectation is that climate change will have an overall negative effect on water quality, increase the need for water quality protection especially from nonpoint pollution, and increase the challenges of nonpoint pollution control. For example, a recent analysis for the Chesapeake Bay found that wetter, warmer weather increases nutrient and sediment loads to the Bay due to both land and riverine processes (Bhat et al. 2020). Policy measures to limit pollution loads from agriculture and other sources beyond those currently planned are likely to be required to meet established water quality goals.

5.7 Summary

This chapter begins with a discussion of the characteristics of agricultural pollution that create unique and significant challenges for pollution policy design. One characteristic is that pollution loads from individual fields and farms cannot be metered and controlled deterministically. This characteristic has fundamental implications for the determination of appropriate policy goals and choice of policy instruments. But other factors are also definitive. One is the scope of the problem. Agriculture is a pervasive source of water pollution affecting water resources small and large. Agricultural sources are widely distributed spatially and many in number. Another is the technological complexity and heterogeneity of the problem. Farms are complex and highly varied enterprises. Uniform technological fixes do not exist. Variations in soil resources and climate lead to wide variations in the pollution propensities of agricultural lands and appropriate technologies even at small scales.

The inability to meter nonpoint emissions removes the possibility of applying the emissions-based instruments preferred for cost-effective pollution control. Plausible alternatives for the compliance base generally directly target farm decisions affecting pollution (e.g., polluting inputs, best management practices, crops, land allocations), or emissions proxies or performance indicators constructed from observations of farm decisions. Regulatory mechanisms applied to these compliance bases may involve regulatory mandates or economic incentives of various forms. The dominant approach used in practice is the use of voluntary compliance instruments that rely on information programs to encourage producers to adopt pollution control practices and participate in programs that provide subsidies in various forms to facilitate adoption. As discussed in the previous chapters, experience demonstrates that this approach is generally not providing needed progress toward water quality goal and is unnecessarily expensive.

We summarize basic options for voluntary compliance approaches in theory and practice, and the main themes emerging from economic research evaluating the performance and potential of voluntary compliance programs. Given that transformational institutional change is not easy, it would be nice if at the end of this chapter we could conclude that there are some simple fixes for voluntary compliance programs that could substantially improve their effectiveness and efficiency, but this is not the case. There are steps that could improve the ecological, economic, and budgetary performance of the voluntary compliance approach. These include better spatial targeting, an emphasis on performance rather than practices, utilization of payment schemes that reduce information rents, and greater utilization of water science and economics, such as loading functions and incentive design theory, in policy design to induce and guide adjustments in input intensities, best management practice implementation, crop selection, and land allocation toward socially preferred outcomes. These are not, however, simple fixes as these steps would generally require major transformations in established programs. Furthermore, there are also fundamental problems for voluntary programs that are inherent to the approach. These include additionality, slippage, reliance on producer self-selection, and a dependence on public budgets to purchase reductions.

It would be nice if we could conclude that the best path forward using mandatory programs was well established. Unfortunately, this is also not the case. Economic research has provided a suite of alternative instruments that are in the Pigouvian tradition of internalizing externalities through policies requiring mandatory compliance. These instruments have received considerable attention in theory, but actual experience is insufficient to provide significant lessons. The problem is one of many imperfect options. First-best policy instruments are easy in theory. First-best theory provides explicit guidance about what should be managed and how to price or regulate it. But as we describe, first-best instruments are implausible. Social choices must be made about what to regulate among the set of relevant private choices that can be observed or calculated, the compliance base in our instrument taxonomy, and how to regulate those private choices given the regulatory mechanisms available. For any choice of compliance bases and regulatory mechanisms, many details (e.g., spatial differentiation) are required for implementation. Alternative combinations and mixes of compliances bases, regulatory mechanisms, and significant details provide many options to consider. A general finding in research on environmental externalities is that when policy choices are limited to second-best instruments, policy performance can be improved by using multiple instruments (Bennear and Stavins 2007, Weersink et al. 1998). Various mixtures may be considered to improve the overall performance. For example, including liability rules within a policy mix may improve overall policy performance when monitoring producers' choices is incomplete as the risk of liability for damages creates incentives for producers to manage unobserved or difficult-to-observe variables (Braden and Segerson 1993). Analysis can eliminate some choices and provide some clear guidance for good choices (e.g., performance based, spatial targeting, spatial differentiation, no opt outs for large polluters, small information rents, high additionality) but the best of the feasible set is ultimately an empirical issue.

In the next three chapters we examine specific policy approaches that are drawing substantial interest as approaches to policy reform. These are water quality trading (Chap. 6), water quality auctions (Chap. 7), and credit stacking (Chap. 8). Each has potentials for addressing many of the limitations of conventional approaches, including consideration of local conditions.

Appendix

Selecting BMP Types and Locations to Minimize Costs

BMPs vary greatly in their cost-effectiveness in reducing pollutants. Their effectiveness and costs also depend on where they are undertaken. The importance of the selection of BMP types and locations is illustrated at multiple scales by two studies of the costs to agriculture of the Chesapeake Bay Total Maximum Daily Load (TMDL). TMDLs are required in the US by the Clean Water Act when a water body does not meet established water quality goals. The USEPA established a TMDL for the Chesapeake Bay in 2010 that requires loading reductions into the Bay of 25% of nitrogen, 24% of phosphorus, and 20% of sediment by 2025. Load reduction targets are set for both point and nonpoint sources distributed across the six states (Delaware, Maryland, New York, Pennsylvania, Virginia, West Virginia) and the District of Columbia (i.e., Washington D.C.) which are located in whole or in part in the Chesapeake Bay watershed.

The TMDL established a phased process for environmental regulators in each state to develop Watershed Implementation Plans (WIPs) to meet the TMDL requirements. Phase I and II WIPs, submitted in 2010 and 2012, respectively, both describe actions and controls to be implemented by 2017 and 2025 to achieve applicable water quality standards. For agricultural load reductions, the WIPs consisted of a list of BMPs that the states planned to implement to meet the goals. The USEPA "tested" the WIPs by simulating the impacts on loads using the Chesapeake Bay Watershed Model (CBWM) Version 5.3.2 (USEPA 2010) to assure that they would satisfy the required load reductions. The CBWM is a hydrological model that simulates pollutant loads delivered to the Bay from sources throughout the watershed. The CBWM divides the watershed into management areas referred to as land-river segments. In Version 5.3.2, counties in the six states formed the basis for defining the segments (Martucci et al. 2006; USEPA 2010, section 3). Some counties were further divided based on within-county nutrient flows.[9] The number of management segments that included some agricultural land was 2426. Virginia had the most management areas with 881,

[9] Phase 5 CBWM has been superseded by Phase 6, which includes a new segmentation (Easton et al. 2017).

followed by Maryland with 671, and Pennsylvania with 582 (Kaufman et al. 2014).

The WIPs I and II provided goals for the levels of implementation needed at the land-river segment level (e.g., hectares in various BMP categories) but limited detail on the spatial location. To assess the impacts of the WIPs on loads using the CBWM, the Chesapeake Bay Program placed BMPs in applicable land areas within segments on a *pro rata* basis. For example, if a WIP called for 10% of the agricultural land area in a state to be in a BMP, the BMP was placed on 10% of the agricultural land area in each model segment with agricultural land (Kaufman et al. 2014). This approach takes no advantage of spatial variation in the effectiveness.

A Basin-Wide Agricultural Cost Analysis

Kaufman et al. (2014) estimated the Bay-basin-wide costs of the Phase I and II WIPs submitted by the states. They also estimated the cost of achieving the load reductions using "Cost-Effective BMP Portfolios." The portfolios were constructed by selecting practices within and across segments within each state to minimize the costs under two scenarios. One scenario did not allow removal of agricultural land from production to achieve the load reduction targets, but the second scenario (Scenario 2) allowed 25% land retirement. The procedure essentially begins by ordering BMPs by cost-effectiveness as determined by both the costs of implementation and efficiency in reducing delivered pollution. The latter is strongly affected by location. The procedure eliminated BMPs that scored poorly on cost-effectiveness so that the set of BMPs in the Cost-Effective Portfolio became a subset of the state WIP BMPs. Their results for Scenario 2 are shown in Table 5.5. It is apparent that significant cost savings compared to the Phase I and II WIPs could result from implementation plans that focus on cost-effectiveness in the selection of BMP types and locations. In Scenario 2 (25% land retirement), Bay-wide cost savings are 60% compared to the WIPs. Among jurisdictions, cost savings relative to the WIPs range from 27% in Virginia to 86% in New York and West Virginia. Kaufman et al. (2014) note that BMP cost data available to the study had little or no variation across segments within jurisdictions. The spatial variations in cost-effectiveness within jurisdictions is therefore primarily reflective of variations in delivery ratios from the segments to the Bay.

Table 5.5 Agriculture cost comparisons: Watershed Implementation Plans versus cost-effective BMP portfolios

	Annualized cost (millions USD)		*% Cost savings from cost-effective portfolio*
State	*Watershed Implementation Plan*	*Cost-effective portfolio*	
Delaware	19.4	3.5	82
Maryland	83.0	12.9	84
New York	71.2	10.1	86
Pennsylvania	378.3	101.6	73
Virginia	307.4	223.6	27
West Virginia	43.9	6.0	86
Chesapeake Bay	903.2	357.7	60

Source: Kaufman et al. (2014). Table modified from Table 5.3

Small Watershed-Scale Cost Analysis

Another study that examines costs savings from attention to the cost-effectiveness of BMP types and locations is Amin et al. (2020). The study begins with the observation that variability in BMP effectiveness is often due to landscape differences that lead to differences in hydrologic and biogeochemical responses. For example, as described in Chap. 4, runoff mitigation practices (e.g., no-till) are more effective in some soils than in others, and cover crops are effective in addressing sediment-bound phosphorus (P) loss in some settings but exacerbate dissolved P loss in areas where erosion is not a primary concern. Top-down planning processes over large areas, like the Chesapeake Bay TMDL WIPs, easily overlook fine-resolution spatial details and site-specific conditions that can significantly affect BMP performance. Furthermore, large basin-scale models like the CBWM are not implemented with the fine-resolution temporal and spatial capabilities necessary for cost-effective management at the farm and small watershed scales at which decisions are made. Amin et al. (2020) use a fine-resolution simulation model to compare the effectiveness and cost of top-down Chesapeake Bay WIPs to a "Smart Strategy" plan for the Spring Creek watershed located in Pennsylvania. The study was subsequently updated, and the method applied to the Conewago and Mahantango

Table 5.6 Annualized agricultural best management practice costs in three Pennsylvania watersheds: Watershed Implementation Plans versus Smart Strategy

Watershed	*Surface Area*	*Watershed Implementation Plan*	*Smart Strategy*	*% Costs Savings from Smart Strategy*
	km²	*1000s USD*		
Conewago	136	328	184	44
Mahantango	223	1025	298	71
Spring Creek	370	562	372	34

Source: Choi et al. (2020). Modified from Table 5.2

watersheds, also in Pennsylvania. All three are small tributaries to the Susquehanna River, which is the main source of freshwater for the Chesapeake Bay. The results are reported in Choi, Ready, and Shortle (2020) and summarized in Table 5.6.

Using the Topo-SWAT hydrological model to simulate agricultural management at the scale of individual fields and hydrologically active areas at the subfield level, Amin and colleagues interpreted the Pennsylvania WIP as it would apply to the three watersheds for reducing N, P, and sediment loads to the Bay by the targeted 25%, 24%, and 20%, respectively. The costs and pollution load reductions for the WIP and for the Smart Strategy plans used field-scale information of the effectiveness of BMP types and locations. The results are as follows:

- Both the top-down WIP and Smart Strategy plans nearly met or exceeded, at the watershed outlet, the Bay-level TMDL target load reductions for N, P, and sediment entering the Bay.
- The Smart Strategy plan used fewer BMPs than those included in the top-down WIP to achieve equal or better pollution reductions at substantially lower costs. Cost savings ranged from 34% to 71% of the WIP costs. The cost savings were the result of targeting practices to high-risk nutrient transport areas and selecting BMPs according to watershed-specific cost-effectiveness.

Watershed Selection

The studies discussed above target spatially based on cost. Another approach to targeting is to consider the variations in benefit-cost ratios of pollution reductions across watersheds. For instance, given limited public resources for pay-for-practice or pay-for-performance programs, how should available resources be devoted to different watersheds? Choi, Ready, and Shortle (2020) estimate benefit-cost ratios for BMP implementations in the three Pennsylvania watersheds discussed above. A unique feature of their study is the consideration of the agglomeration of water quality benefits as pollutants move downstream from their origin. Reducing nutrients and sediment loads in waters upstream from the Bay provides benefits in the location where they are released, in downstream locations between the source and the Bay, and in the Bay. Due to attenuation of pollutants as they move downstream, the water quality impact of the upstream pollution reduction diminishes the further downstream it is from the source. However, the benefits depend not only on the pollution reduction along the stream system, but also on the number of people who enjoy the benefits and where they are located relative to the location of the water quality improvements. The study used a suite of modeling methods and data sources to estimate the impacts of the simulated BMP implementations in each of the three source watersheds, on water quality and water economic benefits in each watershed, in downstream locations between the source watersheds and the Bay, and ultimately in the Bay. The water quality simulations and BMP costs were developed by Amin and colleagues for the Conewago, Mahantango, and Spring Creek watersheds for the WIP and Smart Strategy BMP scenarios. The benefit estimates are based on data and procedures presented in Choi, Ready, and Shortle (2020) and Choi and Ready (2019).

Their results for the Smart Strategy, which produces the largest benefit-cost ratios, are summarized in Table 5.7. The Smart Strategy BMPs generate very large, positive net benefits and high benefit-cost ratios. This is interesting for several reasons. One is that the high benefit-cost ratios are at odds with the overall low benefit-cost ratios for water quality protection in the US discussed in Chap. 2. Although there may be many reasons, the fact that this case involves low-cost agricultural nonpoint pollution controls rather than the high-cost point source controls that are the dominant approach to water quality protection in

Table 5.7 Annualized benefits and costs of agricultural best management practices in selected Pennsylvania watersheds

Benefits/costs	*Watersheds*		
	Spring Creek	*Conewago Creek*	*Mahantango Creek*
	1000s USD		
Benefits from improved stream water quality[a]	6085	14,963	11,339
Benefits from improved Bay water quality[b]	6038	7416	70,864
BMP implementation costs	372	184	298
Net benefits	11,727	22,195	81,905
Benefit-cost ratio	32.6	121.6	275.8

Source: Choi et al. (2020). Table modified from Table 5.2

[a]This category is the benefits received from improved water quality in affected streams above the point where the Susquehanna joins the Chesapeake Bay

[b]This category is the benefits received from improved water quality in the Chesapeake Bay due to the upstream water pollution reductions

the US is significant. Another interesting aspect noted by the authors is that the net benefits would be substantial and the benefit-cost ratios high even if only the benefits realized upstream of the Chesapeake Bay were considered. While the Chesapeake Bay TMDL is the major policy driver of pollution control for agriculture in the region, it is apparent that substantial net benefits accrue in this case study to beneficiaries upstream of the Bay. Most of these upstream benefits accrue in Pennsylvania, indicating the BMPs would provide a regional win-win in net benefits.

Benefit-cost ratios greater than one indicate that the cost-effective BMP implementations achieved by the Smart Strategy approach would be good economic investments in any of these watersheds. However, the variation in the benefit-cost ratio across the three watershed is striking and indicates the potential for very larger differences in the societal returns from BMP investments. Factors found to influence the relative net benefits in the case study include overall pollution reduction, the location of the watershed to populated areas, and the location relative to the Bay.

References

Abler, D., J. Shortle, J. Carmichael, and R. Horan. 2002. Climate change, agriculture, and water quality in the Chesapeake Bay region. *Climatic Change* 55 (3): 339–359.

Allcott, H., and T. Rogers. 2014. The short-run and long-run effects of behavioral interventions: Experimental evidence from energy conservation. *American Economic Review* 104 (10): 3003–3037.

Amin, M.M., T.L. Veith, J.S. Shortle, H.D. Karsten, and P.J. Kleinman. 2019. Addressing the spatial disconnect between national-scale total maximum daily loads and localized land management decisions. *Journal of Environmental Quality* 49: 613–627.

Arheimer, B., J. Dahné, and C. Donnelly. 2012. Climate change impact on riverine nutrient load and land-based remedial measures of the Baltic Sea Action Plan. *Ambio* 41 (6): 600–612.

Babcock, B.A., P.G. Lakshminarayan, J. Wu, and D. Zilberman. 1997. Targeting tools for the purchase of environmental amenities. *Land Economics* 73 (3): 325–339.

Beavis, B., and M. Walker. 1983. Achieving environmental standards with stochastic discharges. *Journal of Environmental Economics and Management* 10 (2): 103–111.

Bechmann, M.E., P.J. Kleinman, A.N. Sharpley, and L.S. Saporito. 2005. Freeze–thaw effects on phosphorus loss in runoff from manured and catch-cropped soils. *Journal of Environmental Quality* 34 (6): 2301–2309.

Bennear, L.S., and R.N. Stavins. 2007. Second-best theory and the use of multiple policy instruments. *Environmental and Resource Economics* 37 (1): 111–129.

Bhat, G., L. Linker, G. Shenk, I. Bertani, J. Rigelman, and P. Claggett. 2020. *Chesapeake Bay program phase 6 climate change model initial findings.* Chesapeake Bay Program. https://www.chesapeakebay.net/channel_files/40316/20200121_-_bhatt_-_uswg_-_chesapeake_bay_program_phase_6_climate_change_model_initial_findings.pdf. Accessed 11 Oct 2020.

Bouraoui, F., L. Galbiati, and G. Bidoglio. 2002. Climate change impacts on nutrient loads in the Yorkshire Ouse catchment (UK).

Braden, J.B., and K. Segerson. 1993. Information problems in the design of nonpoint-source pollution policy. In *Theory, modeling, and experience in the management of nonpoint-source pollution*, 1–36. Boston: Springer.

Braden, J.B., E.E. Herricks, and R.S. Larson. 1989. Economic targeting of nonpoint pollution abatement for fish habitat protection. *Water Resources Research* 25 (12): 2399–2405.

Bromley, D.W. 1978. Property rules, liability rules, and environmental economics. *Journal of Economic Issues* 12 (1): 43–60.

Bromley, D.W., and I. Hodge. 1990. Private property rights and presumptive policy entitlements: Reconsidering the premises of rural policy. *European Review of Agricultural Economics* 17 (2): 197–214.

Choi, D.S., and R. Ready. 2019. Measuring benefits from spatially-explicit surface water quality improvements: The roles of distance, scope, scale, and size. *Resource and Energy Economics*: 101108.

Choi, D.S., R.C. Ready, and J.S. Shortle. 2020. Valuing water quality benefits from adopting best management practices: A spatial approach. *Journal of Environmental Quality* 49 (3): 582–592.

Claassen, R. 2012. The future of environmental compliance incentives in US agriculture: The role of commodity, conservation, and crop insurance programs. *USDA-ERS Economic Information Bulletin* (94).

Claassen, R., V.E. Breneman, S. Bucholtz, A. Cattaneo, R.C. Johansson, and M.J. Morehart. 2004. *Environmental compliance in US agricultural policy: Past performance and future potential* (No. 1473-2016-120702).

Claassen, R., A. Cattaneo, and R. Johansson. 2008. Cost-effective design of agri-environmental payment programs: US experience in theory and practice. *Ecological Economics* 65 (4): 737–752.

Claassen, R., E.N. Duquette, and D.J. Smith. 2018. Additionality in US agricultural conservation programs. *Land Economics* 94 (1): 19–35.

Cooper, J.C., 2003. A joint framework for analysis of agri-environmental payment programs. *American Journal of Agricultural Economics* 85 (4): 976–987.

Diebel, M.W., J.T. Maxted, P.J. Nowak, and M.J. Vander Zanden. 2008. Landscape planning for agricultural nonpoint source pollution reduction I: A geographical allocation framework. *Environmental Management* 42 (5): 789–802.

Dorfman, J.H. 1996. Modeling multiple adoption decisions in a joint framework. *American Journal of Agricultural Economics* 78 (3): 547–557.

Easton, Z., D. Scavia, R. Alexander, L. Band, K. Boomer, P. Kleinman, J. Martin, A. Miller, J. Pizzuto, D. Smith, and C. Welty. 2017. Scientific and Technical Advisory Committee Chesapeake Bay Watershed Model Phase 6 Review.

Ferraro, P.J., and M.K. Price. 2013. Using nonpecuniary strategies to influence behavior: Evidence from a large-scale field experiment. *Review of Economics and Statistics* 95 (1): 64–73.

Ferraro, P.J., and R.D. Simpson. 2002. The cost-effectiveness of conservation payments. *Land Economics* 78 (3): 339–353.

Ferraro, P., K.D. Messer, and S. Wu. 2017. Applying behavioral insights to improve water security. *Choices* 32 (4): 1–6.

Fleming, P. 2017. Agricultural cost sharing and water quality in the Chesapeake Bay: Estimating indirect effects of environmental payments. *American Journal of Agricultural Economics* 99 (5): 1208–1227.

Fleming, P., E. Lichtenberg, and D.A. Newburn. 2018. Evaluating impacts of agricultural cost sharing on water quality: Additionality, crowding In, and slippage. *Journal of Environmental Economics and Management* 92: 1–19.

Fraser, R. 2002. Moral hazard and risk management in agri-environmental policy. *Journal of Agricultural Economics* 53 (3): 475–487.

Fraser, B. 2004. On the Use of Targeting to Reduce Moral Hazard in Agri-environmental Schemes. *Journal of Agricultural Economics* 55 (3): 525–540.

Gall, H.E., J. Park, C.J. Harman, J.W. Jawitz, and P.S.C. Rao. 2013. Landscape filtering of hydrologic and biogeochemical responses in managed catchments. *Landscape Ecology* 28 (4): 651–664.

Gburek, W.J., A.N. Sharpley, L. Heathwaite, and G.J. Folmar. 2000. Phosphorus management at the watershed scale: A modification of the phosphorus index. *Journal of Environmental Quality* 29 (1): 130–144.

Goulder, L.H., and I.W. Parry. 2008. Instrument choice in environmental policy. *Review of Environmental Economics and Policy* 2 (2): 152–174.

Griffin, R.C., and D.W. Bromley. 1982. Agricultural runoff as a nonpoint externality: A theoretical development. *American Journal of Agricultural Economics* 64 (3): 547–552.

Hanley, N., R. Faichney, A. Munro, and J.S. Shortle. 1998. Economic and environmental modelling for pollution control in an estuary. *Journal of Environmental Management* 52 (3): 211–225.

Hansen, L., and D. Hellerstein. 2006. Conservation program design: Better targeting, better outcomes. United States Department of Agriculture Economic Research Service *ERS Econ. Brief, 2.*

Hertel, T.W., and C.Z. de Lima. 2020. Climate impacts on agriculture: Searching for keys under the streetlight. *Food Policy* 95: 101954.

Hertel, T.W., and D.B. Lobell. 2014. Agricultural adaptation to climate change in rich and poor countries: Current modeling practice and potential for empirical contributions. *Energy Economics* 46: 562–575.

Horan, R.D. 2001. Differences in social and public risk perceptions and conflicting impacts on point/nonpoint trading ratios. *American Journal of Agricultural Economics* 83 (4): 934–941.

Horan, R.D., and J.S. Shortle. 2017. Endogenous risk and point-nonpoint uncertainty trading ratios. *American Journal of Agricultural Economics* 99 (2): 427–446.

Horan, R.D., J.S. Shortle, and D.G. Abler. 1998. Ambient taxes when polluters have multiple choices. *Journal of Environmental Economics and Management* 36 (2): 186–199.

———. 2004. The coordination and design of point-nonpoint trading programs and agri-environmental policies. *Agricultural and Resource Economics Review* 33 (1): 61–78.

Jackson-Smith, D.B., M. Halling, E. de la Hoz, J.P. McEvoy, and J.S. Horsburgh. 2010. Measuring conservation program best management practice implementation and maintenance at the watershed scale. *Journal of Soil and Water Conservation* 65 (6): 413–423.

Kaufman, Z., D. Abler, J. Shortle, J. Harper, J. Hamlett, and P. Feather. 2014. Agricultural costs of the Chesapeake Bay total maximum daily load. *Environmental Science & Technology* 48 (24): 14131–14138.

Khanna, M. 2001. Sequential adoption of site-specific technologies and its implications for nitrogen productivity: A double selectivity model. *American Journal of Agricultural Economics* 83 (1): 35–51.

Khanna, M., M. Isik, and D. Zilberman. 2002. Cost-effectiveness of alternative green payment policies for conservation technology adoption with heterogeneous land quality. *Agricultural Economics* 27 (2): 157–174.

Koponen, H.T., and P.J. Martikainen. 2004. Soil water content and freezing temperature affect freeze–thaw related N 2 O production in organic soil. *Nutrient Cycling in Agroecosystems* 69 (3): 213–219.

Lankoski, J., and M. Ollikainen. 2003. Agri-environmental externalities: A framework for designing targeted policies. *European Review of Agricultural Economics* 30 (1): 51–75.

———. 2013. Counterfactual approach to assessing agri-environmental policy: The case of the Finnish water protection policy. *Review of Agricultural and Environmental Economics* 94 (2): 165–193.

Lankoski J., E. Lichtenberg, and M. Ollikainen. 2010. Agri-environmental Program Compliance in a Heterogeneous Landscape. *Environmental and Resource Economics* 47 (1): 1–22.

Laukkanen, M., and C. Nauges. 2014. Evaluating greening farm policies: A structural model for assessing agri-environmental subsidies. *Land Economics* 90 (3): 458–481.

Li, Y., B.M. Chen, Z.G. Wang, and S.L. Peng. 2011. Effects of temperature change on water discharge, and sediment and nutrient loading in the lower Pearl River basin based on SWAT modelling. *Hydrological Sciences Journal* 56 (1): 68–83.

Lichtenberg, E. 2004a. Some hard truths about agriculture and the environment. *Agricultural and Resource Economics Review* 33 (1): 24–33.

———. 2004b. Cost-responsiveness of conservation practice adoption: A revealed preference approach. *Journal of Agricultural and Resource Economics* 29 (3): 420–435.

———. 2014. Conservation, the Farm Bill, and US Agri-environmental policy. *Choices* 29 (3): 1–6.

———. 2018. Conservation programs in the 2018 farm bill. *AEI Paper & Studies*: 3D.

Lichtenberg, E., and R. Smith-Ramirez. 2011. Slippage in conservation cost sharing. *American Journal of Agricultural Economics* 93 (1): 113–129.

Lichtenberg, E., and D. Zilberman. 1988. Efficient regulation of environmental health risks. *The Quarterly Journal of Economics* 103 (1): 167–178.

Lintern, A., L. McPhillips, B. Winfrey, J. Duncan, and C. Grady. 2020. Best management practices for diffuse nutrient pollution: Wicked problems across urban and agricultural watersheds. *Environmental Science & Technology* 54 (15): 9159–9174.

Marshall, E., M. Aillery, M. Ribaudo, N. Key, S. Sneeringer, L. Hansen, S. Malcolm, and A. Riddle. 2018. *Reducing nutrient losses from cropland in the Mississippi/Atchafalaya River Basin: Cost efficiency and regional distribution*. United States Department of Agriculture Economic Research Service No. 1477-2018-5724.

Martucci, S.K., J.L. Krstolic, J.P. Raffensperger, and K.J. Hopkins. 2006. *Development of land segmentation, stream-reach network, and watersheds in support of hydrologic simulation program-fortran (HSPF) modeling, Chesapeake Bay Watershed, and adjacent parts of Maryland, Delaware, and Virginia*. Reston: U.S. Geological Survey.

McSweeny, W.T., and J.S. Shortle. 1990. Probabilistic cost effectiveness in agricultural nonpoint pollution control. *Journal of Agricultural and Applied Economics* 22 (1): 95–104.

Meier, H.M., K. Eilola, and E. Almroth. 2011. Climate-related changes in marine ecosystems simulated with a 3-dimensional coupled physical-biogeochemical model of the Baltic Sea. *Climate Research* 48 (1): 31–55.

Moxey, A., B. White., and A. Ozanne. 1999. Efficient contract design for agri-environment policy. *Journal of Agricultural Economics* 50 (2): 187–202.

Najjar, R.G., C.R. Pyke, M.B. Adams, D. Breitburg, C. Hershner, M. Kemp, R. Howarth, M.R. Mulholland, M. Paolisso, D. Secor, and K. Sellner. 2010. Potential climate-change impacts on the Chesapeake Bay. *Estuarine, Coastal and Shelf Science* 86 (1): 1–20.

National Research Council. 2012. *Exposure science in the 21st century: A vision and a strategy*. National Academies Press.

Nguyen, N.P., J.S. Shortle, P.M. Reed, and T.T. Nguyen. 2013. Water quality trading with asymmetric information, uncertainty and transaction costs: A stochastic agent-based simulation. *Resource and Energy Economics* 35 (1): 60–90.

Nowak, P., S. Bowen, and P.E. Cabot. 2006. Disproportionality as a framework for linking social and biophysical systems. *Society and Natural Resources* 19 (2): 153–173.

Ollikainen, M., B. Hasler, K. Elofsson, A. Iho, H.E. Andersen, M. Czajkowski, and K. Peterson. 2019. Toward the Baltic Sea socioeconomic action plan. *Ambio* 48 (11): 1377–1388.

Opalinski, N.F., D.P. Schultz, H.E. Gall, and M. Royer. 2016. Development of a decision-making framework for BMP design to reduce loads during "Hot

Moments". In *2016 ASABE annual international meeting*, 1. American Society of Agricultural and Biological Engineers.

Organization for Economic Cooperation and Development (OECD). 2012. *Water quality and agriculture: Meeting the policy challenge*. OECD Publishing.

Ozanne, A., T. Hogan, and D. Colman. 2001. Moral hazard, risk aversion and compliance monitoring in agri-environmental policy. *European Review of Agricultural Economics* 28 (3): 329–348.

Pannell, D.J., and R. Claassen. 2020. The roles of adoption and behavior change in agricultural policy. *Applied Economic Perspectives and Policy* 42 (1): 31–41.

Pannell, D.J., G.R. Marshall, N. Barr, A. Curtis, F. Vanclay, and R. Wilkinson. 2006. Understanding and promoting adoption of conservation practices by rural landholders. *Australian Journal of Experimental Agriculture* 46 (11): 1407–1424.

Preisendanz, H.E., T.L. Veith, Q. Zhang, and J. Shortle. 2020. Temporal inequality of nutrient and sediment transport: a decision-making framework for temporal targeting of load reduction goals. *Environmental Research Letters 16* (1): 014005.

Qiu, Z., T. Prato, and F. McCamley. 2001. Evaluating Environmental Risks Using Safety-First Constraints. *American Journal of Agricultural Economics* 83 (2): 402–413.

Ribaudo, M.O. 1989. Targeting the conservation reserve program to maximize water quality benefits. *Land Economics* 65 (4): 320–332.

Ribaudo, M., and J. Shortle. 2019. Reflections on 40 Years of Applied Economics Research on Agriculture and Water Quality. *Agricultural and Resource Economics Review* 48 (3): 519–530.

Ribaudo, M., R.D. Horan, and M.E. Smith. 1999. *Economics of water quality protection from nonpoint sources: Theory and practice* (No. 1473-2016-120718).

Ribaudo, M., J. Savage, and M. Aillery. 2014. An economic assessment of policy options to reduce agricultural pollutants in the Chesapeake Bay. *USDA-ERS Economic Research Report* 166.

Roberts, M.J., and S. Bucholz. 2006. Slippage in the Conservation Reserve Program or spurious correlation? A rejoinder. *American Journal of Agricultural Economics* 88 (2): 512–514.

Savage, J., and M. Ribaudo. 2016. Improving the efficiency of voluntary water quality conservation programs. *Land Economics* 92 (1): 148–166.

Segerson, K. 1988. Uncertainty and incentives for nonpoint pollution control. *Journal of Environmental Economics and Management* 15 (1): 87–98.

———. 2013. Voluntary approaches to environmental protection and resource management. *Annual Review of Resource Economics* 5 (1): 161–180.

Segerson, K., and J. Wu. 2006. Nonpoint pollution control: Inducing first-best outcomes through the use of threats. *Journal of Environmental Economics and Management* 51 (2): 165–184.

Shortle, J.S. 1990. The allocative efficiency implications of water pollution abatement cost comparisons. *Water Resources Research* 26 (5): 793–797.

Shortle, J.S., and J.W. Dunn. 1986. The relative efficiency of agricultural source water pollution control policies. *American Journal of Agricultural Economics* 68 (3): 668–677.

Shortle, J.S., and R.D. Horan. 2001. The economics of nonpoint pollution control. *Journal of Economic Surveys* 15 (3): 255–289.

Shortle, J., and R.D. Horan. 2013. Policy instruments for water quality protection. *Annual Review of Resource Economics* 5 (1): 111–138.

———. 2017. Nutrient pollution: A wicked challenge for economic instruments. *Water Economics and Policy* 3 (02): 1650033.

Shortle, J.S., and T. Uetake. 2015. *Agri-environmental policy in the United States.* OECD Publishing.

Shortle, J.S., D.G. Abler, and R.D. Horan. 1998. Research issues in nonpoint pollution control. *Environmental and Resource Economics* 11 (3–4): 571–585.

Shortle, J.S., M. Ribaudo, R.D. Horan, and D. Blandford. 2012. Reforming agricultural nonpoint pollution policy in an increasingly budget-constrained environment. *Environmental Science and Technology* 46 (3): 1316–1325.

Shrestha, J., P.A. Niklaus, E. Frossard, E. Samaritani, B. Huber, R.L. Barnard, P. Schleppi, K. Tockner, and J. Luster. 2012. Soil nitrogen dynamics in a river floodplain mosaic. *Journal of Environmental Quality* 41 (6): 2033–2045.

Simmelsgaard, S. 1991. Estimation of nitrogen leakage functions – Nitrogen leakage as a function of nitrogen applications for different crops on sand and clay soils. In *Nitrogen fertilizers in Danish Agriculture – Present and future application and leaching*, Institute of Agricultural Economics Report 62 (in Danish: Kvaelstofgödning i landbruget – behov og udvasking nu og i fremtiden). English summary, ed. S. Rude, 135–150. Copenhagen.

Talberth, J., M. Selman, S. Walker, and E. Gray. 2015. Pay for performance: Optimizing public investments in agricultural best management practices in the Chesapeake Bay Watershed. *Ecological Economics* 118: 252–261.

Thaler, R.H., and C.R. Sunstein. 2008. Nudge: Improving decisions about health. In *Wealth, and happiness.* Yale University Press.

US Department of Agriculture Economic Research Service (USDA-ERS). 2020. *Conservation Programs.* https://www.ers.usda.gov/topics/natural-resources-environment/conservation-programs.aspx

US Environmental Protection Agency (USEPA). 2010. Chesapeake Bay Phase 5.3 Community Watershed Model, EPA 903S10002-CBP/TRS-303-10. Annapolis: USEPA, 2010.

———. 2020. *Overview of total maximum daily load.* Accessed 8 June 2020.

US General Accounting Office (USGAO). 2017. USDA's environmental quality incentives program could be improved to optimize benefits GAO-17-225: Published: April 13, 2017.

Wainger, L.A., G. Van Houtven, R. Loomis, J. Messer, R. Beach, and M. Deerhake. 2013. Tradeoffs among ecosystem services, performance certainty, and cost-efficiency in implementation of the Chesapeake Bay total maximum daily load. *Agricultural and Resource Economics Review* 42 (1): 196–224.

Weersink, A., J. Livernois, J.F. Shogren, and J.S. Shortle. 1998. Economic instruments and environmental policy in agriculture. *Canadian Public Policy/Analyse de Politiques*: 309–327.

Wu, J. 2000. Slippage effects of the conservation reserve program. *American Journal of Agricultural Economics* 82 (4): 979–992.

———. 2005. Slippage effects of the conservation reserve program: Reply. *American Journal of Agricultural Economics* 87 (1): 251–254.

Wu, J., and B.A. Babcock. 1998. The choice of tillage, rotation, and soil testing practices: Economic and environmental implications. *American Journal of Agricultural Economics* 80 (3): 494–511.

Wu, J., D. Zilberman, and B.A. Babcock. 2001. Environmental and distributional impacts of conservation targeting strategies. *Journal of Environmental Economics and Management* 41 (3): 333–350.

Yano, Y., and D. Blandford. 2009. Use of compliance rewards in agri-environmental schemes. *Journal of Agricultural Economics* 60 (3): 530–545.

Zimmermann, A., and W. Britz. 2016. European farms' participation in agri-environmental measures. *Land Use Policy* 50: 214–228.

CHAPTER 6

Water Quality Trading

6.1 Introduction

Emissions trading was first proposed as a pollution control policy instrument by economists Tom Crocker (1966) and John Dales (1968) in the 1960s in the form now known as cap-and-trade. Emissions trading began to move from concept to practice in the US in the mid-1970s with programs that allowed companies to trade air emissions between within-company sources (e.g., between smokestacks in a facility) (Tietenberg 1990). The first true cap-and-trade program was the US Acid Rain program established by the Clean Air Act Amendments of 1990 to control sulfur dioxide emissions. Since then more have followed at local, regional, and national scales for a range of air pollutant types. Assessments of air emission trading conclude that well-designed programs can limit emissions to target levels and reduce abatement control costs compared to traditional regulatory approaches (Goulder 2013; Schmalensee and Stavins 2017; Tietenberg 2010).

Emissions trading programs for water pollution control, commonly referred to as water quality trading, are rare. Trading has not had the scope of applications nor the compelling successes that are found in air applications (Fisher-Vanden and Olmstead 2013). Reasons for this difference include physical aspects of water pollution control that pose challenges for

J. Shortle et al., *Water Quality and Agriculture*, Palgrave Studies in Agricultural Economics and Food Policy,
https://doi.org/10.1007/978-3-030-47087-6_6

market design, flaws in market design and development, and institutional restrictions on markets. Yet there is substantial interest in, and good reasons for, considering water quality trading as a component in the policy mix for water quality protection.

Water quality trading has the potential to address many of the shortcomings of the voluntary compliance approach to agriculture without adopting the shortcomings of the regulatory approaches used for point sources. It is performance rather than practice based. Credible implementation requires the use of scientific understanding about the relationships between farm practice and pollution loads. It can be implemented with spatial targeting. It can be used to integrate the management of point and nonpoint sources to improve the overall effectiveness and efficiency of water quality protection. The use of markets enables polluters to use their private information to discover cost-effective allocations and provides desirable innovation incentives.

Research on the design of emissions trading programs is extensive but almost exclusively addressed to trading between conventional point sources. In this case emissions are observable and largely nonstochastic. To extend trading to include agricultural nonpoint sources is to bring the challenges of unobservable and stochastic emissions to market design. The archetypal water quality trading program is a point-nonpoint trading in which wastewater treatment plants can trade with each other or with agricultural or other nonpoint sources.

We devote the first part of this chapter to the basic theory of emissions trading. The basic theory is a theory of trading between point sources, generally referred to as point-point trading in the water quality domain. The basic theory is crucial to understanding both the merits of trading for efficient pollution control and the challenges of realizing those merits when trading with nonpoint sources. We then turn to methods for including agricultural nonpoint sources in trading programs and design issues for managing nonpoint risk. The remainder of the chapter presents a review and assessment of real-world trading programs. The number of active water quality trading programs is small. Recent inventories find several dozen initiatives in a variety of developmental stages. These include exploratory studies, pilot and demonstration projects, programs with established rules that are open for trading, and pilots or once active programs that have been suspended (Fisher-Vanden and Olmstead 2013; Stephenson and Shabman 2019). Almost all are within the US. Trading programs with agriculture are found in Canada and New Zealand. Of the 26 active programs in the US, 24 permit trading between point and non-point sources, the latter mainly agricultural.

6.2 Basic Trading Models and the Merits of Markets

Emissions trading takes two basic forms: cap-and-trade and baseline-and-credit (Tietenberg 2003). Cap-and-trade programs set an explicit limit (or cap) on the total level of emissions from regulated polluters. In cap-and-trade, polluters are required to hold allowances (permits or licenses) if they pollute. Baseline-and-credit programs impose regulatory limits (baselines) on individual regulated polluters' emissions. Polluters who reduce their emissions below their baseline earn credits that can be used by other polluters to meet their regulatory limits. Baseline-and-credit trading for air emissions was the earliest form of trading in the US (Tietenberg 1990). Cap-and-trade is now the most common trading form for air emissions (Goulder 2013). Water quality trading is almost exclusively baseline-and-credit trading.

6.2.1 The Basic Cap-and-Trade Model

We introduce the basic cap-and-trade model with a simple hypothetical case: a single pollutant is to be managed for a single receptor, polluters control their emissions deterministically, and emissions are metered accurately by source and are uniformly mixed. Under these assumptions, cost minimization for any given aggregate emission target requires equalization of marginal abatement costs (per the equi-marginal principle for uniformly mixed pollutants developed in Chap. 2). The regulator sets a total emissions cap equal to the target emissions and issues a supply of tradeable emissions allowances equal to the cap. The allowances may be distributed free of charge or by auction.[1] We refer to this basic model as an *emissions permit system*.

[1] When issued free of charge one approach to the initial allocation, referred as "grandparenting," is based on historical emissions. Another approach used in the European Union Emissions Trading System, referred to as benchmarking, distributes allowances that cover the emissions of sources with emissions well below the industry average (the lowest 10% in any sector). Sources that do not meet the benchmark receive smaller initial allocations.

Once the initial allocation has been distributed, polluters may buy allowances if they seek to emit more than the initial allocation or sell if they seek to profit from selling allowances. The economic results depend on the behavior of market participants. The standard for considering the efficiency of trading is a perfectly competitive market equilibrium. In perfectly competitive markets, price-taking buyers and sellers exchange standardized commodities at an established market price. In an emissions permit system, the commodity is standardized emissions allowances. Standardization means that demanders do not differentiate between suppliers based on commodity attributes as all commodities are the same. Competition removes the power of individual suppliers or demanders to control price and results in a uniform market equilibrium price (the one price rule).

Consider an individual polluter, say a food processing plant. Denote the plant's discharges during the compliance period as e. The plant's initial allowance allocation for the compliance period is denoted $\bar{e}$. The plant's baseline (pre-permitting) emissions level is e_0. The plant manager has three interconnected decisions: the plant's emissions (or equivalently abatement) level under the trading program (e), whether to participate in the market, and how to participate in the market (as a buyer or seller).

Consider first the manager's optimal emissions (abatement) decision. A market price for allowances creates an opportunity cost for emissions as polluters can benefit from abatement to generate revenues from selling allowances if they have allowances to sell. Table 6.1 shows abatement ($a = e_0 - e$), marginal abatement costs (MAC), and the difference between the allowance price and the MAC where the allowance price is $p_e = 12.5$ per allowance.[2] If the plant is seeking to sell allowances, the allowance price is the marginal revenue from allowance sales. The difference between the price and MAC is the marginal profit from selling allowances.

[2] The price would be expressed in monetary units such as USD or EUR per allowance. For the hypothetical example, the reader may select their preferred currency.

From Table 6.1, the allowance price exceeds MAC up to $a = 5$, beyond which MAC exceeds the price. Correspondingly, the marginal profit from selling allowances is positive when $a < 5$, zero when $a = 5$, and negative when $a > 5$. The plant can add more to revenues than costs by increasing abatement to sell allowances up to $a = 5$ if its initial allocation enables selling 5 or more allowances. The plant would add more to cost than revenue if it reduced emissions further. The profit-maximizing abatement for the plant if selling allowances is therefore $a = 5$ at which $p_e = MAC$.

But the plant will also choose $a = 5$ if it buys allowances. If the plant's initial allocation is zero, it must buy allowances to operate. The benefit of buying is the avoided abatement costs as it increases emissions. The cost is the expenditures on the allowances. At the margin, the plant compares the marginal benefit of additional emissions (i.e., avoided abatement costs) to the marginal cost of buying additional allowances. The former is given by the MAC and the latter is the allowance price. In Table 6.1 the difference between the allowance price and the MAC is the negative of the marginal net benefit (MAC - 12.5) of buying allowances to increase emissions. It pays to buy additional allowances if $MAC > 12.5$ as the marginal benefit exceeds the marginal cost. Buying additional allowances is uneconomic if $MAC < 12.5$. Accordingly, the profit-maximizing abatement of the plant if buying allowances is also $a = 5$. From this analysis the manager's optimal emissions choice depends on the allowance price and the marginal

Table 6.1 Trading market participation

Abatement (kg)	*Marginal abatement cost ($)*	*Total abatement cost ($)*	*Marginal profit ($) (12.5 - MAC)*
0.00	0.00	0.00	12.50
1.00	0.50	0.17	12.00
2.00	2.00	1.33	10.50
3.00	4.50	4.50	8.00
4.00	8.00	10.67	4.50
5.00	12.50	20.83	0.00
6.00	18.00	36.00	-5.50
7.00	24.50	57.17	-12.00
8.00	32.00	85.33	-19.50
9.00	40.50	121.50	-28.00
10.00	50.00	166.67	-37.50

Source: Authors' creation

abatement cost but not the initial allocation. The optimal emissions level in either case corresponds to the abatement level at which $p_e = MAC$. Given that all polluters face the same price, the result will be equalization of MACs across polluters and therefore a least-cost allocation for the aggregate pollution target. Specifically, given n pollution sources, the market will satisfy the cap and result in an allocation of abatement (or emissions) with the following property:

$$p_e = MAC_1 = MAC_2 = \ldots = MAC_n$$

Whether the polluters participate in the market as buyers or sellers depends on whether their initial allowance allocation is greater or smaller than the optimal emissions. If the initial allocation is greater than the plant's optimal emissions, then it is optimal to sell. If the initial allocation is less than optimal emissions, then it is optimal to buy. The amount bought or sold will be the difference between the initial allocation and optimal emissions. Polluters who participate in the market as sellers profit from the existence of the market. Polluters who participate in the market as buyers have lower profits than they would in the absence of regulation but higher profits than they would have if their emissions were limited to their initial allocations. Buyers therefore also profit from the existence of the market. Box 6.1 describes the emissions permit system equilibrium algebraically and graphically.

Box 6.1 Emission Permit Market Equilibrium

In Fig. 6.1, the x-axis is total emissions from n polluters. Where e_i denotes the emissions from polluter i, $i = 1, 2, \ldots n$, and total emissions are written $\sum_{i=1}^{n} e_i$. Denoting e_i^0 as polluter i's pre-market emissions, $E = \sum_{i=1}^{n} e_i^0$. The x-axis intercept, E, is total pollution from n polluters prior to establishment of the permit market. T is the emissions cap set by the environmental agency. Denoting $\overline{e}_i$ as polluter's initial allowance, an initial permit allocation equal to the cap means $\sum_{i=1}^{n} \overline{e}_i = T$. The vertical line at T depicts the *total market supply curve*

(continued)

Box 6.1 (continued)

for allowances. Total pollution abatement can be read along the x-axis as the reduction in total emissions from the pre-market baseline. Defining $a_i = e_i^0 - e_i$ as polluter *i*'s abatement, total abatement is then $\sum_{i=1}^{n} a_i = \sum_{i=1}^{n} \left(e_i^0 - e_i\right)$. Read from left to right, the *market demand curve for allowances* shows the total quantity of allowances that would be demanded by polluters at any allowance price to cover their emissions. Read from right to left, the demand curve is the *marginal abatement cost curve for total abatement.* The demand curve is derived from the MACs of polluters analogous to those shown in Table 6.1. Specifically, polluters choose their abatement levels to equate MACs with the allowance price. Subtracting the total abatement provided by polluters at any allowance price p_e on the y-axis from the pre-market baseline emissions E gives the quantity of emissions allowances demanded at the price. The *market equilibrium price* p_e^* is the price at which the demand for allowance equals the supply of allowances. Denote s_i as the allowances sold and b_i as the allowances bought by polluter i in the market equilibrium. Polluters' emissions are restricted to the allowances they own, so $e_i \le \overline{e}_i + b_i - s_i$ for any i. In a market equilibrium, the sum of purchases must equal the sum of sales, so $\sum_{i=1}^{n} \left(b_i - s_i\right) = 0$. This implies that total emissions will not exceed the cap: $\sum_{i=1}^{n} e_i \le \sum_{i=1}^{n} \overline{e}_i = T$.

6.2.2 *Trading Versus Emission Standards*

The principal case for trading is the capacity of markets to achieve pollution targets at a lower cost than conventional regulatory mechanisms. An environmental agency with perfect information about polluters' costs and the capacity to compute the least-cost solution could assign emissions levels to polluters that are cost minimizing. Possessing this knowledge might not be problematic for a small number of publicly owned wastewater treatment plants or simple common industrial processes. But in regulatory

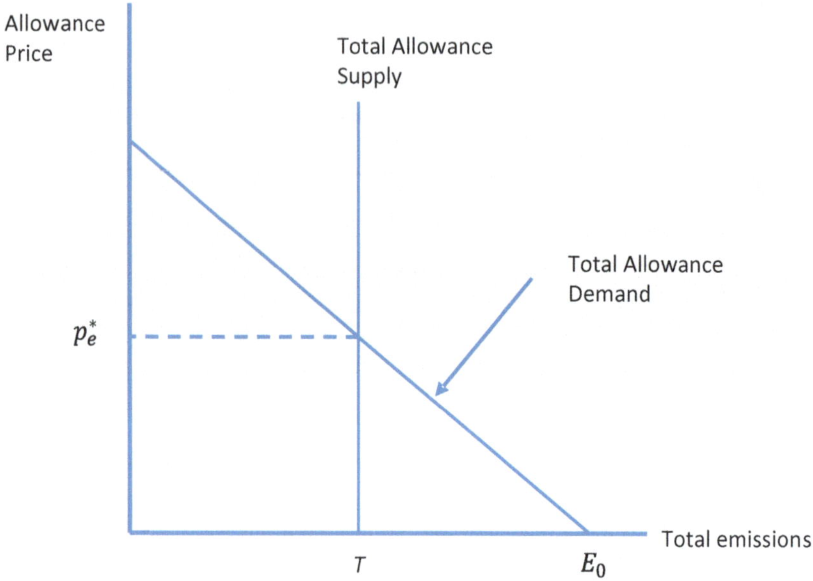

Fig. 6.1 Emissions permit system market equilibrium. (Source: Authors' creation)

situations with more than a few private polluters operating in dynamic economic environments, regulatory processes to discover and implement low-cost solutions will be expensive and imperfect. In contrast, information revelation and economic coordination are tasks that well-functioning markets do and can do well. If a market can achieve policy makers' environmental objectives at lower social cost, societal economic and environmental objectives are better served by relying on the market.

6.2.3 Emissions Permits Versus Emissions Taxes

As discussed in Chap. 2, cost minimization under the basic assumptions of the emissions permit system could also be achieved with a uniform tax or charge on emissions where the tax rate per unit of emissions is set equal to the market equilibrium allowance price p_e. Following the marginal decision analysis method above, we would find that optimizing behavior by individual polluters would lead them to set their emissions levels to equate their MACs with the administratively set tax price, thus resulting in equalization of MACs. In this case, there is no efficiency-based reason to prefer an emissions permit system to a tax or a tax to an emissions permit system.

But for this equivalence, the environmental agency must know the allowance price that equilibrates the market. Given increasing MACs, total emissions will vary inversely with the tax rate. A tax rate that is lower or higher than the market equilibrium allowance price would result in total emissions greater or lesser than the cap. For an equivalent result, the agency must be able to perfectly predict the market equilibrium price. This requires the regulator knowing polluters' MACs. Achieving the target at least cost with the market requires no such knowledge. In the economic, institutional, and physical environments we have assumed thus far, the agency need only set up the market and set the aggregate permit supply cap to obtain the cap at least cost.[3] This is the fundamental insight of Crocker and Dales.

6.2.4 *Nonuniform Mixing and Multiple Receptors*

The equi-marginal principle for least-cost abatement requires emissions from different sources to be perfect substitutes in determining ambient environmental conditions. Emissions from different sources are perfect environmental substitutes if a one-for-one exchange would leave ambient environmental conditions unaffected. Uniformly mixed pollutants are perfect substitutes.

Uniform mixing is not generally the case in water pollution problems. Furthermore, it is often the case that water pollutants entering water resources at one location in a water system cause damages at multiple downstream locations. Effective and cost-efficient pollution control must take account of such spatial patterns of discharges and their environmental

[3] Several factors can affect the relative performance of taxes and trading in application that may qualify this conclusion, including differences in transaction costs between the mechanisms (Stavins 1995). Furthermore, if the regulator is not committed to achieving a specific target but will instead allow some variability in the emissions, then there are circumstances under which taxes may be economically preferred to markets (Weitzman 1974). The cap in cap-and-trade places an upper limit on emissions. Changes in the demand to pollute will cause variations in the price of allowances but the pollutions do not vary. An emissions tax fixes the price of emissions but allows emissions to fluctuate with changes in the demand to pollute. Depending on the characteristics of the benefits and costs of pollution, there are circumstances in which the cap is economically preferred to the tax and vice versa (Horan and Shortle 2021).

impacts. One approach to addressing nonuniform mixing and multiple receptors is an *ambient permit system* in which allowances are specified in terms of contributions to delivered pollution loads at discrete downstream locations (Montgomery 1972). The ambient permit system is a cap-and-trade in pollution loads delivered downstream rather than in emissions. The key design elements of the ambient permit system are receptor-specific caps set to achieve receptor-specific water quality goals and receptor-specific allowances denominated in terms of pollution loads delivered to the receptors. The essential information needed to translate emissions to contributions to downstream-delivered pollution are delivery ratios that give the fraction of emissions released at one location that are delivered to downstream receptors (see Box 6.2).

If the management problem is one with a single receptor but nonuniform mixing, cost minimization requires distributing abatement across sources to minimize the costs of achieving the cap on the delivered load for the one receptor. The condition established for this case in Chap. 2 is equalization of the MACs of delivered pollution across sources. Because there is only one receptor, the ambient permit system will require only one market for trading allowances for delivered pollution as was the case with the emissions permit system. A perfectly competitive market would

Box 6.2 Trade Ratios

Trade ratios are a common feature of water quality programs (Shortle 2013; Fisher-Vanden and Olmstead 2013). A trade ratio is a rule defining the rate at which an allowance or credit for one pollutant (e.g., nitrogen or phosphorous), source (e.g., point or nonpoint), or location type can be exchanged for another. The primary purpose of trade ratios is to address imperfect substitution between different types of polluters (e.g., point, nonpoint) and different locations. Trade ratios for this purpose are intended to translate emissions from different source types, locations, or pollutant types into water quality impact equivalents (e.g., USEPA 2007). Trade ratios are sometime used for other purposes. For example, the Greater Miami River of Ohio nitrogen credit trading program developed a trade ratio scheme intended to speed trading by establishing a time schedule for trade ratios that provided better terms of trade for credit buyers for credits purchased prior to a specified date (Shortle 2013).

establish a single price (p_a) per allowance. As was the case with the emissions permit system, it will generally pay polluters to participate in the market as buyers or sellers. Analogous to the emissions permit system market, profit maximization will lead participants to choose abatement levels and corresponding buying or selling activity resulting in the equalization of the marginal cost of abating delivered pollution and the market price of allowances.

As described in Chap. 2, the MAC for delivered pollution is the MAC at the point of discharge divided by the delivery ratio. Given that all participants are choosing their abatement levels to equate the MAC of delivered pollution abatement to the ambient pollution price, the result will be to minimize the cost of achieving the ambient cap. Specifically, the market would satisfy the cap and result in the following relationship between prices and marginal abatement costs:

$$p_a = \frac{MAC_1}{\beta_1} = \frac{MAC_2}{\beta_2} = \ldots = \frac{MC_n}{\beta_n}$$

where $0 \leq \beta_i \leq 1$ is the delivery ratio to the receptor from source, i, $i = 1, 2, \ldots, n$.

The ambient permit system as described requires the correct delivery ratio for each polluter. If this information is not available, information intensity can be is reduced by using zonal approaches (Hung and Shaw 2005; Johnson 1967). In practical applications, the implementation of an ambient permit system for water pollution with a single receptor would divide the upstream source locations into zones (e.g., stream reaches, subwatersheds) and establish delivery ratios for the zones. Analytically equivalent to the ambient permit system in the single receptor case with pollution sources divided into zones would be an emission allowance trading with terms of trade governed by trade ratios based on zonal delivery ratios as described in Box 6.2 (Montgomery 1972; Tietenberg 1985). Trade ratios of this type are used in several water quality trading programs (Fisher-Vanden and Olmstead 2013; USEPA 2007).

Multiple receptors with caps on delivered loads set for each entail a significant increase in the complexity of efficient market design. A cost-minimizing allocation with multiple receptors minimizes aggregate abatement costs in the water system subject to satisfying all receptor-specific caps on delivered pollution loads. As noted in Chap. 2, the efficiency conditions for multiple receptors is mathematically complex as there is in general a condition for each source that considers the impacts of all affected downstream receptors (see, e.g., Hung and Shaw 2005; Montgomery 1972).

A cost-minimizing ambient permit system with multiple receptors would require multiple allowance markets, one for each receptor. Polluters would be required to hold a portfolio of allowances for all the receptors they affect. Since there are multiple receptor markets, a perfectly competitive equilibrium in each would generate an ambient allowance price for each. Polluters would again benefit from market participation as buyers or sellers.

A perfectly competitive ambient permit market equilibrium will minimize the social costs of achieving policy-determined caps for the multiple receptors provided that the initial allowance allocation satisfies certain technical conditions (Krupnick et al. 1983; Montgomery 1972). However, the ambient permit system has limited real-world appeal for multiple receptors. One reason is the significant complexity and potentially high transaction costs for polluters participating in multiple markets. A second is that the technical conditions for initial endowments generally cannot be satisfied (Krupnick et al. 1983). This has led to interest in pragmatic alternatives (Farrow et al. 2005; Hung and Shaw 2005; Konishi et al. 2015; Krupnick et al. 1983; Rodríguez 2000).

One compelling design for water quality markets with multiple receptors is the trade ratio system proposed by Hung and Shaw (2005). The trade ratio system is an emissions permit system with modifications to address nonuniform mixing and multiple receptors. As above, the trade ratio system divides the regulated water system into zones. As in the ambient permit system, pollution load caps are set for each zone based on zonal water quality objectives. In contrast to the ambient permit system, emissions caps are also set for each zone. The total load cap applies to the delivered pollution to a zone from all sources. The emissions load cap applies to emissions within the zone. The procedure for determining the emissions caps is as follows: starting at the top of the water system and moving downstream, the zonal emissions caps are set by subtracting the allowable load moving into the zone from upstream zones from the total allowable load for the zone. Tradeable allowances are issued for each zone equal to the zonal emissions cap. Table 6.2 illustrates the determination of zonal emissions caps for a hypothetical case with four zones under the assumption that the delivery ratio from one zone to the next is 0.5 for all four zones. Table 6.3 shows the trade ratios.

Polluters seeking to increase their emissions can purchase allowances from others within their zone or from polluters in other zones. Polluters

Table 6.2 Trade ratio system: zonal total load and emissions caps

Zone	*Zonal load cap*[a]	*Zonal emissions cap*[b]
1	10	10
2	15	10
3	20	12.5
4	10	1.125

Source: Authors' creation

[a]The zonal load cap is a limit on total pollution in a zone from all sources – upstream polluters or polluters within the zone. This is set to meet zonal water quality objectives

[b]The zonal emissions cap applies to polluters within the zone. It is derived by subtracting pollution delivered from upstream sources from the zonal load cap. In this example the load delivered from upstream source is computed using the zonal load cap and the delivery ratios in Table 6.3

Table 6.3 Trade ratio system: delivery/trade ratios

Source zone	*Receiving zone*			
	1	*2*	*3*	*4*
1	1.000	0.500	0.250	0.125
2	0.000	1.000	0.500	0.250
3	0.000	0.000	1.000	0.500
4	0.000	0.000	0.000	1.000

Source: Authors' creation

seeking to profit from reducing emissions can similarly sell allowances defined for their zone to buyers in their own zone or other zones. Allowances trade within a zone at a ratio of one to one. Allowance trades between zones would trade at rates based on delivery ratios between the zones.

The method used to set the zonal emissions caps for each zone essentially bundles upstream polluters with rights to contribute to downstream pollution. Polluters need not assemble a portfolio of allowances for all zones that are affected by their effluents. Specifically, a polluter seeking to increase emissions will rationally purchase rights either from polluters in the same zone or from upstream polluter(s). This is because purchases of downstream allowances cannot be used to meet upstream zonal emissions caps.

An alternative to the Hung and Shaw trade ratio system is the damage-denominated trade ratio system proposed by Farrow et al. (2005).[4] Analogous to the trade ratio system, a river system is divided into zones with trade ratios used to define terms of trade between zones. Unlike the trade ratio system, the damage-denominated trade ratio is not designed to achieve water quality goals at specific points in the system. It is instead designed to limit the aggregate social cost of pollution across the zones. Allocations of allowances to zones and trade ratios between zones are set with this objective. A comparison of the two approaches by various criteria based on simulation experiments of temperature trading on the Vermillion River in Minnesota (US) demonstrate that each approach has advantages and disadvantages with respect to economic and water quality criteria (Konishi et al. 2015). The chief conclusion is that empirical evaluation of trading designs is crucial to optimizing designs in complex spatial environments.

6.2.5 Cap-and-Trade vs. Baseline-and-Credit

The discussion above has been presented in the context of cap-and-trade. The issues of nonuniform mixing and multiple receptors apply whether the type is cap-and-trade or baseline-and-credit. Baseline-and-credit trading is essentially a modification of traditional emissions standards that increases polluters' options for regulatory compliance. Traditional standards impose periodic (e.g., annual) maximum limits on emissions from specific sources (e.g., smokestacks, sewer outfalls), and require the limits to be met at the source. The requirement that limits are met at the source prevents emissions reductions from one source being used to meet the limits of another. Prohibiting the use of emissions reductions from one source to offset emissions from another serves no environmental purpose if environmental conditions are unaffected by the offset. However, the inability to use offsets increases the costs of pollution control when the incremental cost of pollution abatement differs between sources. With baseline-and-credit trading, a source may meet the limit in part or in whole (depending on trading rules) by acquiring offsetting emission reductions from other sources.

The use of markets to improve policy performance when command-and-control regulatory programs fail is a common thread in the history of environmental economics. The transition from regulations to markets is

[4] The damage-denominated trade ratio label is due to Konishi et al. (2015).

sometimes facilitated by small steps. Baseline-and-credit trading served this purpose in air emissions trading (Johnstone and Tietenberg 2004; Tietenberg 1990). However, while baseline-and-credit can be economically advantageous compared to command-and-control, it is not the form of trading that economic research recommends if market design can begin with a clean sheet. Cap-and-trade is the preferred form.

Cap-and-trade entails explicit caps on aggregate emissions, whereas baseline-and-credit does not. A baseline-and-credit market could obtain the same economic and environmental outcomes as the emission permits system described above for a uniformly mixed pollutant and single receptor if the regulatory baselines of the individual sources add up to the cap and if trading activity within the baseline-and-credit market equalizes marginal costs. The first condition would require that procedures for setting and resetting baselines for individual sources are designed to achieve and maintain an explicit aggregate emissions cap. This would be no easy task in a dynamic economy in which the number, size, and structure of polluting facilities among which the cap is to be divided up are routinely changing. The problem would be multiplied with multiple receptors.

Another difference between baseline-and-credit and cap-and-trade is significant regulatory discretion over allowable trades in the former. Cap-and-trade defines standardized commodities in the form of pollution allowances with enforcement entailing comparison of actual emission against allowable emissions at the end of the compliance period. Baseline-and-credit trading typically involves complex and cumbersome processes to certify that actions to generate proposed credits comply with the various rules, verify that actions to produce proposed credits have been undertaken, and register the resulting credits as available for use by demanders. In some instances, each trade requires public scrutiny. Research indicates that such procedures adversely affect market participation and the overall efficiency of trading (Boom and Dijkstra 2009; Dewees 2001; Buckley et al. 2007; Ghosh et al. 2011; Ribaudo and Savage 2014; Ribaudo et al. 2014; Woerdman and Nentjes 2019). Similarly, cumbersome administrative procedures burden regulators.

With one exception, water quality trading programs that include agriculture are a hybrid form of baseline-and-credit trading. This reflects to a large degree the institutional environment in which water quality trading with agriculture has been developed rather than the merit of the method. Specifics are discussed in a Sect. 6.4.2.

6.3 Trading with Agricultural Nonpoint Sources: Conceptual Issues

The assumption that emissions are accurately observed at little or no cost and are completely under the control of the polluter in cap-and-trade or baseline-and-credit programs is a very important one. When this assumption holds, conditions for efficient allocations are defined in terms of allocations of observed emissions that are deterministically controlled. Tradeable allowances are correspondingly defined in terms of emissions or measures that are derived from emissions (i.e., contributions to downstream pollution loads). The violation of this assumption by agricultural nonpoint sources is the most consequential for the application of trading to pollution control from agriculture. Two issues stand out. One is the appropriate definition of the traded commodity. The second is the appropriate structure of the cap. With respect to the commodity, the inability to observe actual emissions means that the preferred commodities for trading, emissions allowances, or emissions reduction credits are not available because emissions cannot be measured. The obvious question then is what to trade. With respect to the cap, because nonpoint pollution cannot be controlled deterministically, a strict cap on aggregate total emissions or emissions delivered to receptors may be implausible or uneconomic because stochastic variations in emissions would result in periodic cap violations. The question then is what kind of cap is appropriate to manage uncertain emissions. These two questions are interrelated because a cap applies to aggregations of the traded commodity (Horan and Shortle 2011).

6.3.1 Agricultural Nonpoint Commodities

The inability to monitor agricultural nonpoint emissions leads pollution control programs to focus on observable farming inputs and practices that affect their emissions – e.g., tillage practices, fertilizer applications rates and application practices, pesticide choices and application practices, erosion and runoff control practices, vegetative land cover types, livestock densities, and manure management practices. Observable inputs offer potential commodities for environmental markets. For example, a cap-and-trade market to limit aggregate fertilizer applications could be created by placing a cap on total fertilizer applications in a geographic region and issuing tradeable fertilizer application allowances to producers to limit applications to the cap. However, regulatory markets managing farm input use that can spatially fine-tune farming practices to achieve desired water quality outcomes cost-effectively are implausible (Shortle et al. 1998).

Better tradeable commodities than specific inputs are estimates of nonpoint emissions, provided that reasonably valid and reliable estimation methods are available and implementable at acceptable cost for the regulatory application. The use of nonpoint estimates as basis for nonpoint commodities is the approach taken in most water quality trading programs involving agriculture (Willamette Partnership 2015). Estimation methods vary greatly in their complexity. At the simple end are predetermined factors that specify the emissions reductions that would be credited to specific Best Management Practices (BMPs) (Willamette Partnership 2015). The factors are typically estimates of steady-state annual average percentage reductions in nitrogen, phosphorus, sediment, or other pollutants that are expected from specific BMPs. Site conditions, farming practices, and other potentially relevant factors are not considered. The reduction would be typically based on literature reviews and expert opinions. With this approach the trading authority provides market participants with lists of approved BMPs and the credits they provide.

More advanced methods use models to estimate emissions or emission reductions using site-data and mathematical algorithms. The degree of model complexity increases with the number and spatial detail of variables describing physical site conditions (soils, topography, climate), the number of BMPs included, capacity to analyze combinations of interacting BMPs, details on farming practices (crops, rotations, nutrient management history), sophistication of the algorithm used to estimate outcomes, and the specificity of emission estimates in time (e.g., steady-state long term average vs. shorter timescales) or space (edge-of-field, local watershed outlet, downstream locations) (Arabi et al. 2012; Loucks and van Beek 2017). At the simpler end of models are spreadsheet calculators that require minimal site-data and use simple generalized algorithms. An example of this type is the calculator for estimating nitrogen and phosphorus load reduction to the Chesapeake Bay from reduced fertilizer and manure applications developed for Pennsylvania's nutrient credit trading program (PADEP 2020). The calculator determines the long-term annual average change in nutrients available in soils for removal to water resources based on crop type, soil type, nutrient application history as it effects soil nutrient availability, and crop nutrient uptake using a simple nutrient balance model. The nitrogen and phosphorus reductions are multiplied by additional factors to estimate reduction in nutrients delivered to the

Chesapeake Bay using factors from the Chesapeake Bay Watershed Model (CBP 2017).

Toward the complex end is the Nutrient Tracking Tool developed by the US Department of Agriculture's Natural Resource Conservation Service (Saleh et al. 2015). The tool is an interface that facilitates the use of an advanced agri-environmental model (the Agricultural Policy Environmental eXtender (APEX)) to assess the effects of individual or combinations of BMPs on edge-of-field emissions (Gassman et al. 2009). The Nutrient Tracking Tool is used for nutrient-reduction credit calculations for trading programs for the Ohio River and the portion of the Chesapeake Bay located in the state of Maryland (Willamette Partnership 2015). Because the Nutrient Tracking Tool models edge-of-field emissions, applications to determine delivered emissions require use of other models to determine the delivery of pollutants to receptors. Like Pennsylvania, the Maryland trading program uses factors from the Chesapeake Bay Watershed model to translate edge-of-field reductions into delivered pollution load reductions. As is evident from this discussion, the use of multiple models to connect water quality management options to water quality outcomes is common in water quality modeling (Loucks and van Beek 2017).

An especially significant application of models is found New Zealand's Lake Taupo nitrogen cap-and-trade program described further below. The trading program uses the OVERSEER@ model to estimate nutrient flows to farm boundaries.[5] The model is used as a decision tool to support producers' trading decisions and to producer compliance.

An advantage of simpler techniques, either in the form of predetermined rules or simple spreadsheet models, are the comparatively low development and implementation costs and complexity. Cost and complexity are barriers to trading program development, to participation in trading markets, and to trading activity. Thus, simpler models appeal to reduce market development and administration costs, and to facilitate market activity. The appeal of more sophisticated models is the expectation of greater accuracy and reliability, greater likelihood of success in achieving water quality targets, and therefore greater program credibility. The presumption that more sophisticated models are more accurate is not always warranted. As discussed in Chap. 3, prediction errors from

[5] The model is described at https://www.overseer.org.nz/

contemporary water quality models are substantial. Implementation and interpretation errors can be another source of errors (Beven 2005, 2006; Moriasi et al. 2016).

The presence of uncertainty about the effects of agricultural practices on water quality should not lead to abandonment of trading as an instrument for agricultural nonpoint pollution management. The same uncertainty poses challenges for any policy instrument intended to achieve environmental goals at minimal cost. The question is how best to address uncertainty and whether the data and tools to do so are available and implementable at acceptable cost. This question will generally be context dependent. As water quality models are also used by performance-based instruments we discuss in Chaps. 7 and 8, we return to the issue of models in Chap. 9.

6.3.2 Stochastic Cap Definition

The choice of commodity and estimation method should not be considered separately from the structure of the cap. Indeed, the cap is an essential tool for addressing model uncertainty and natural variability. With uncertainty comes the probability that a fixed emissions cap will be periodically violated unless radical and highly costly measures to reduce agricultural nonpoint pollution are implemented. This brings us back to the topic of how to express policy goals for stochastic emissions raised in Chap. 5.

One approach presented there was safety rules that limit the frequency of violations of the water quality goal to an acceptable level. While conceptually appealing, implementing the approach is practically daunting. This is because managing the frequency of violations requires managing the mean, variance, and possibly other statistical moments of the probability distribution of emissions. The statistical moments of pollution cannot be managed except crudely by nonpoint commodities based on estimated emissions. Managing risk efficiently would require highly complex commodities that provide explicit incentives to manage the relevant moments. What those might be is an open question as the state of water quality and economic modeling and capacities of water quality regulators have not evolved sufficiently to say.

As described in Chap. 5, explicit management of risk in water pollution control is rare and generally simplistic. We gave the example there of the use of a *margin of safety* in the construction of Total Maximum Daily

Loads in US water quality management. This approach offers a practical, if crude, approach to risk management. The cap in a cap-and-trade with agricultural sources and point sources in which the agricultural commodity is estimated, emissions would be limited to the sum of the actual point source emissions and the estimated nonpoint emissions. The cap would be set to limit the load to a level sufficient to achieve the water quality objective with a margin of safety. This approach does not explicitly address statistical attributes of the load other than the mean. Trading estimated emissions provides no direct incentives to choose practices that reduce variance. Producers choosing practices for compliance with allowances will choose practices that reduce estimated emissions cheaply. The probability of violating the target is simply reduced by depressing the mean to shift the probability distribution to the left.

6.3.3 *Point-Nonpoint Uncertainty Trade Ratios*

Another approach to addressing risk, and one that has become standard in water quality trading programs with nonpoint sources, is the use of uncertainty trade ratios. The concept underlying an uncertainty trade ratio is that uncertain nonpoint source emissions are not perfect substitutes for certain point source emissions and that the two should not therefore trade one to one. A point-nonpoint uncertainty trade ratio is typically defined as the quantity of estimated nonpoint pollution required to offset a unit of point source pollution.[6] In trading programs with point-nonpoint trade ratios, the ratio is typically greater than one. If the trade ratio is 2:1, a common ratio in actual programs, 2 units of estimated nonpoint emission are required to offset one unit of actual point source emissions. A unit of estimated nonpoint abatement is essentially considered to be half as effective as a unit of actual point source abatement.

Some implications of trade ratios for the allocation of pollution between point and nonpoint sources can be demonstrated by considering the structure of the cap in a cap-and-trade program in which estimated nonpoint emissions can trade with actual point source emissions. Assuming perfect mixing, the cap limits aggregate emissions from point sources, denoted e, and total estimated emissions from nonpoint sources, denoted μ. The equation describing how loads can be allocated between point and

[6] Heterogeneity in marginal damage costs can provide another rationale for point-nonpoint trade (Lankoski et al. 2008).

nonpoint sources to meet the cap when an uncertainty trade ratio is used can be expressed as $e + \tau\mu = T$, where τ is the uncertainty trade ratio and T is the total pollution cap. If $\tau = 1$ and implying one-to-one trading, the equation for the cap is $e + \mu = T$. In this case, any combination of point source emissions and estimated nonpoint emissions that adds to T satisfies the cap. The cap could be met, for example, by allocations with $e = 0$ and $\mu = T$, or $e = T$ and $\mu = 0$. If the ratio is 2:1, so $\tau = 2$, the cap equation is $e + 2\mu = T$. In this case, the cap could again be met by $e = T$ and $\mu = 0$ but not by $e = 0$ and $\mu = T$. The maximum allowable nonpoint load under that cap is $T/2$. Consequently, with the 2:1 trade ratio the cap would allow only half as much nonpoint pollution as with the 1:1 ratio. The use of uncertainty trade ratios greater than one therefore serve to limit the extent to which agricultural nonpoint pollution can be used to meet a cap.

Discounting nonpoint pollution reductions relative to point sources using point-nonpoint trade ratios has an intuitive appeal in that it would seem to provide a greater probability of achieving water quality goals when uncertain nonpoint pollution reductions can be substituted via trading for certain point source pollution reductions. But the approach may be wrong headed. Trade ratios greater than one are essentially a tax on point source purchases of nonpoint pollution reductions that will reduce substitution of nonpoint pollution control for point source pollution control. Given the significant variability of nonpoint pollution (e.g., recall Fig. 5.1), the effect may be to preserve rather than decrease overall uncertainty about pollution loads. Specifically, in assessing a trade between a point source and a nonpoint source, the question should be whether the probability distribution of emission is better after the trade. This cannot be determined by the changes in the mean pollution. It also requires consideration of the effect on the spread or variance of emissions. If BMP choices that reduce estimated nonpoint emissions also reduce the emission variance, there is a double dividend of sorts – a reduction in mean and variance. Research on uncertainty trade ratios in fully capped point-nonpoint trading markets show that optimal trade ratios may well be less than one (Hennessy and Feng and 2008; Horan and Shortle 2005, 2011, 2017; Malik et al. 1993; Shortle 1987, 1990). Trade ratios greater than one can be risk increasing rather than risk reducing!

6.4 Water Quality Trading in Practice: United States

Water quality trading emerged in the US not as a stand-alone mechanism to control water pollution but as a means to reduce the costs of compliance with the Clean Water Act (CWA) (see Chap. 2). The primary mechanism for regulating pollution under the CWA is technology-based effluent standards that all permitted point sources are required to meet. These standards can be augmented by water quality-based effluent standards when additional controls are needed to meet water quality goals. Discharge permits specify which standards apply. When a regulated water does not meet its water quality standards, the CWA requires states to establish a Total Maximum Daily Load (TMDL) (Chaps. 2 and 5) specifying maximum allowable pollution loads for the water body. TMDLs must also specify how the allowable load is to be allocated among source classes (e.g., wastewater treatment plants, agricultural nonpoint sources).

Given the regulatory asymmetry between point sources (subject to laws and administrative mechanisms enabling strict regulation) and agricultural nonpoint sources (largely unregulated), the first course of action for water quality authorities to address unmet water quality goals has been to modify point source permits granted under the National Pollutant Discharge Elimination System (Ribaudo and Shortle 2019). The result is to push point source marginal abatement costs ever higher relative to agricultural nonpoint sources and to limit the water quality gains that could be achieved by additional restrictions on nonpoint sources (Bingham et al. 2000). A mechanism that could increase agricultural nonpoint pollution control without the political and administrative challenges of direct regulation of agriculture would have significant appeal in this environment. Supporting this view is a US Environmental Protection Agency (USEPA) (USEPA 2001) study estimating that water quality trading between point and nonpoint sources could reduce TMDL compliance costs by $1 billion or more annually between 2000 and 2015. Most of the water quality trading development in the US has occurred since the mid-1990s, prompted by the interest of federal and state water quality authorities in cost-effective approaches to TMDL compliance (Fisher-Vanden and Olmstead 2013; Shortle 2013; Stephenson and Shabman 2011).

The USEPA established a draft framework for trading in 1996 and issued policy to guide the development of trading programs by state water quality authorities in 2003 (Fisher-Vanden and Olmstead 2013). The policy required that trading take place within a watershed or a defined area for which a TMDL has been approved. It indicated that the USEPA would support trading of nutrients and sediment, but that trading of other pollutants would trigger increased scrutiny and require prior approval. The guidelines also prohibited the use of trading by point sources to meet their technology-based effluent standards. Trading was therefore limited to meeting water quality-based standards. This rule has been relaxed in some cases (Fisher-Vanden and Olmstead 2013). Trading program development has been subsequently encouraged and supported by the USEPA and the US Department of Agriculture (USDA), including technical and financial support for water quality trading projects, the development of tools like the Nutrient Tracking Tool, and the development of guidelines for permit writers. The USEPA issued a policy memorandum in 2019 reaffirming the agency's support for trading and providing principles to guide program development. Parallel to the development of federal and state trading policies has been the formation of public-private partnerships to promote the concept and to provide tools for program development (Willamette Partnership 2012, 2015; USEPA-USDA 2006).

6.4.1 US Trading Program Overview

Experiments with trading in the US date from the mid-1980s. Prompted by TMDL compliance challenges and encouraged by federal policy support, most of the 26 programs that are currently active were initiated in the latter half of the 1990s into the latter half of the 2000s (Fisher-Vanden and Olmstead 2013; Ribaudo and Gottlieb 2011; Shortle 2012; Stephenson and Shabman 2019). The following summary of active programs is primarily from the 2019 inventory by Stephenson and Shabman (2019).

Pollutants Traded Almost all active programs (22 of 26) trade nutrients (nitrogen and/or phosphorus). Other pollutants are temperature, biological oxygen demand, and urban stormwater runoff.

Pollution Sources Almost all programs (24 of 26) allow trading between point and nonpoint sources, the latter mainly agricultural. The two programs limited to point sources are for phosphorus on the Minnesota River (Minnesota River Basin Trading) and nitrogen on the Connecticut River (Long Island Sound Nitrogen Credit Exchange).

Offset Programs Six programs involve one-time offset agreements in which a permitted point source meets a portion of its effluent standard with off-site discharge reductions. This is typically done by contracting with agricultural or other nonpoint sources to install BMPs. Pioneering cases were for industrial sources on the Minnesota River (Rahr Malting Company (1997) and Southern Minnesota Beet Sugar Cooperative (1999)) to regulate biological oxygen and phosphorus. More recent examples indicating the potential scope are for temperature pollution caused by municipal wastewater treatment plants discharging to the Rogue (2010) and Tualatin (2005) Rivers in Oregon.

Rules-Based Trading Programs One-time offsets do not involve the systematic, multi-buyer, multi-seller trading economists prescribe to fully realize expected efficiency gains from trading. Stephenson and Shabman categorize 18 of the active programs as rules based, enabling trading to satisfy enforceable discharge limits on a set of discharge sources subject to rules that determine when and how those sources can acquire reductions from other sources. The regulated sources are point sources in 12 of the rules-based programs. Urban land development was the regulated source in 5 of the 18. Agricultural nonpoint sources are not subject to limits in any.

Nearly two thirds of the rules-based programs are for nitrogen and/or phosphorus in east coast or near east coast states. These include programs in three Chesapeake Bay basin states (Maryland, Pennsylvania, Virginia), North Carolina (Neuse River basin and Tar-Pamlico Sound), and the Long Island Sound Nitrogen Credit Exchange (Connecticut). Rules-based programs are also found in Colorado (Bear Creek, Chatfield Reservoir, Dillon Reservoir), Florida (Lower St. Johns River Basin), Minnesota (Minnesota River Basin Trading), and Wisconsin (Red Cedar River Nutrient Trading).

Demonstration Projects Two of the 26 active programs are demonstration projects in the Ohio River Basin (Great Miami River Watershed Trading Pilot, Ohio River Basin Trading Project) that target agricultural nonpoint sources. These demonstration programs operate without any mandatory caps on any source. Research grants and donations are used to purchase nitrogen and phosphorus credit-generating activities from agricultural sources.

Geographic Scope Active trading programs address specific pollutants in specific water bodies. The scope varies greatly, from small, localized water quality problems to large or portions of large basins. The former is exemplified by one-time offset programs. An example of the latter is the Pennsylvania Nutrient Credit Trading program for nitrogen and phosphorus loads that reach the Chesapeake Bay from the Potomac and Susquehanna River basin portions located in Pennsylvania. The Susquehanna basin in Pennsylvania is 20,960 square miles (54,286 square kilometers). The Potomac basin in Pennsylvania is 1584 square miles (4103 square kilometers). The Ohio River Basin Trading Project is an interstate demonstration program for the portions of Indiana, Kentucky, and Ohio located in the Ohio River basin.

Trading Types Trading programs that include agriculture are a hybrid baseline-and-credit type. These programs are generally intended to allow point sources to achieve abatement cost savings by acquiring credits from low-cost agricultural nonpoint sources. Program rules define eligibility requirements to be met by agricultural sources if they are to produce and sell credits, BMPs that can be used to produce credits, in some programs restrictions on the use of cost-shared BMPs, methods for computing credited pollution reductions, credit duration, credit verification and certification procedures, and trading rules, including trade ratios. Rules vary from program to program (Willamette Partnership 2015). Based on USEPA policy guidelines and program-specific objectives, rules are generally designed to have agricultural credit trades more than compensate for the avoided reductions in point source discharges, resulting in an overall net reduction in pollution and overall water quality improvement (Willamette Partnership 2015).

Delivery ratios are commonly used to address attenuation of nutrients as they move through water systems. Point-nonpoint trade ratios substantially more than 1:1 are common. Large ratios are intended to produce net pollution reductions and to address the relative uncertainty of nonpoint reductions. Some programs use equivalence ratios to account for different forms of the same pollutant, such as biologically available phosphorus and bound phosphorus. Some trade ratios are intended solely to assure that trading produces net pollution reductions (e.g., retirement ratios) or to provide reserves to compensate for nonpoint project failures (e.g., reserve ratios).

Market Structures The theory presented in Sect. 6.4 assumed perfectly competitive markets but did not go into details of how markets are organized. Three types of market structures are found: bilateral, clearinghouse, and exchange markets (Fisher-Vanden and Olmstead 2013; Selman et al. 2009; Woodward et al. 2002). Bilateral trading entails the least formal organization of the market and is generally expected to result in the highest transaction costs per trade. Transaction costs include search costs for finding, choosing, and recruiting trading partners, negotiating terms, contracting, and monitoring and enforcement of agreements. In bilateral trading, participants discover partners through informal processes and negotiate trades with no or limited information about the market. Exchange markets provide formal mechanisms for matching buyers and sellers, providing price information to market participants, and setting prices. Exchange markets are the structure implied by textbook trading models and the market structure found in high trading-volume air emissions markets. Exchange markets are expected to minimize transaction costs per trade, but they are not suited to situations with few market participants and infrequent trading. Clearinghouse markets fall between the informality of bilateral negotiations and the formality of exchange markets. In clearinghouse markets one or more intermediaries recruit and match participants.

Consistent with the limited nature of the exchange activity, one-time offset trades are organized through bilateral negotiations in all but one case (Fisher-Vanden and Olmstead 2013). Pennsylvania operates the only exchange market for water quality trading in the US (see Box 6.3). Most rules-based trading is organized through bilateral negotiations or clearinghouses, and in some, both structures are operative (Fisher-Vanden and Olmstead 2013). Within the clearinghouse category there is considerable variation in how buyers and sellers are connected, prices determined, and the extent to which genuine market exchanges occur between willing buyers and sellers (Shortle 2013). Some illustrations of clearinghouse market structures follow:

- The Minnesota River Basin Pollutant Trading program operates under a group permit for phosphorus discharges applicable to 47 sources (Fisher-Vanden and Olmstead 2013). The program facilities trading additional phosphorus abatement required by a TMDL on the Minnesota River. Trades are organized by bilateral negotiations.

Box 6.3 Pennsylvania Nutrient Credit Trading Program

The Pennsylvania Nutrient Credit Trading Program was developed in 2005 as one element in a set of initiatives designed to meet its obligations under the Chesapeake Bay Program agreements and in anticipation of the TMDLs in 2010. It is a partially capped baseline-and-credit trading in which agricultural and other sources can sell nitrogen and phosphorus credits to regulated point sources. Point sources are also eligible to generate credits. The program is administered by Pennsylvania's Department of Environmental Protection (PADEP) and operates in the portions of the Susquehanna and Potomac River Basins, both tributaries to the Chesapeake Bay, located in Pennsylvania.

Agricultural and other nonpoint sources can earn credits by implementing BMPs. To be eligible to produce and sell credits they must first meet a set of regulatory compliance and farming practice requirements. Analogous to a zonal ambient trading system, the river basins are divided into source areas referred to as land-river segments based on the Chesapeake Bay Watershed Model (CBWM) discussed in Chap. 3. Credits are calculated as reductions in delivered pollution to the Bay. The calculations use relationships from the CBWM and farm management data. PADEP takes 10% of each credit generated to fund a credit reserve intended to provide a buffer against defaults by credit suppliers. The program applies analogous rules to point sources that seek to generate credits by reducing emissions below the regulatory requirements.

Producers and credit buyers can participate and compete directly in the market. The state provides an online tool for calculating and registering credits to facilitate their participation and beginning in 2010 provides a nutrient credit clearinghouse that is managed by the Pennsylvania Infrastructure Investment Authority, a state agency (known as PENNVEST) that traditionally was charged with financing water infrastructure investments. Credit exchanges can be made through bilateral negotiations, but PENNVEST also conducts periodic double auctions to buy and sell credits.

(*continued*)

Box 6.3 (continued)

The program is impressive as a policy architecture for facilitating market-based trading (Shortle 2013). However, like other US trading programs, the partially capped baseline-and-credit model along with regulatory barriers and point source abatement preferences limit demand for agricultural credits with the result that there is little trading activity with agriculture (Stephenson and Shabman 2019; Shortle 2012, 2013).

- The Tar-Pamlico Nutrient Trading program is a point-point, point-nonpoint nutrient trading program (Nguyen et al. 2006). Analogous to the Minnesota program, participants are subject to an aggregate cap on nutrient discharges under a group permit. The permit allows trading among the 15 members of the Tar-Pamlico Basin Association. Trades are organized through bilateral negotiations. Unlike the Minnesota program, the Tar-Pamlico program also enables trading with nonpoint sources. The Association funds agricultural nonpoint reductions to offset loads beyond the allowed levels. The nonpoint reductions are organized by a clearinghouse managed by the North Carolina Agricultural Cost Share program. Agricultural BMPs are cost shared at 75%. The Neuse River Basin Trading program is similar. Nonpoint reductions used in the program are sourced from wetlands restoration projects.
- The Great Miami River Watershed Trading Pilot (Ohio) operates as a conservation auction (Shortle 2013; WCS 2005). The intermediary is the Water Conservation Subdistrict of the Miami Conservancy District, a quasi-governmental agency. The Subdistrict buys nutrient credits from agricultural sources. The credits are purchased from producers through periodic reverse auctions. To be eligible, producers' bids are required to be developed and submitted in collaboration with participating Soil and Water Conservation Districts.[7] The purchases are funded by donations from local municipal wastewater

[7] US Soil and Water Conservation Districts are independent subdivisions of state governments organized typically by county. They were originally developed in the 1930s to address soil and water conservation problems.

treatment plants and federal grants. The supply side of the market is very market-like. The demand side is not driven by regulatory compliance but public contributions. The program demonstrates an institutional and market design that can induce producer participation and result in BMP implementation for water quality protection. There is, however, research indicating that the additionality of the program is modest (Mezzatesta et al. 2013). The program also illustrates the essentiality of a regulatory or other mechanism to create demand for credits. Restrictions on nutrient loads from point sources expected to drive the market were never developed. The program was dependent on demonstration funding and has become inactive as the supporting programs have become inactive.

- The Long Island Sound Nutrient Credit Exchange is commonly categorized as a trading system but is more accurately described as an effluent reduction charge/subsidy program (Shortle 2013). The intermediary is the Connecticut Department of Environmental Protection. It assigns effluent limits to 79 wastewater treatment plants. Plants that over-comply are awarded credits for which they receive payments from the agency managing the program. Sources that under-comply are debited credits and are required to make payments to the agency for the shortage. The payment rate for credits and debits is not set by the market but by an advisory group appointed by the state legislature. Dischargers do not exchange credits, and the supply and demand for credits are not required to balance.

6.4.2 *The Partially Capped Baseline-and-Credit Model*

In the typical point-nonpoint program, potential traders are wastewater treatment plants and farms. The wastewater plants are subject to explicit regulatory limits on their emissions. The farms are not. Wastewater treatment plants can purchase credits from farms or from other wastewater plants to use toward meeting their regulatory limits. Wastewater plants may therefore participate as credit buyers or sellers. Farms, not being subject to emissions limits, have no economic motive to purchase credits, participate in the market only as credit sellers. The hybrid baseline-and-credit programs that include agriculture are therefore voluntary compliance pay-for-performance programs for agricultural or other nonpoint sources. They are not stand-alone water quality programs as are cap-and-trade but an add-on to a regulatory structure that essentially serves as a tool for reducing compliance costs for point sources.

The limit on total regulated emissions in the hybrid baseline-and-credit programs with agriculture is determined by the regulatory limits on the wastewater treatment plants. The programs are sometimes described as being partially capped as regulated point sources are subject to regulatory limits, but agricultural sources are not. Baseline-and-credit trading with agricultural sources do not have total pollution caps comparable to the caps of cap-and-trade.

Due to the absence of explicit regulatory limits for agricultural sources, a key task in the design of a baseline-and-credit program is the definition of the baselines used for calculating agricultural credits. One approach would be to have baselines reflect actual business-as-usual emissions for individual farms. In this case, credits would measure reductions in estimated emissions relative to what occurred prior to the implementation of the trading program. Other things equal, this is the preferred baseline for ecological integrity and economic efficiency. For ecological integrity (see Box 6.4) of credit trading, credits generated should correspond to actual pollution reductions. For economic efficiency, the marginal costs of generating credits should correspond to the marginal costs of abatement so that allocations are guided by relative marginal abatement costs. This will not be the case if the baselines differ from business-as-usual emissions.

Implementation of a business-as-usual baseline would require a history of farming practices over an averaging period and estimates of the

Box 6.4 Ecological Integrity

Ecological integrity is a key concern when credits produced by lightly regulated or unregulated nonpoint sources can be used for regulatory compliance by point sources. Do credited nonpoint reductions truly offset avoided point source reductions in time and place? This is probably too stringent a requirement for ecological integrity given the inherent variability of nonpoint pollution. More pragmatically, do nonpoint credits on average produce the same water quality outcomes as the avoided point source reductions?

One issue when considering ecological integrity is the degree to which point and nonpoint pollutants are environmental substitutes. For eutrophication problems, especially when point and nonpoint

(*continued*)

Box 6.4 (continued)

sources mix to determine loads in downstream waterbodies, loads delivered contemporaneously are close environmental substitutes. Long-term averages may be as well. A second issue and one that receives much attention is the accuracy of the models used to estimate nonpoint pollution reductions. Minimally they should have a sufficiently high correlation with actual performance and be considered by policy makers and stakeholders as acceptable for policy.

Behavioral issues are equally important. In Chap. 5 we introduced the concepts of non-additionality and slippage that affect the results of conventional voluntary compliance pay-for-practice or land retirement programs. These behavioral phenomena are also ecological integrity issues for pay-for-performance programs like water quality trading or water quality auctions we introduce in Chap. 7. In baseline-and-credit trading programs, producers adopt practices to produce credits according to program rules, including specification of eligible practices. Eligibility rules that allow credits for practices that producers are likely to adopt for their own private benefit invite non-additionality. Solving this problem requires consideration of likely net private benefits of practices when defining eligible practices. Those that are likely to be profitable ought to be excluded. Planners cannot know net private benefits for all practices in all contexts, but standard enterprise budgeting methods can generally provide useful sorting.

Slippage and other within-farm spillover effects described in this chapter and Chap. 5 result from partial accounting of farm emissions. Whole-farm environmental accounting is essential to a policy that is effective and ecologically valid. Policy makers must know not only what producers do on fields or pastures enrolled in pay-for-practice or pay-for-performance programs, but also what will be done in other parts of the farm. Ideally, farms should be held for the overall environmental performance, not just performance in fields participating in pay-for-performance or pay-for-practice program.

Cap-and-trade programs inherently require whole-farm caps. Baseline-and-credit programs impose caps when regulatory baselines are applicable to all allowed polluters. The current hybrid baseline-and-credit program used for agricultural sources has fundamental integrity issues because of the absence of whole-farm caps.

corresponding emissions. Constructing this baseline would be no small task given the data required and the need to use models to estimate pollution loads. But then, estimating reductions in pollution is fundamental to the concept of trading with agricultural sources. This is not, however, the typical approach. Agricultural baselines are selected, for example, to reflect minimal standards of "good" farming practice, regulatory principles that producers are expected to meet before being eligible to profit from selling credits, and concerns for fairness in the treatment of producers with different conservation practice histories and with farms relative to regulated point sources (Ghosh et al. 2011; Ribaudo et al. 2014; Willamette Partnership 2015). Research concludes that typical approaches diminish the cost-effectiveness of baseline-credit markets (Ghosh et al. 2011; Ribaudo et al. 2014).

The absence of explicit caps on agricultural sources and the pay-for-performance structure lead to problems for trading with agriculture analogous to other voluntary compliance programs (Chap. 5). One problem is the additionality of the credited pollution reductions (Miller and Duke 2012; Ribaudo et al. 2014; Stephenson and Shabman 2019; Woodward et al. 2016). Credits awarded to unregulated agricultural or other nonpoint sources for implementing control practices that they would adopt even without a credit sale do not reduce pollution relative to what it would otherwise be. Such a credit is nonadditional. This issue would be of no consequence in baseline-and-credit programs with agriculture if agricultural BMPs that are eligible credits were unprofitable without trade. But some BMPs are profitable and their adoption economically warranted without additional incentives from credit sales or subsidies. A second problem when credits are allowed from uncapped sources is slippage and other spillovers that can result in unaccounted increases in pollution due to credit-producing activities (Fleming et al. 2019; Stephenson and Shabman 2017, 2019; Stephenson et al. 2009). Slippage is a concern when there is no comprehensive accounting of the total emissions from unregulated sources. Procedures for certifying credits from unregulated agricultural sources are typically done on a project-by-project basis without considering the total emissions from the entire farm.

Non-additionality and slippage reduce the effectiveness and efficiency of trading in baseline-and-credit systems that admit trading with uncapped sources. Such induced economic responses also create and exacerbate uncertainty about the economic and environmental consequences of trading. The disincentive to trade with nonpoint sources created by high uncertainty trade ratios will be even more problematic in baseline-and-credit programs that permit trading with unregulated nonpoint sources as it works against bringing the risky uncapped sources under control. This is a perverse outcome.

6.5 Water Quality Trading in Practice: Non-US

Significant trading programs outside of the US are found in Australia, Canada, and New Zealand (Fisher-Vanden and Olmstead 2013). The Hunter River Salinity Trading Program in New South Wales (Australia), implemented in 2002, trades salinity discharges from coal mines and power plants (New South Wales DECCW 2006). It is managed by the New South Wales Office of Environment and Heritage. The regulated section of the river is divided into segments, which are monitored for flow and salinity levels. Discharges are prohibited under low flow conditions and unrestricted under flood conditions unless salinity concentrations exceed a limit. Discharges are limited during high flows with the cap contingent on salinity concentrations. The program is essentially a cap-and-trade program, but allowances are called credits. Forty-three entities were included in the initial credit allocation. Credit durations are limited but new credits are issued by periodic auctions. An online trading platform was developed for exchanging credits with prices and credit transactions negotiated by buyers and sellers. A small (three source) localized point-point trading program exists to manage nutrients discharged to South Creek in the greater Sydney metropolitan area.

The South Nation River Watershed Trading program in Ontario (Canada) was established in 2000 to allow new and expanding dischargers of industrial and municipal wastewater to meet stricter phosphorus limits by purchasing agricultural credits at a ratio of 4:1 (Shortle 2013). The program was developed and is managed by South Nation Conservation (SNC), a regional watershed management agency created by Province of Ontario. Producers do not participate directly in trading; South Nation Conservation negotiates the trades. Dischargers pay an established credit price based on the average cost of producing agricultural credits. Payments to the SNC are deposited in the SNC Clean Water Fund, which is used to finance credit-generating agricultural projects SNC credit sales augment other funding sources that finance BMP implementation in the watershed. A phosphorus offset program is being piloted for urban phosphorus run-off to Lake Simcoe, also in Ontario.

The most interesting trading programs for agriculture is the Lake Taupo nitrogen cap-and-trade program in Waikato (New Zealand) established in 2009 by the provincial environmental authority (now the Waikato Regional Council). The program is unique as the world's only agricultural nonpoint source cap-and-trade program (Duhon et al. 2015; Shortle 2013) (see Box 6.5). The program established a catchment-wide cap on nitrogen losses from agricultural activities, and created enforceable nitrogen discharge allowances (NDAs) for agricultural sources. Producers can trade (via bilateral negotiations) allowances among themselves or sell allowances to a public fund (Lake Taupo Protection Trust (LTPT)).

Box 6.5 California Grassland Area Selenium Trading Program

The Lake Taupo trading program is the only true cap-and-trade program for agriculture now in operation. It is, however, not the first cap-and-trade program to be applied to agriculture. That distinction belongs to the now inoperative trading program operated by the Grassland Area Farmers (GAF) in the San Joaquin Valley of California (Shortle 2013).

The GAF is an association of seven irrigation districts that was established in 1996 to control the amount of selenium discharged to waterways from agricultural lands. Drainage systems installed to remove excess irrigation water from farm fields transported selenium to a wildlife refuge. Selenium is naturally occurring in the soils but harmful to waterfowl when concentrated to toxic levels. Among the strategies developed by the GAF to reduce selenium loads was a trading program. The program established a cap on the total allowable regional selenium load that could be discharged into a canal carrying drainage water to the refuge and allocated the cap among the member irrigation districts. The districts could meet their load allocations through selenium reductions within the district and/or by purchasing selenium offsets from other districts. Trades were negotiated by the individual districts directly. The program produced 39 trades and met the drainage area's water quality goals during the two years in which it operated. Trading was suspended after installation of a collective drain water treatment facility.

(*continued*)

Box 6.5 (continued)

Several features of the program are interesting. Its cap on agricultural sources is atypical of WQT models that include agriculture. Another significant and related feature is the "commodity" traded. Trading programs that involve agricultural land generally do not measure agricultural emissions because the nonpoint nature of those emissions makes routine accurate measurements infeasible. Instead, such markets use calculations of agricultural nonpoint pollution emissions. Trading in the GAF was based on actual emissions. This was possible because the trading was conducted between irrigation districts rather than between farmers and because the irrigation districts collected drainage water in sumps before pumping into the drainage canal.

When established, the Lake Taupo program set the nitrogen cap at 20% below 2001–2005 discharge levels. The cap was to be achieved in two stages. In the first stage, producers were issued NDAs at no charge. The allocations were based on past production. In the second stage, the LTPT, funded by government grants, is responsible for purchasing and permanently retiring NDAs from producers to meet the cap. Trading rules allow producers to exceed their initial NDA allocation if they offset the excess by securing an equivalent decrease in nitrogen losses elsewhere in the catchment. This enables trading. The program uses a nutrient budgeting tool, OVERSEER® (Overseer 2020) to estimate nitrogen loss from pastoral land using data on location, land use practices (e.g., animal shelters and feed pads, effluent management, animal species and their management and stocking rate, supplements, nitrogen inhibitors, irrigation, fertilizer use) and landscape and climatic characteristics (e.g., wetland areas, topography, climate, soil type) (Duhon et al. 2015). The model is used both for determining initial allocations and compliance monitoring.

6.6 *Ex Ante* and *Ex Post* Assessments

Given that the fundamental objective of water quality trading is to achieve or exceed water quality goals at minimum cost or at lower costs than traditional policies, key considerations in *ex ante* or *ex post* assessments are performance with respect to water quality goals and control costs. Comprehensive assessments may consider broader environmental, economic, and social outcomes. For instance, a broader economic and ecological interest in water quality trading with agriculture is ancillary environmental benefits as BMPs used to improve water quality as they can also affect carbon fluxes, wildlife habitat, and other ecosystem services (Johnstone and Tietenberg 2004).

6.6.1 Ex Ante *Assessment*

Theory provides a compelling case for emissions trading when the assumptions about the water environment of the basic model hold reasonably well. Theory is of less use for predicting economic and ecological outcomes for agricultural nonpoint pollution because the problem violates crucial assumptions. Empirical simulations using integrated economic-biophysical models are especially useful in this case. Applications include the Baltic Sea (e.g., Hautakangas and Ollikainen 2019), the Chesapeake Bay (e.g., Hanson and McConnell 2008; Kaufman et al. 2014; Van Houtven et al. 2012; Wainger et al. 2013), and the Gulf of Mexico (e.g., Ribaudo et al. 2005). The research generally supports an expectation that well-designed water quality trading programs that include agricultural nonpoint sources can increase the efficiency of water quality protection by comparison to traditional water quality protection policies in appropriate applications.[8]

A common indicator of potential gains from trade with agriculture is comparison of the marginal costs of pollution abatement in agriculture versus regulated sectors. It is often the case that marginal costs of reducing expected agricultural loads are substantially lower than those for reducing

[8] A fair criticism of the *ex ante* empirical simulation research is that studies do not embrace the full physical and economic complexity of the problem, which may result in overly optimistic expectations of the potential benefits. For instance, it is common to dispense with explicit modeling of participant behaviors in markets and simply assume that trading will result in the least-cost allocation for the set of trading rules. The trading equilibrium is then computed as the cost-minimizing allocation. This approach excludes explicit consideration of how participant behavior within markets affects outcomes. In one study addressing this issue Nguyen et al. (2013) use an agent-based model to simulate boundedly rational agents operating in bilateral and clearinghouse markets. They find that market-based trading in either the bilateral or clearinghouse form provides a better outcome than traditional regulations, but they also find that cost minimization within a set of market rules is not a good prediction of the trading outcome.

point source or urban nonpoint source loads. For example, as noted in Chap. 5, the Chesapeake Bay Program provides estimates of the cost-effectiveness for nitrogen and phosphorus control for 247 Best Management Practices for agriculture, urban, and other land uses in the Chesapeake Bay watershed. The average cost-effectiveness of agricultural BMPs for reducing nitrogen is $266 per lb. of reduction and for phosphorus is $519 per lb. of reduction (2018 USD) (CBP 2020). The comparable numbers for reductions from developed land (e.g., urban stormwater) are for nitrogen $5208 and for phosphorus $41,133. Differences in cost-effectiveness of this kind suggest substantial potential gains from trade.

6.6.2 Ex Post *Assessment: Non-US Programs*

Australia's Hunter River point-point salinity trading program is a stand-out success for water quality trading. Unlike most North American water quality trading programs, it is not an add-on intended to enhance the efficiency and effectiveness of a preexisting regulatory system but a stand-alone cap-and-trade regulatory mechanism. Trades are organized through an online exchange market with clear rules and limited regulatory supervision of trading beyond assurance that credits cover discharges. It is a well-monitored and well-understood river system. These features align well with the assumptions of the basic textbook trading model. Assessments indicate that the program has achieved established targets for water quality at a lower cost than the prior regulatory scheme and allowed expansion of economic activity that otherwise might not have occurred (Collins 2005; Kraemer et al. 2002; New South Wales DECCW 2006).

New Zealand's Lake Taupo trading model is also a stand-out water quality trading program as a demonstration that the cap and-trade model can be applied to effectively control agricultural nonpoint pollution. Assessments of the program indicate that the cap is reducing nitrogen losses from farms, and that producers are participating in trading necessary for the market to serve the economic purposes of trading (Duhon et al. 2015). The program called for a cumulative 20% nitrogen load reduction via purchases of nitrogen allowances from landowners by the LTPT. This was achieved by 2015 in a total of 25 trades. Water quality in the lake is reported to be improving. These allowances were removed in perpetuity. Ongoing trading occurs between private landowners. With the cap having been met by 2015 via the LTPT's purchases and removal of allowances, subsequent trading does not reduce the overall load but reallocates allowances among landowners. Between the initiation of the program in 2011 and fall 2020, there have been a total of 55 trades between landowners in the Lake Taupo catchment.[9] Nitrogen prices in fall 2020 were in the neighborhood of 400–500NZ per kilogram. Recommendations for improving the economic performance of the market focus on reducing transaction costs that depress trading activity by eliminating cumbersome requirements for *ex ante* trade approvals and extra monitoring requirements for producers who participate in trades, and increasing market coordination by making allowance price information available to producers (Duhon et al. 2015).

Canada's South Nation River point-nonpoint trading program is a successful demonstration of the potential for a relatively simple hybrid baseline-and-credit program to reduce the costs of point source pollution control and produce net pollution reductions when the regulatory environment makes trading with agricultural sources economically beneficial. After the inception of the program, new or expanding facilities discharging phosphorus in the program area opted to meet phosphorus limits by purchasing agricultural credits at the steep trade ratio of 4:1 rather than incur the expense of upgrading their treatment facilities (O'Grady 2011; Shortle 2013). Agricultural projects funded by the program are estimated to have reduced phosphorus discharges (O'Grady 2011).

6.6.3 Ex Post *Assessment: Active US Programs*

The most typical setting for US trading programs is one in which regulators face unpleasant choices because water quality standards are not met by point source regulations, additional point source effluent control is infeasible or economically harmful, and nonpoint sources are significant pollution sources (Shabman and Stephenson 2007). *Ex ante* studies of point-nonpoint trading suggest that point-nonpoint trading programs should involve meaningful trading activity and point source cost savings. The key issue then, given the uncertainty about the impacts of BMP adoption on water quality, is the actual water quality outcomes. Yet the most noted feature of water quality trading programs involving agricultural sources is the lack of trading activity. In their 2019 review of US trading programs, Stephenson and Shabman (2019) find that active rule-based

[9] Figures on trades and prices are from personal communication with Jacqui Bolton, Resource Officer, Waikato Regional Council, 10/26/2020.

trading programs that allow point sources to trade with agricultural nonpoint sources rarely produce even a single agricultural trade. When agricultural trades do occur they are largely confined to stormwater runoff management programs for new urban developments in which land developers can avoid some on-site urban stormwater controls by purchasing agricultural credits generated from land conversion activities (e.g., converting agricultural land to forest) or landscape restoration activities (e.g., stream restoration or riparian buffer expansion).

The lack of point-nonpoint trading activity has led assessments to focus on trading institutions and the economic and regulatory environments in which they are implemented as they may affect participation and trading activity by credit suppliers and demanders. Institutional design elements that have been identified as potential barriers to producer participation and credit supply include steep baseline participation requirements that require costly investments before a producer can be eligible to generate credits, restrictions on BMPs approved for generating credits or inflexible BMP design standards that increase producers' out-of-pocket and opportunity costs of BMP installation, requirements that practices that generate credits be implemented before credits can be sold, and the significant regulatory involvement associated with the generation, certification, and verification of credits, and credit trade approvals (Ghosh et al. 2011; Motallebi et al. 2017; Ribaudo and Gottlieb 2011; Ribaudo and Savage 2014; Ribaudo et al. 2014; Fisher-Vanden and Olmstead 2013; Stephenson and Shabman 2019).

Another design element that has been identified as a barrier to trade activity is trade ratios. Trade ratios are appropriate components of economically efficient trading programs to address issues of imperfect substitution between discharges from locations (delivery ratios), of different pollutant forms (equivalence ratios), or source types (point-nonpoint). Inefficiencies arise when the ratio settings do not align with the functional purpose. This is particularly the case for point-nonpoint trade ratios (Ribaudo and Gottlieb 2011; Fisher-Vanden and Olmstead 2013; Stephenson and Shabman 2019). Point-nonpoint trade ratios more than unity are like a tax on trades with nonpoint sources. With trade ratios of 2:1 or higher, the tax is substantial and would be expected to depress transactions unless credit demand is highly inelastic. Yet, as discussed previously, economic research on trade ratios suggest that large ratios can have perverse effects on water quality risk and when optimally specified may be less than one. Large ratios may be set under the expectation that strong demand for credits will offset the disincentive effect to produce large net pollution reductions. The evidence suggests otherwise.

While institutional designs increase the costs of supplying agricultural nonpoint credits (e.g., steep baseline requirements) and further increase the costs of point sources utilizing them (high point-nonpoint trade ratios), the near absence of trades across programs, many with an adequately large number of potential buyers and sellers to support trading activity, suggests that it is the demand for credits that is at the core of the problem. Fundamentally, producers can be expected to sell credits if the demand for credits makes it profitable. Stephenson and Shabman (2019) make the case that it is the missing demand that is the problem for US programs. They tie weak demand primarily to the structure of trading rules and the institutional structure of the CWA and derivative regulations that limit choices and influence the preferences of regulated sources. In contradiction of the expressed interest in trading to improve the economic efficiency, disincentives and conditions emerging from the complex permitting structures and multiple regulatory conditions of regulatory enforcement can radically dampen the willingness and ability of regulated sources to pay for nonpoint source credits (Stephenson and Shabman 2011, 2017, 2019). Furthermore, regulated point sources demonstrate a willingness to bear additional compliance costs to utilize technologies and control practices under their direct control, forgoing the use of trading even when credit prices are low (Stephenson and Shabman 2019). Legal liability risks that exist under existing law and mismatches between the scope and duration of agricultural control practices relative to point sources are factors in this preference (Jones and Vossler 2014; Suter et al. 2013; Stephenson and Shabman 2019). Regulated point sources face long-term obligations for relatively large volumes of wastewater while agricultural BMPs are often short-term management practices and generate a relatively small number of credits. Permit compliance through agricultural nonpoint source credits would require multiple contracts with producers/landowners, covering thousands of acres with frequent renewals (Stephenson et al. 2010).

6.7 Conclusions

The concept of trading in markets conjures up the notion of trading the rights to something tangible. In the case of standard emissions trading, the rights traded are rights to release measured emissions. The unobservable and stochastic nature of nonpoint pollution does not seem compatible with the concept of trading a tangible good. Yet, theorists and policy innovators have suggested that the merits of emission trading could be brought to nonpoint pollution control by defining the traded good in terms of estimated pollution. Policy makers have, in dozens of local and regional cases, accepted the concept, resulting in planning initiatives, pilot programs, and the implementation of trading programs that include agricultural sources for local and regional problems. In the US, where almost all water quality trading programs are located, the USEPA and USDA have positively endorsed the concept of trading for nutrients and other pollutants from agriculture and seek to facilitate trading through policy, technical support, and the development of tools for estimating agricultural emissions.

The partially capped baseline-and-credit that has been implemented in the US is generally a disappointment. Where implemented, it is, for agriculture, another voluntary compliance program that is of the pay-for-performance type. With voluntary participation, there is no assurance of reductions in agricultural pollution loads. The primary economic appeal of the trading model is to reduce point source compliance costs by shifting high-abatement costs from high-cost point sources to low-cost nonpoint sources. Reducing point source costs in this way is a worthy goal given the substantial inefficiency of the US water pollution controls, but net reductions in pollution require agricultural reductions to more than offset the allowed point source pollution, which in turn leads to trading rules that discourage market participation and trading activity. In practice, US point-nonpoint programs have generally produced neither significant point source cost savings nor net pollution reductions because participation in trading programs and trading volumes have been negligible. The approach depends on a substantial demand from point sources to purchase agricultural credits. The economic and regulatory environment in the US does not deliver the demand and seems unlikely to without fundamental changes in the water pollution policy architecture.

We do not yet believe that the experience of trading is definitive about the merits of trading. The review of applications of water quality trading indicates the instrument has not yet lived up to expectations and remains to be proven. However, while there are no great successes in agricultural water quality trading, several experiments suggest that the mechanism has potential as a component of better water quality policy (Shortle 2013). The Great Miami River (US), Lake Taupo (New Zealand), and South

Nation River (Canada) have elements that are encouraging for the potential of water quality trading programs that include agriculture. In the Great Miami River, these elements include the use of an auction mechanism to price and exchange credits and the use of producer-trusted organizations (Soil and Water Conservation Districts) to provide technical assistance and facilitate producer participation. The South Nation River is also noteworthy for the development of a market structure that facilitates exchange using an organization producers know and trust.[10] The Lake Taupo is a promising application of cap-and-trade. The Pennsylvania trading program simultaneously demonstrates the feasibility of developing a plausible architecture for a large-scale trading program but also highlights the importance of demand to trading activity. As in other cases, the Pennsylvania program exists within an overall regulatory architecture under the Clean Water Act that limits effective demand for agricultural credits.

An important aspect of each of these initiatives is that they emerged as local or regional solutions to local or regional water quality problems and were developed with active engagement of local or regional stakeholders. In the Great Miami River and Ontario, the initiatives were led by long-established conservation agencies. In New Zealand, it was the Waikato Regional Council (formerly Environment Waikato). Leadership in water quality trading program development, both point-point and point-nonpoint, has come from innovators at the state (provincial) levels or below.

Water quality trading skeptics often focus on the challenges imposed by nonpoint uncertainty and other aspects of the water environment that pose challenges to functioning markets. But the experience to date suggests that the institutional environment is of crucial importance. Despite the lack of success to date, our view is that water quality trading has significant appeal as a mechanism for improving the efficiency and effectiveness of managing nutrients and some other pollutants from agriculture for reasons mentioned in the introduction. One source of this appeal is that trading in the cap-and-trade form is a systematic watershed-based approach that explicitly recognizes and integrates the management of multiple pollution sources toward an explicit water quality goal. A second is the

[10] The South Nation River program in Ontario is the hybrid baseline-and-credit type. The success of the program is the result of an economic, regulatory, and institutional environment that delivered the requisite point source demand and facilitated trading, albeit on a very small scale.

expectation that trading can achieve water quality goals at lower cost than traditional regulatory approaches. Realizing these potential benefits requires market designs that have ecological integrity but that also facilitate trading activity (Horan and Shortle 2011). Policy makers need to pay as much attention to design elements that affect demand, market participation, and the costs of trading activity as they do to elements that affect ecological integrity. In the US, two fundamental reforms are needed to make substantial use of trading. One would be to establish regulatory restrictions on agricultural emissions that are below business-as-usual emission levels and utilize these for baseline definition. Agriculture would then be subject to regulatory caps that are now missing the hybrid baseline-and-credit approach. Another would be to shift from the baseline-and-credit to cap-and-trade as in the New Zealand model. Whether in the US or elsewhere, the experience in both point-point and point-nonpoint trading to date suggests that local and regional initiative to address local and regional problems is a promising path, both to produce programs best suited to the context and to provide the experience needed to advance the method.

Two topics we have not taken up in this chapter are credit stacking and policy coordination of trading programs with agricultural conservation and pollution control programs. We return to these topics in Chap. 8.

References

Arabi, M., D.W. Meals, and D.L.K. Hoag. 2012. Watershed modelling: National institute of food and agriculture – Conservation effects assessment project. In *How to build better agricultural conservation programs to protect water quality: National Institute of Food and Agriculture conservation effects assessment project*, ed. D.L. Osmond, D.W. Meals, D.L.K. Hoag, and M. Arabi, 84–119. Ankeny: Soil Conservation Society of America.

Beven, K. 2005. On the concept of model structural error. *Water Science and Technology* 52 (6): 167–175.

———. 2006. A manifesto for the equifinality thesis. *Journal of Hydrology* 320 (1–2): 18–36.

Bingham, T.H., T.R. Bondelid, B.M. Depro, R.C. Figueroa, A.B. Hauber, S.J. Unger, and G.L. Van Houtven. 2000. *A benefits assessment of water pollution control programs since 1972: Part 1, The benefits of point source controls for conventional pollutants in rivers and streams.* EPA Contract Number 68-C6-0021. Research Triangle Park: Research Triangle Institute.

Boom, J.T., and B.R. Dijkstra. 2009. Permit trading and credit trading: A comparison of ca based and rate-based emissions trading under perfect and imperfect competition. *Environmental and Resource Economics* 44 (1): 107–136.

Buckley, N.J., S. Mestelman, and R.A. Muller. 2007. Baseline-and-credit emission permit trading: Experimental evidence under variable output capacity. In *Experimental methods in environmental economics*, ed. T. Cherry, S. Kroll, and J. Shogren. New York: Routledge Press.

Chesapeake Bay Program (CBP). 2020. Chesapeake Assessment and Scenario Tool (CAST) Version 2017d. Chesapeake Bay Program Office. Last accessed 18 June 2020.

Collins, D. 2005. Australian experiences with water quality trading. Presented at the workshop on water quality trading in Canada, Old Chelsea, Québec, September 19–20. Available at www.policyresearch.gc.ca/doclib/Water0905_DrewCollins_E.pdf

Crocker, T.D. 1966. The structuring of atmospheric pollution control systems. In *The economics of air pollution*, ed. H. Wolozin, 61–86. New York: W.W. Norton.

Dales, J. 1968. *Pollution, property and prices*. Toronto: University of Toronto Press.

Dewees, D.N. 2001. Emissions trading: ERCs or allowances? *Land Economics* 77 (4): 513–526.

Duhon, M., H. McDonald, and S. Kerr. 2015. *Nitrogen trading in Lake Taupo: An analysis and evaluation of an innovative water management policy*. Available at SSRN 2653472.

Farrow, R.S., M.T. Schultz, P. Celikkol, and G.L. Van Houtven. 2005. Pollution trading in water quality limited areas: Use of benefits assessment and cost-effective trading ratios. *Land Economics* 81 (2): 191–205.

Fisher-Vanden, K., and S. Olmstead. 2013. Moving pollution trading from air to water: Potential, problems, and prognosis. *Journal of Economic Perspectives* 27 (1): 147–171.

Fleming, P., E. Lichtenberg, and D.A. Newburn. 2019. *Water quality trading in the presence of existing cost share programs*. Available at SSRN 3368559.

Gassman, P.W., Williams, J.R., Wang, X., Saleh, A., Osei, E., Hauck, L.M., Izaurralde, R.C. and Flowers, J., 2009. The Agricultural Policy Environmental Extender (APEX) model: An emerging tool for landscape and watershed environmental analyses.

Ghosh, G., M. Ribaudo, and J. Shortle. 2011. Baseline requirements can hinder trades in water quality trading programs: Evidence from the Conestoga watershed. *Journal of Environmental Management* 92 (8): 2076–2084.

Goulder, L.H. 2013. Markets for pollution allowances: What are the (new) lessons? *Journal of Economic Perspectives* 27 (1): 87–102.

Hanson, J.C., and K.E. McConnell. 2008. Simulated trading for Maryland's nitrogen loadings in the Chesapeake Bay. *Agricultural and Resource Economics Review* 37 (2): 211–226.

Hautakangas, S., and M. Ollikainenm. 2019. Nutrient trading between wastewater treatment plants in the Baltic Sea Region. *Environmental and Resource Economics* 73 (2): 533–556.

Hennessy, D.A., and H. Feng. 2008. When should uncertain nonpoint emissions be penalized in a trading program? *American Journal of Agricultural Economics* 90 (1): 249–255.

Horan, R.D., and J.S. Shortle. 2005. When two wrongs make a right: Second best point nonpoint trading. *American Journal of Agricultural Economics* 87 (2): 340–352.

———. 2011. Economic and ecological rules for water quality trading. *Journal of the American Water Resources Association* 47 (1): 59–69.

———. 2017. Endogenous risk and point-nonpoint uncertainty trading ratios. *American Journal of Agricultural Economics* 99 (2): 427–446.

———. 2021. The song remains not the same: Correlated intercept and slope uncertainties matter to prices vs quantities. *Journal of the Association of Environmental and Resource Economists.* Forthcoming.

Hung, M.F., and D. Shaw. 2005. A trading-ratio system for trading water pollution discharge permits. *Journal of Environmental Economics and Management* 49 (1): 83–102.

Johnstone, N., and T. Tietenberg. 2004. *Ex post evaluation of tradeable permits*, 9–44. Paris: OECD Publishing.

Jones, L.R., and C. Vossler. 2014. Experimental tests of water quality trading markets. *Journal of Environmental Economics and Management* 68 (3): 449–462.

Kaufman, Z., D. Abler, J. Shortle, J. Harper, J. Hamlett, and P. Feather. 2014. Agricultural costs of the Chesapeake Bay total maximum daily load. *Environmental Science and Technology* 48 (24): 14131–14138.

Konishi, Y., J. Coggins, and B. Wang. 2015. Water quality trading: Can we get the prices of pollution right? *Water Resources Research* 51 (5): 3126–3144.

Kraemer, A., E. Interwies, and E. Kampa. 2002. Tradeable permits in water resource protection and management: A review of experience and lessons learned. In *Implementing domestic tradeable permits: Recent developments and future challenges.* Paris: Workshop proceedings, Organization for Economic Cooperation and Development.

Krupnick, A.J., W.E. Oates, and E. Van De Verg. 1983. On marketable air-pollution permits: The case for a system of pollution offsets. *Journal of Environmental Economics and Management* 10 (3): 233–247.

Lankoski, J., E. Lichtenberg, and M. Ollikainen. 2008. Point/nonpoint effluent trading with spatial heterogeneity. *American Journal of Agricultural Economics* 90 (4): 1044–1058.

Loucks, D.P., and E. van Beek. 2017. Water quality modeling and prediction. In *Water resource systems planning and management.* Springer.

Malik, A.S., D. Letson, and S.R. Crutchfield. 1993. Point/nonpoint source trading of pollution abatement: Choosing the right trading ratio. *American Journal of Agricultural Economics* 75 (4): 959–967.

Mezzatesta, M., D.A. Newburn, and R.T. Woodward. 2013. Additionality and the adoption of farm conservation practices. *Land Economics* 89 (4): 722–742.

Miller, K., and J.M. Duke. 2012. Additionality and water quality trading: Institutional analysis of nutrient trading in the Chesapeake Bay watershed. *Georgetown International Environmental Law Review* 25: 521.

Montgomery, W.D. 1972. Markets in licenses and efficient pollution control programs. *Journal of Economic Theory* 5 (3): 395–418.

Moriasi, D.N., K.W. King, D.D. Bosch, D.L. Bjorneberg, S. Teet, J.A. Guzman, and M.R. Williams. 2016. Framework to parameterize and validate APEX to support deployment of the nutrient tracking tool. *Agricultural Water Management* 177: 146–164.

Motallebi, M., D.L. Hoag, A. Tasdighi, M. Arabi, and D.L. Osmond. 2017. An economic inquisition of water quality trading programs, with a case study of Jordan Lake, NC. *Journal of Environmental Management* 193 (15): 483–490.

New South Wales Department of Environment and Conservation. 2006. *Hunter river salinity trading scheme working together to protect river quality and sustain economic development*. Department of Environment and Conservation, Sydney, Australia.

Nguyen, N., R.T. Woodward, M. Matlock, A. Denzer, and M. Selman. 2006. *A guide to market-based approaches to water quality*. July 2006.

Nguyen, N.P., J.S. Shortle, P.M. Reed, and T.T. Nguyen. 2013. Water quality trading with asymmetric information, uncertainty, and transaction costs: A stochastic agent-based model. *Resource and Energy Economics* 35 (1): 60–90.

O'Grady, D. 2011. Sociopolitical conditions for successful water quality trading in the South River Nation watershed, Ontario, Canada. *Journal of the American Water Resources Association* 47 (1): 39–51.

Overseer. 2020. Overseer Scientific Model. Online resource. Available at https://www.overseer.org.nz/our-model Accessed October 19th 2020.

PADEP. 2020. *Nonpoint nutrient calculation spreadsheet*. https://www.dep.pa.gov/Business/Water/CleanWater/NutrientTrading/Pages/Trading Resources.aspx. Accessed 1 Mar 2020.

Ribaudo, M.O., and J. Gottlieb. 2011. Point-nonpoint trading – Can it work? *Journal of the American Water Resources Association* 47 (1): 5–14.

Ribaudo, M.O., and J. Savage. 2014. Controlling non-additional credits from nutrient management in water quality trading programs through eligibility baseline stringency. *Ecological Economics* 105: 233–233.

Ribaudo, M., and J. Shortle. 2019. Reflections on 40 years of applied economics research on agriculture and water quality. *Agricultural and Resource Economics Review* 48 (3): 519–530.

Ribaudo, M.O., R. Heimlich, and M. Peters. 2005. Nitrogen sources and Gulf hypoxia: Potential for environmental credit trading. *Ecological Economics* 52 (2): 159–168.

Ribaudo, M., J. Savage, and J. Talberth. 2014. Encouraging reductions in nonpoint source pollution through point-nonpoint trading: The roles of baseline choice and practice subsidies. *Applied Economic Perspectives and Policy* 36 (3): 560–576.

Rodríguez, F. 2000. On the use of exchange rates as trading rules in a bilateral system of transferable discharge permits. *Environmental and Resource Economics* 15 (4): 379–395.

Saleh, A., O. Gallego, and E. Osei. 2015. Evaluating nutrient tracking tool and simulated conservation practices. *Journal of Soil and Water Conservation* 70 (5): 115A–120A.

Schmalensee, R., and R.N. Stavins. 2017. The design of environmental markets: What have we learned from experience with cap and trade? *Oxford Review of Economic Policy* 33 (4): 572–588.

Selman, M., S. Greenhalgh, E. Branosky, C. Jones, and K. Guiling. 2009. *Water quality trading programs: An international overview*. Washington, DC: World Resources Institute.

Shabman, L., and K. Stephenson. 2007. Achieving nutrient water quality goals: Bringing market-like principles to water quality management. *Journal of the American Water Resources Association* 43 (4): 1076–1089.

Shortle, J.S. 1987. The allocative implications of comparisons between the marginal costs of point and nonpoint source pollution abatement. *Northeastern Journal of Agricultural and Resource Economics* 17 (1): 17–23.

———. 1990. The allocative efficiency implications of abatement cost comparisons. *Water Resources Research* 26 (5): 793–797.

———. 2012. *Water quality trading in agriculture*. Paris: OECD Publishing.

———. 2013. Economics and environmental markets: Lessons from water-quality trading. *Agricultural and Resource Economics Review* 42 (1): 57–74.

Shortle, J.S., D.G. Abler, and R.D. Horan. 1998. Research issues in nonpoint pollution control. *Environmental and Resource Economics* 11 (3–4): 571–585.

Stavins, R.N. 1995. Transaction costs and tradeable permits. *Journal of Environmental Economics and Management* 29 (2): 133–148.

Stephenson, K., and L. Shabman. 2011. Rhetoric and reality of water quality trading and the potential for market-like reform. *Journal of the American Water Resources Association* 47 (1): 15–28.

Stephenson, Kurt, and Leonard Shabman. 2017. Nutrient assimilation services for water quality credit trading programs: A comparative analysis with nonpoint source credits. *Coastal Management* 45 (1): 24–43. https://doi.org/10.1080/08920753.2017.1237240.

Stephenson, K., and L. Shabman. 2019. Can water quality trading fix the agricultural nonpoint source problem? *Annual Review of Resource Economics* 9: 95–116.

Stephenson, S.K., D. Parker, C. Abdalla, L. Shabman, J. Shortle, C. Jones, B. Angstadt, B. Rose, D. King, and D. Hansen. 2009. Evaluation framework for water quality trading programs in the Chesapeake Bay watershed. *Evaluation framework for water quality trading programs in the Chesapeake Bay Watershed.* Chesapeake Bay Science and Technical Advisory Committee.

Stephenson, K., Aultman, S., Metcalfe, T. and Miller, A., 2010. An evaluation of nutrient nonpoint offset trading in Virginia: A role for agricultural nonpoint sources? *Water Resources Research* 46 (4).

Suter, J.F., J.M. Spraggon, and G.L. Poe. 2013. Thin and lumpy: An experimental investigation of water quality trading. *Water Resources and Economics* 1: 36–60.

Tietenberg, T. 2003. The tradable-permits approach to protecting the commons: Lessons for climate change. *Oxford Review of Economic Policy* 19 (3): 400–419.

———. 1985. *Emissions trading, an exercise in reforming pollution policy.* Resources for the Future.

———. 1990. Economic instruments for environmental regulation. *Oxford Review of Economic Policy* 6 (1): 17–33.

———. 2010. Cap-and-trade: The evolution of an economic idea. *Agricultural and Resource Economics Review* 39 (3): 359–367.

US Environmental Protection Agency (USEPA). 2001. *The national costs of the total maximum daily load.* Program. EPA-841-D-01-003. Washington, DC: Environmental Protection Agency.

———. 2007. *WQT toolkit for permit writers.* Office of Wastewater Management, Water Permits Division. EPA 833-R-07-004.

US Environmental Protection Agency and US Department of Agriculture (USEPA and USDA) Offices of Inspector General. 2006. *Inspector general evaluation report: Saving the Chesapeake Bay Watershed requires better coordination of environmental and agricultural resources.* EPA OIG Report No. 2007-P-00004 USDA OIG Report No. 50601-10-Hq, November 20, 2006.

Van Houtven, G., R. Loomis, J. Baker, R. Beach, and S. Casey. 2012. *Nutrient credit trading for the Chesapeake Bay: An economic study.* Annapolis: Chesapeake Bay Commission.

Wainger, L.A., G. Van Houtven, R. Messer, J. Breach, and M. Deerhake. 2013. Tradeoffs among ecosystem services, performance certainty, and cost-efficiency in implementation of the Chesapeake Bay total maximum daily load. *Agricultural and Resource Economics Review* 42 (1): 196–224.

Water Conservation Subdistrict of the Miami Conservancy District (WCS). 2005. *Great Miami River watershed water quality credit trading program operations manual.* Miami Conservancy District, Dayton, Ohio, USA.

Weitzman, M.L. 1974. Prices vs. quantities. *The Review of Economic Studies* 41 (4): 477–491.

Willamette Partnership, World Resources Institute, and the National Network on Water Quality Trading. 2015. *Building a water quality trading program: Options and considerations.* http://willamettepartnership.org/wp-content/uploads/2015/06/BuildingaWQT Program-NNWQT.pdf

Woerdman, E., and A. Nentjes. 2019. Emissions trading hybrids: The case of the EU ETS. *Review of Law and Economics* 15 (1).

Woodward, R.T., R.A. Kaiser, and A.M.B. Wicks. 2002. The structure and practice of water quality trading markets. *Journal of the American Water Resources Association* 38 (4): 967–979.

Woodward, R.T., D.S. Newburn, and M. Mezzatesta. 2016. Additionality and reverse crowding out of pollution offsets in water quality trading. *Ecological Economics* 128: 224–231.

CHAPTER 7

Water Quality Auctions

7.1 Introduction

Water quality trading as described in Chap. 6 is a performance-based approach to pollution management that in pure form relies on voluntary exchange of property rights to set prices. From Chap. 6 it is apparent that the dominant baseline-and-credit trading model in which agriculture is uncapped is a limited mechanism for reducing agricultural loads. It can serve as a relief valve for point source costs, but absent a sufficient point source demand for agricultural credits and trading rules that result in net reductions, the model is not a good choice for achieving substantial reductions in agricultural loads. A cap-and-trade model is a promising alternative for some agricultural pollutants but real-world experience demonstrating feasibility and effectiveness of the approach is very limited. Cap-and-trade would also be a radical departure from the dominant pay-the-polluter paradigm. Where this departure is not politically feasible, cap-and-trade is not a viable option. In this case, another performance-based pricing mechanism for purchasing pollution reductions to consider is a water quality auction. Auctions retain the traditional pay-the-polluter approach and can be designed to target problem areas and to elicit spatially efficient pollution controls. Auctions also have significant proof-of-concept.

J. Shortle et al., *Water Quality and Agriculture*, Palgrave Studies in Agricultural Economics and Food Policy,
https://doi.org/10.1007/978-3-030-47087-6_7

A water quality auction can be considered as a form of *conservation auction*, which is used to purchase various types of resource conservation or protection activities on privately owned land. Conservation auctions are applications of auction types known as reverse auctions, procurement auctions, or tendering systems. Conservation auctions have received significant attention in economic research and in applications, as they can facilitate two goals of conservation programs with voluntary participation: environmental effectiveness and budgetary cost-efficiency. Auctions induce landowners or managers whose properties can provide environmental benefits effectively and at low cost to participate in voluntary conservation programs. Recruiting such landowners via auctions maximizes the environmental benefits from a fixed conservation budget. These attributes are clearly desirable for water quality programs relying on payments to producers and voluntary participation. Water quality auctions have a strong theoretical background, as the economic theory of auctions and conservation auctions are well developed (Klemperer 2002; Latacz-Lohmann and van der Hamsvoort 1997, 1998; McAfee and McMillan 1987; Rolfe et al. 2017; Stoneham et al. 2003). There is also a long and extensive history of using auctions that provide enormous practical experience.

The origins of auctions as a resource allocation mechanism are in antiquity. Auctions were used in ancient Egypt and Rome to allocate ordinary goods and even more. Perhaps the strangest auction was the one used to select the Emperor of Rome Marcus Didius Severus Julianus. The Emperor paid 25,000 sesterces to each soldier in the Pretorian Guard and won the ex tempore auction over Flavius Sulpicianus. The story has a sad ending though. A Roman general, Septimus Severus, who was leading Roman legions in Pannonia (current Hungary) declared himself the Emperor and brought his legions to Rome. Under the threat of Severus' legions, the Pretorian Guard abandoned Julianus and the Senate of Rome sentenced him to death. Marcus Didius Severus Julianus was Emperor 66 days. This gives a unique meaning to the "winner's curse." Most auctions have happier endings.

Conservation auctions have been used in several countries to pursue environmental objectives, including biodiversity conservation and water quality improvements in agriculture, and biodiversity conservation in forestry since the 1980s. The first and largest is the US Department of Agriculture's Conservation Reserve Program (CRP) initiated in 1985 to remove cropland from production. The long and varied experience provides empirical evidence on auctions as environmental policy tools.

Auctions offer possibilities for targeting, spatial differentiation, and practice selection that serve the effectiveness and cost-efficiency of environmental protection. Specifically, they can be used to help overcome the four fundamental deficiencies of traditional voluntary agri-environmental schemes: weak targeting of environmental outcomes, weak targeting of locations that can be most effective and efficient in providing desired environmental outcomes, a focus on farm practices rather that the targeted environmental goods, and asymmetric information concerning the producers' true conservation costs.

In this chapter we introduce the concept and methods of conservation auctions, review applications in various countries, and discuss the merits of auctions compared to traditional voluntary compliance pay-for-practices approaches. To date there has been little use of auctions specifically targeting water quality problems. We discuss how the conservation auction model can be extended to targeted water quality protection and appropriate applications.

7.2 Conservation Auctions: The Basics

Conservation auctions are used by governments or other entities (e.g., environmental nonprofits) to purchase actions that produce environmental benefits. To simplify the discussion, we assume the prospective buyer is a government agency. Potential suppliers are typically producers or other rural landowners whose lands can provide the desired benefits. The agency solicits bids from eligible landowners. The bids consist of the proposed conservation actions and the payment requested as compensation. The term bid refers to the combination of these two components. Bids are selected according to environmental performance criteria and cost. Landowners want to cover their cost at minimum, and they may seek higher compensation than the minimum amount they would be willing to accept. This behavior is called rent-seeking. Profits (payments above costs) obtained from rent-seeking in auctions are called "information rents." In pursuing rents, bidders must consider the likelihood that their bid will be rejected in favor of a lower bid from a competing landowner. Competitive bidding moderates rent-seeking, reducing the agency's costs and increasing the effectiveness of public expenditures.

7.2.1 *Basic Auction Design*

The basic components of a conservation auction are: (1) specifications for the conservation or environmental actions to be procured, (2) the bid rule, (3) the bid ranking (selection) rule, (4) the reservation price (bid cap) rule, and (5) whether the auction is "one shot" or repeated. Standard auction theory assumes the transaction is at "arm's length," meaning buyers and sellers act in their own self-interest independently without one party influencing the other. Some conservation auction implementations have eschewed the "arm's length principle" with buyers, or organizations cooperating with the buyers, working with potential sellers to encourage participation and provide technical assistance in preparing bids.

A common application of conservation auctions, highlighted by the CRP, is to remove environmentally sensitive land from crop production and install resource-conserving practices (e.g., vegetative covers, riparian buffers, wetlands). Procurement specifications define eligible land types and locations, allowable vegetative covers, annual maintenance, allowable non-crop uses, and contract duration. Specification for an auction to procure water pollution reductions would similarly specify eligible land types, locations, and actions. Removing cropland from production and installing riparian buffers, wetlands, or permanent grasses would be options. For working lands, pollution reductions are achieved with changes in pollution input use and implementation of Best Management Practices (BMPs). Auction rules would define allowable practices and standards for their implementation to assure additionality and other dimensions of ecological integrity (Box 6.3).

The *bid rule* determines how winning producers are compensated. The two main alternatives are "pay according to the bid," or "pay the same compensation to all." In the terminology of auction theory, "pay according to bid" is a first-price *discriminatory pricing* auction, as every bid accepted receives a different compensation. "Pay the same compensation to all" is a *unit price auction* (also known as a *uniform price auction)*. Discriminatory pricing is the norm in conservation auctions. An exception is a uniform price auction for environmentally sensitive land conservation in Canada (Brown et al. 2011). Discriminatory price auctions are more easily understood by bidders and eliminate uncertainty about what they will receive if their bid is accepted. Under uniform pricing a bid only determines the chance of winning but not the payment that the bidder receives. These factors may lead to better participation by bidders in discriminatory price auctions than in unit price auctions (Brown et al. 2011; Latacz-Lohmann and Schilizzi 2005).

The *bid ranking rule* defines the agency's method for ranking bids. A simple bid ranking rule is to select bids according to the requested price. Budget efficiency can be improved by more advanced approaches. These include benefit-cost ratios or cost-effectiveness ratios that rank bids according to the ratio of costs to environmental outcomes. The CRP, for example, uses an Environmental Benefits Index to rank bids. The index was initially used to construct a benefit-cost ratio but was subsequently amended to rank bids based on an adjusted cost-concept (Hellerstein 2017).

A *reservation price rule,* or a bid cap, may be used to exclude very high bids. The bid cap may or may not be made public. Setting a bid cap publicly may increase competition among producers, moderating bids and increasing budgetary efficiency. However, announcing a bid cap can also lead to increased bids if the cap becomes a target for bidders, reducing budgetary cost-efficiency. Whether announced or not, a risk from setting a cap is that if set too low, some high-impact but also high-cost projects may be excluded (Hellerstein 2017). The same effect may happen if the conservation budget is announced to the public (Latacz-Lohmann and van der Hamsvoort 1997; Myerson 1981).

The choice between a sequential or single round auction depends on the agency's objectives and resources. Suppose that the agency has a specific target for the environmental good, and that it has a sufficient budget to achieve the objective with a single auction. If the auction is successful, the single round is adequate. Alternatively, suppose the agency has a specific target but its annual budget is limited. It may then choose a sequence of auctions to add to the supply of the environmental good over time. Research shows, however, that participants learn from experience from repeated auctions and adjust their bidding behavior in ways that increase information rents and reduce the budget efficiency of subsequent auctions (e.g., Kirwan et al. 2005).

7.2.2 *The Economics of Bidding*

To develop a formal understanding of bidding behavior, consider a producer generating a bid for a nutrient pollution control project. Following the typical conservation auction, consider bid selection in a discriminatory pricing auction where winners receive the payment specified in their bid.

In determining a bid, the producer seeks to earn rents, but the rent-seeking will be tempered by the expectation that a high bid will not be accepted. The bids will therefore be determined based on the producer's subjective understanding of the effect of the bid on the probability of acceptance.

The producer's profit from normal operation in the absence of the auction is π^0. Undertaking the nutrient controls entail costs that would reduce farm profit to π^A. Compensation according to the bid, denoted by ω, offsets this loss by the amount of the bid. The producer's subjective probability that bid is rejected increases with the size of the bid.[1] This probability is denoted $F(\omega)$, an increasing function in ω: the higher the compensation asked, the higher the probability of the bid being rejected. The probability of the bid being accepted is simply one minus the probability of rejection $(1 - F(\omega))$. The bid that maximizes the producer's expected profit from participation in the auction is given by Eq. (7.1)

$$\omega^* = \left(\pi^0 - \pi^A\right) + \frac{1 - F\left(\omega^*\right)}{f\left(\omega^*\right)}. \tag{7.1}$$

The profit difference in parentheses is the producer's cost of the nutrient-reduction project (i.e., the reduction in profits). From the buyer's perspective, the ideal bid is the producer's cost, $\omega^* = (\pi^0 - \pi^A)$. This is the first term on the right-hand side of the equation. If the agency knew the producer's cost, it could reject bids exceeding the costs. In consequence, Eq. (7.1) includes the second term on the right-hand side. This term is the probability of the bid being selected divided by the change in the probability, $f(\omega)$, due to a marginal increase in the bid. The term is positive, which means that a producer's optimal bid is more than the cost. This term is the information rent the producer receives if the bid is accepted. The origin of concept of information rent in the context of auctions is now evident. If received it is the result of the producer's cost information being private, unknown to the agency.

[1] Subjective probability refers to an individual's expectation of the likelihood of an outcome based on their own judgment and experience.

7.3 Auctions vs. Uniform Price Subsidies

The existence of information rent means that a budget devoted to a water quality auction will not go entirely to covering the costs of providing water quality benefits. But the relevant question when considering alternative subsidy schemes is whether auctions provide better outcomes than using administratively determined payment rates. Given asymmetric information, administratively determined payment rates cannot optimally differentiate across participants. The usual practice is to use uniform or flat-rate payment programs.

As described in Chap. 5, voluntary programs for working lands provide a menu of water quality measures (e.g., buffer strips, gypsum applications, cover crops, reduced tillage). Subsides are offered by fixed payment rates. The budgetary efficiency of auctions compared to uniform payment programs can be demonstrated most directly by disregarding the practice selection, which affects the pollution reduction from land enrolled in the program, and focusing on the selection of land into the program. Assume for simplicity that the auction removes cropland from production for conversion to an environmentally benign use. Assume also for simplicity that all eligible cropland has the same pollution propensity. The agency is thus indifferent between the parcels for environmental effectiveness. The agency's objective is served by enrolling as many parcels as possible within the budget. The uniform payment mechanism pays producers the same amount per parcel enrolled. The auction mechanism ranks parcels by producers' bids and pays winning producers according to their bids.

The following analysis is from Ollikainen et al. (2019). Let the annual conservation budget be G and denote the area payment per ha by s. To keep the graphical illustration as clear as possible, it is assumed that each participating producer supplies a field parcel of one hectare to the program. The cost to producer i is denoted by c_i.

Figure 7.1 presents parcel selection for a uniform payment given the fixed budget. The horizontal axis lists parcels from the lowest cost to the highest. The vertical axis denotes costs and the flat-rate payment. Producers may enter the program until the budget is exhausted. Producers offering parcels 1 through 4 profit by enrolling the parcels at the uniform payment rate. Enrollment is breakeven for the producer enrolling parcel 5. Producers with parcels 6 and 7 would choose not to enroll even if the

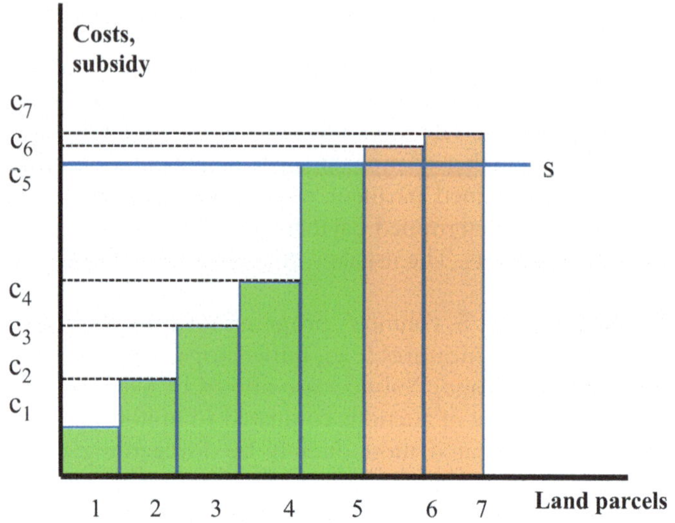

Fig. 7.1 Flat-rate payment and farmers' participation. (Source: Ollikainen et al. 2019)

budget remained available. However, assume that the conservation budget is exhausted with parcel 5 (technically $s*5 = G$). While the producer enrolling parcel 5 receives compensation exactly equal to the costs, other participating producers receive excess compensation. These surpluses are also information rents. Graphically the rents are the area between the pillars and the horizontal flat rate *s*. Because of overcompensation, only five parcels can be chosen as no budget remains for parcels 6 and 7.

Figure 7.2 illustrates parcel selection with the auction. Parcels are enrolled in the program solely according to their bids. The sum of the cost pillar and a new light red pillar indicates the size of the bid for each parcel. The sum of these figures defines the *bid curve* associated with the parcels supplied. The bid curve is located above the true cost curve reflecting the information rent included in the bids as described by Eq. (7.1). However, the bid curve starts well below the uniform payment rate *s*, due to the competition between producers created by auction, and surpasses it at some point. As producers are paid according to their bids, the auction reduces information rents to producers below what they received with the

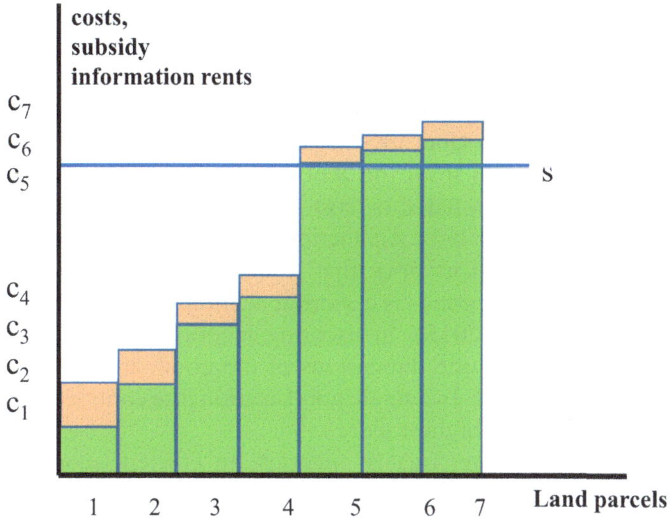

Fig. 7.2 Tendering: conservation budget and farmers' participation. (Source: Ollikainen et al. 2019)

flat-rate subsidy (the reduction of information rent is the area above the bids and below s). As the agency now saves money, it can enroll more parcels in the program. In Fig. 7.2, parcels 6 and 7 are now enrolled in the program, indicating that water protection effort increases with the more efficient use of the agency budget.

Given that the more parcels enrolled, the greater water quality benefits, this analysis illustrates that the auction provides a better outcome for the same budgetary cost as the uniform payment policy: information rents are lower than in the voluntary program with a flat-rate payment, and a larger number of pollution-reducing projects are accepted to the program.

7.4 Bid Ranking for Water Quality Auctions

In the preceding analysis water quality gains were produced by removing land parcels from production. The water quality benefits from all parcels were assumed to be identical. In this case the agency is indifferent between parcels except for the payments. Since this will not generally be the case, methods for considering heterogenous water quality responses related to differences in pollution propensities and pollution control practices are needed. Including environmental benefits in an auction format poses challenges. Bids must provide information (e.g., for working lands, types and locations of BMPs, and land areas treated by BMPs) to determine the environmental outcomes from the actions offered by farmers. Also, bid rankings must be differentiated by cost and the estimated environmental outcomes offered by the bids. An auctions format that includes both the quality of the good and its price that is suitable for purchasing multi-attribute environmental goods is a scoring auction (Asker and Cantillion 2008; Hellerstein et al. 2015). In a scoring auction the buyer announces a scoring rule. Bids include dimensions of the good that describe attributes per the scoring rule. In a single good auction, the contract is awarded to the supplier with the highest score.

In application to water quality auctions, an efficient scoring rule is based on the performance of the bids in reducing pollution loads, requiring indicators of water quality impacts of practice selection. In this section we consider this issue and the effect of the choice on bidding behavior.

7.4.1 *Quantifying the Environmental Good*

A water quality auction requires rules for quantifying the water quality outcome of proposed projects. Following the example of CRP scoring auction, the rules may be expressed as a worksheet that producers are required to use in the construction of their bids. Entries would give levels for eligible actions. The entries would be used to calculate a water quality performance indicator. The indicator might be a point system, an index, or an estimate of the pollution load reduction. For the auction, bids would consist of the calculated performance according to the indicator and the corresponding request for compensation. A comparison of a flat-rate pay-for-performance mechanism to the auction would lead to the same conclusion as found above. The auction mechanism will provide a greater pollution reduction as measured by the agency's indicator than the flat-rate payment for the same budget.

As discussed in Chap. 5, the economically preferred method for ranking water quality projects is by benefit-cost ratios. An example of benefit-cost ratios for agricultural BMPs is in the Appendix of Chap. 5. This method could be used in ranking bids for water quality auctions provided adequate information on benefits is available. Bids would be ranked by the estimated benefits they produce divided by the bid price. This method requires the agency to have data and procedures for providing estimates of pollution reductions and the corresponding benefits that are considered of acceptable quality for the purposes of the auction.

Benefit estimates are valuable for ranking bids when the emissions reductions from eligible bids are not perfect substitutes in producing water quality improvements. For example, consider an auction to purchase reductions in a single pollutant delivered to a water body affected by many sources. In this case, delivered pollution reductions from different sources are perfect substitutes. Assuming individual bids do not exhaust the budget, ranking projects by their cost-effectiveness, defined by price/pollution reduction, will be the same as ranking by benefit-cost ratios. The agency would only need to have data and procedures for providing estimates of delivered pollution reductions that are considered of acceptable quality for the purposes of the auction.

Cost-effectiveness rankings become problematic if projects reduce different pollutants or mixes of pollutants or affect different receptors. Consider, for example, two bids in an auction to purchase reductions in nutrient loads. One bid is $100 for a nitrogen reduction of 10 kg and a phosphorous reduction of 5 kg. The bid/pollution-reduction ratio is $10 per kilogram for nitrogen and $20 for phosphorous. The second bid is also for $100 but with the bid/pollution ratios reversed, which of the projects is better? If benefits can be estimated, project benefit-cost ratios can be calculated to rank bids.

Given the significant uncertainty about environmental benefits or legal or political limitations on the use of benefits in environmental decisions, another ranking tool that can be used is the Environmental Benefits Index (EBI). EBIs were developed originally for the US CRP program to rank parcels for removal from crop production but they can also be applied to working agricultural lands as well (Hellerstein 2017). The purpose of the EBI is to evaluate the performance of projects that provide multiple or multi-attribute environmental goods. For example, auctions are used to purchase environmentally sensitive parcels that provide habitat, air and water quality, and other public goods but not in equal proportions. EBIs apply weights to individual metrics of the environmental goods to produce a single performance indicator. EBIs have not been used for water quality auctions but the concept can be applied to water quality auctions involving single or multiple pollutants when economic benefit estimates are not available or where agency preferences differ.

EBIs might also be useful for evaluating projects for single pollutants if available methods for estimating pollution reductions are not considered sufficiently valid or reliable for use in ranking projects. Expert assessments would provide the bases for ranking (e.g., Iho et al. 2014).

7.4.2 *Bidding with Water Quality Metrics*

Whether benefits, estimated load reductions, or EBIs are used for measuring water quality performance, the bid will consist of the performance measure, the practice that would be implemented to produce the indicated performance, and the compensation. Let Z denote the performance measure. As above, a producer will submit the bid that maximizes the expected return from participation. It can be demonstrated that the optimal bid is

$$\omega^* = \left(\pi^0 - \pi^A\right) + \frac{1 - F\left(\theta_i^*\right)Z}{f\left(\theta_i^*\right)}, \tag{7.2}$$

where $\theta = \omega/Z$ (see Iho et al. 2014). Equation (7.2) is similar in form to Eq. (7.1) but there is a crucial difference. The first term, the conservation cost, is as before. But the information rent term is modified by the inclusion of performance measure. Z enters the probability terms $(1 - F(\theta))$ and $f(\theta))$ and also directly in the numerator of the information rent term. The effect is to increase the size of the optimal bid, and the size of the information rent. This result has positive and negative implications. The "good news" is that the incentives for high-impact (i.e., high Z) producers to participate in the auction is greater than would be the case without the performance indicator. This implies that use of performance measures increases the environmental effectiveness of the auction. The "bad news" is that producers who know they have comparatively high performance will utilize this information to increase their information rents.

The cost-effectiveness of bids is determined by the ratio of compensation to performance indicators (ω^*/Z). The budgetary efficiency and overall cost-effectiveness of the program are served by selecting bids according to cost-effectiveness starting with the most cost-effective until the conservation budget is exhausted.

7.5 Conservation and Water Quality Auctions in Practice

Experience supporting the use of auctions for purchasing environmental goods is compelling. Environmental auctions have been applied successfully in agriculture. Most applications and pilots to date are designed to purchase a suite of environmental benefits and use multi-attribute EBIs. We describe this general experience and the small number of water quality-focused auctions.

7.5.1 US Conservation Reserve Program

As noted earlier, the first and oldest conservation auction is the CRP, established by the US Food Security Act 1985 (see Box 5.6). The auction enrolls landowners into a program to remove highly erodible and otherwise environmentally sensitive land from crop and livestock production (Hellerstein 2017). The long-term goal is to re-establish land cover to improve water quality, reduce soil erosion, and improve wildlife habitat. Participating contracts are typically for 10–15 years. Producers are required to establish and maintain a vegetative cover on the lands during the contract period. As of 2015, the program enrolled 9.8 million hectares at a cost of 1.8 billion USD per year (US Department of Agriculture (USDA) Farm Service Agency 2015). Evaluations find that the CRP has improved a variety of environmental goods, including wildlife habitat and water quality (Hansen 2007; Ribaudo et al. 2001). These studies also indicate that the program's environmental benefits cover its costs.

When first initiated, the CRP was intended primarily as a commodity supply control program to reduce excess production (Hellerstein 2017). The program was subsequently re-envisioned as an environmental program, leading to the development and use of EBI for ranking bids in 1990. As the CRP was intended to serve multiple environmental objectives, the EBI incorporated a variety of factors relevant to ground and surface water quality and other objectives. The EBI was originally constructed as a benefit-cost ratio but modified after 1997 to include several other factors, including costs, making the EBI a cost-adjusted indicator (Hellerstein 2017). This form of the EBI provides program administrators flexibility to favor low-cost or high-benefit lands (by changing the weights), and thus to respond to the set of offers submitted. It also adds ambiguity to the landowners' decision making (Hamilton 2010).

7.5.2 *Australian Conservation Pilots and Programs*

Australia has piloted and implemented several conservation auctions (Rolfe et al. 2017). The first was the Victoria Bush Tender pilot conducted in 2000–2001 to protect areas of the remnant native vegetation on private lands. The pilot provided the foundation for the subsequent Bush Tender program that was implemented to purchase nature conservation contracts over several years (2001–2016) (Stoneham et al. 2003). Under the contracts, producers implement various practices, such as fencing to exclude livestock to promote biodiversity, controlling pests, and minimizing habitat disturbance, that promote the remnant native vegetation for biodiversity. The Auction for Landscape Recovery (ALR) was a pilot conducted in Western Australian over two years (2003–2005) to purchase biodiversity and other environmental benefits (Hajkowicz et al. 2007). The Reef Trust Tender – Wet Tropics program, implemented in 2014 is intended to reduce nitrogen discharges that harm the Great Barrier Reef. The program provides financial incentives to sugarcane producers in the Wet Tropics to improve their nitrogen use efficiency by adopting nitrogen management practices. The competitive tendering ended by 2015.

The Catchment Care auction purchases riparian conservation (Connor et al. 2008). It was established in 2004 to fund watercourse and riparian zone restoration and protection activities in the Onkaparinga catchment in South Australia. The auction was a sealed bid, first-price discriminatory auction. The bids by landowners were assessed against an EBI and the required compensation. The EBIs were calculated for each site using three criteria: rarity and significance of the geomorphological type, the degree of hydrological disturbance, and the condition, connectivity, and conservation significance of the remnant vegetation. This specification of the EBI combines habitat and water quality aspects. An integral part of the auction was that program administrators visited landowners to provide them with information about management options and to help them construct their bids. Bids were accepted for consideration continuously during an enrollment period and selections made noncompetitively based on the information in the bid. Accordingly, there was not a systematic simultaneous ranking and evaluation of all sites (Connor et al. 2008). Because of this format and the engagement between the buyer and sellers, the Catchment Care auction is not entirely "by the book" for auction theory as it is not at arm's length.

7.5.3 *Conservation Auctions: Other Countries*

Auctions have been piloted in Finland, Germany, Scotland, and other countries. The Grassland Conservation Pilot Tender (2003–2005) in Germany sponsored several pilot auctions to purchase biodiversity and conserve grassland and to encourage broader participation in agri-environmental programs (Groth 2008). Groth (2008) found improvements in cost-effectiveness of up to 36% compared to flat-rate policy designs.

Scotland and Finland have piloted auctions to promote forest ecosystem services and biodiversity. Scotland established the Challenge Fund Forests in 1977 and Finland has the Forest Biodiversity Program for Southern Finland (METSO), which organized a pilot auction called Trading in Nature Values in 2003–2006. This program rented forest with high biodiversity values with 10 years contracts outside wood production in Southwest Finland (Juutinen and Ollikainen 2010). While the Scottish auction followed the standard "buyer hands off" auction procedure, the Finnish auction had much in common with the Australian Catchment Care auction, being of much help to landowners considering participation in the auction.

7.5.4 *US Water Quality Auctions*

The US provides two examples of water quality auctions. One was developed as a mechanism for purchasing nutrient-reduction credits in Ohio's Great Miami River Water quality trading program. The second is an auction for buying and selling nutrient-reduction credits in Pennsylvania's water quality trading program.

The Great Miami River auction is a definitive example of a reverse auction for purchasing pollution reductions from producers (Newburn and Woodward 2012; Shortle 2013). The auction was developed in 2005 by a regional water conservation agency (the Water Conservation Subdistrict (WCS), Miami Conservancy District) as an incentive to accelerate water quality improvements in the Great Miami Watershed in southwestern Ohio. It was developed to provide regulated point sources with the opportunity to purchase nutrient-reduction credits from agricultural sources. The WCS serves as a clearinghouse, buying pollution-reduction credits from agricultural sources through the auction and transferring them to point sources.

In the auction, producers provide bids and propose BMPs to generate load reductions. The proposed Best Management Practices (BMPs) must be newly adopted and cannot receive funding from federal or state cost-share programs. To be eligible to participate, producers were required to develop and submit their bids with participating County-based Soil and Water Conservation Districts. The Districts were subsequently responsible for monitoring activities. A worksheet was developed for calculating credits for field-based and livestock BMPs. The field-based BMPs credit calculations make use of factors in the Universal Soil Loss Equation, such as vegetative cover type, slope, and soil erodibility. The livestock worksheets consider manure management practices and the treatment of milk house water. The key factors are proximity to waterways, covered surface, animal weight and number, and management technology (Newburn and Woodward 2012). The total number of credits is calculated as a sum of annual nitrogen and phosphorus load reduction calculated using the worksheets multiplied by the duration of conservation contracts length, which may be a 5- or 10-year period.

Although developed within the context of a water quality trading program, the program is better characterized as a water quality auction. Funds for the purchase of credits came largely from federal grants to support the program as a demonstration project. In this respect the program was a success with several round of auctions over a period of years purchasing several hundred BMP projects (Shortle 2013). Auctions were discontinued with the completion of the funding. The program was developed with the expectation that anticipated water quality regulations would lead to a demand from wastewater treatment plants for the credits. The regulations were never developed so that the demands essential for trading never materialized. The case illustrates the fundamental importance of both the demand for credits to successful trading programs and the budget for purchasing environmental goods to successful auctions.

An *ex post* evaluation of the auction shows great variability in the types of BMPs included in bids, most being field-based rather than livestock-based bids (Newburn and Woodward 2012). In the first rounds, the reverse auction demonstrated considerable costs savings but after repeated rounds Newburn and Woodward (2012) found signs of increased strategic bidding. Successful auctions resulting in BMP implementation provide support for the potential for water quality auctions. However, this case also indicates that the additionality problem that hinders the effectiveness of programs subsidizing the adoption of BMPs is also a concern for water quality auctions. In an analysis of the program Woodward et al. (2016) find no statistically significant increase in the number of conservation practices due to the trading program.

Pennsylvania implemented a water quality trading program in 2005 to meet its obligations under agreements with the US and other states in the Chesapeake Bay watershed, and in anticipation of the Chesapeake Bay TMDL implemented in 2010 (see Box 6.3). Pennsylvania's Department of Environmental Protection administers the program, which enables regulated point sources to meet effluent limits using nitrogen and phosphorus credits acquired from agricultural nonpoint sources. To earn and sell credits from nutrient reductions, producers are required to meet a set of baseline farming practice requirements. Credits are calculated following procedures that estimate the long-term annual average nutrient reductions in the Chesapeake Bay resulting from BMP adoption. The procedures essentially determine the on-farm nutrient reduction which is multiplied by a set of factors that determine the reduction in delivered pollution to the Bay.

In 2010, the state launched a nutrient credit clearinghouse operated by the state's public infrastructure investment authority, PENNVEST. Credit exchanges can be made through bilateral negotiations, but the clearing house also conducts periodic double auctions to buy and sell credits. Equilibrium in auctions is a settlement price that uses the weighted average of the lowest bid and the highest offer prices (PENNVEST 2014). As described in Chap. 6, trading activity has been limited, likely due to a low demand for credits.

7.5.5 *Finnish Water Quality Pilot Auction*

A pilot water quality auction conducted in Finland provides an example of the construction and uses of an EBI for phosphorus loading (Iho et al. 2014). The purpose of the pilot auction was to reduce phosphorus loading to the Baltic Sea from working agricultural lands, and test producers' willingness to participate in an auction mechanism.

A unique feature of the Finish pilot was the development of a phosphorus EBI for ranking bids. The EBI focuses on algae-available phosphorus in the forms of dissolved reactive phosphorus and particulate phosphorus. The loading of dissolved phosphorus is closely related to soil phosphorus levels, while the loading of particulate fractions is affected by erosion susceptibility, captured by field slopes in the EBI (Iho et al. 2014). Delivery of both forms is inversely related to the distance of the parcels to waterways. In the EBI, phosphorus fractions were weighed according to their algal availability. The dissolved fractions were 100% algal available while 16% of the particulate phosphorus loading was counted in (Ekholm et al. 2005). This means that 6.25 kilograms of particulate phosphorus loading is equivalent to a kilogram of dissolved phosphorus. This relationship is translated into a continuous index value between 0–100. The conservation measure offered in the auction was a gypsum application of four tons per hectare (for the impacts of gypsum on phosphorus loading, see Ekholm et al. 2012).

The starting point for the EBI is the nutrient runoff potential: the greater the potential load from a given field parcel, the greater the load reduction achieved by applying gypsum to the parcel. Fields with high potential loads will therefore be assigned a high EBI. Potential loads were determined for a total of 75 parcel types. To make the index more tractable for potential bidders, certain slope and all production technology classes were combined to form several different parcel types. Table 7.1 shows how the specific values of the above three features were translated to index values.

Table 7.1 illustrates the EBI in the same format as it was presented to local producers in the pilot. The EBI gives the highest values on parcels with a P-status above 14 mg/l adjacent to surface waters and had a slope steeper than 6%. For example, a parcel adjacent to a ditch, with a slope between 1.5% and 6% and a P-status between 8 mg/l and 14 mg/l received an EBI value of 42.

Table 7.1 Phosphorus benefit index: Finnish Agriculture Auction Pilot

P-status	*Slope < 1.5%*			*Slope 1.5–6%*			*Slope > 6%*		
	Abutting			*Abutting*			*Abutting*		
	Land	*Ditch*	*Surface water*	*Land*	*Ditch*	*Surface water*	*Land*	*Ditch*	*Surface water*
<8	8	17	19	14	30	34	29	66	73
8–14	13	28	31	18	42	46	34	77	85
>14	18	41	46	24	55	61	40	90	100

Source: Iho et al. (2014)

The pilot suggests that the auction format with the EBI performed well. It also demonstrated that producers were able to make their bids based on the EBI, which facilitated a cost-efficient selection field in the pilot. Importantly, the auction format also enrolled some of the most environmentally sensitive parcels in the area, and the information rents in the pilot were low (Iho et al. 2014).

7.6 What Role for Water Quality Auctions?

We see significant experience in the use of conservation auctions. Some, such as the Great Miami River and the Reef Trust Tender – Wet Tropics program, are solely for purchasing water quality protection from agriculture in targeted locations. Some, such as the Conservation Reserve Program and Catchment Care Program, purchase agricultural activities that, though not specifically targeted to water quality, can contribute to water quality as one outcome in the suite of environmental goals. In this and the next section we ask how well conservation auctions have performed and what role they might play in addressing water quality problems cause by agriculture.

7.6.1 Assessing the Conservation Auction Programs

The previous discussion suggests at least four aspects that should be considered when assessing the performance of conservation auctions: environmental effectiveness; size of information rents, which affect budget efficiency; success in recruiting producers into conservation programs; and possible social cost savings relative to policy-relevant alternatives. Producing information on these metrics can be challenging. EBIs or other performance indicators provide predictions of environmental outcomes with varying degrees of specificity with respect to location and affected environmental goods. However, *ex ante* predictions and *ex post* outcomes can be different. Data limitations and multiple uncontrolled sources of causation can make actual environmental effects and other outcomes of auctions difficult to determine. For instance, *ex ante* comparisons of conservation auctions to flat-rate policies can be produced quite easily but determination of information rents *ex post* is difficult. Auction data includes bids, payments, and EBIs or other indicators if used, but information on actual conservation costs of participants and nonparticipants are unknown. This is what only producers know. Thus, to extract information rents one must produce reliable estimates of the conservation costs.

Empirical studies comparing conservation auctions to flat-rate policies find that auctions achieve better environmental outcomes than flat-rate policies for a given budget by reducing information rents to increase the resources devoted to conservation activities (e.g., Latacz-Lohmann and Schilizzi 1998; Stoneham et al. 2003; Connor et al. 2008; Kirwan et al. 2005). Connor et al. (2008) found that high administrative costs associated with farm visits, tutoring regarding the bidding process, and scoring and ranking bids can be reducing auction cost-efficiency relative to a comparable flat-rate policy but that the auction still offers better environmental performance. The consensus of these studies is that conservation auctions are at least as efficient as flat-rate policies and in most cases more efficient by a wide margin.

Turning to information rents, Table 7.2 summarizes findings from four studies. Each entry is for a sealed bid discriminatory auction mechanism. The results suggest that information rents are not negligible but also not very large. Kirwan et al. (2005) find that information rents in the CRP have been gradually increasing, most likely due to the learning effects, as producers gradually get better at the bidding process and actively seek

Table 7.2 Estimated information rents from conservation programs and pilots

Article	*Conservation program*	*Environmental good*	*EBI*	*Information rent*
Latacz-Lohmann and Schilizzi (2005)	Challenge funds	Planting forests for ecosystem services	Biodiversity, recreation	45%
Kirwan et al. (2005)	Conservation reserve program	Biodiversity and water quality	Environment and costs	10–40%
Groth (2008)	Grassland conservation pilot	Agricultural biodiversity	Features of biodiversity	
Juutinen and Ollikainen (2010)	METSO; trading in nature values, a pilot	Forest biodiversity	Biodiversity and costs	20–40%
Iho et al. (2014)	Gypsum pilot for P runoff	Water quality, eutrophication	Features of P runoff	10%

Source: Authors' creation

higher information rents. Juutinen and Ollikainen (2010) end up with the same magnitude for the Finnish Trading in Nature Values, which they partly explain by the fact that this pilot was a mixture of auction and negotiations.

The smallest information rents were found in the Finnish gypsum pilot, most likely because all producers used the same technology with known costs. Known cost limits the likelihood that high bids will be winning bids, thus moderating bids and information rents.

Several implications can be drawn from this research. First, auctions can be used effectively to recruit producers located in high-priority source areas into voluntary compliance programs for water quality protection. Auctions can be expected to work well for targeting pollution control to larger priority source areas where there are enough producers to assure competitive bidding. They can also be expected to work well in environments where several smaller priority source areas contribute to a common water quality problem. Within these contexts, the effectiveness of the auction in reducing water pollution will depend on the rules defining the environmental good. Other things equal, cost-effectiveness is served by competitive bidding, but the extent will depend on the size of the information rents relative to the total budget.

The case for auctions weakens for water quality problem involving disconnected priority source areas with few farmers. Auction theory suggests that it is virtually impossible to establish an auction market for small-scale, local environmental goods (Latacz-Lohmann and van der Halmwoort 1998). For these areas, alternative instruments are required.

As with most environmental policy design, details matter greatly to auction performance (Hellerstein 2017; Rolfe et al. 2017). There is much academic research to understand participant behavior, the effects of design on performance, and to fine-tune design details. Experimental economics methods are now used extensively for this purpose (Cason and Gangadharan 2004; Hellerstein and Higgins 2010; Hellerstein et al. 2015; Latacz-Lohmann and Schilizzi 2005; Schilizzi and Latacz-Lohmann 2007).

7.6.2 *Farm Assistance and Transaction Costs*

We have noted that in several auctions the purchasing agency or a cooperating organization engages directly with producers prior to the submission of bids, providing technical assistance for project selection and guidance in preparing bids. In this, the conduct of conservation auctions differs from the standards of auction theory. In most cases, the technical assistance relates to the selection of practices and the effects of practice selection on the environmental performance indicator.

The potential benefit from outreach and engagement is increased producer participation and better results for producers and support for the programs. Breetz et al. (2005) find, for example, that engaging trustworthy third parties, like Soil and Water Conservation Districts, can reduce producers' reluctance to participate in water quality trading programs. Additionally, institutionalized technical assistance could be programmatically beneficial as a means beyond auction rules to assure ecological integrity.

But there are also trade-offs from institutionalized assistance. One involves program costs. Farm visits and technical advice reduce producers' transaction costs but increase administrative costs. These costs are well documented, for instance, in the Australian Catchment Care program (Connor et al. 2008). Perhaps more significant for budgetary-cost-effectiveness is that such assistance may lead to producers having greater success in their pursuit of information rents. For instance, Woodward and Newburn (2012) conclude that the involvement of Soil and Water Conservation Districts offices in the Great Miami River auctions reduced the potential cost savings from the reverse auction structure as some agents learned decision rules in the auction that they utilized to help producers bid strategically. A similar result was found in the Finnish conservation auction (Juutinen and Ollikainen 2010). Given these trade-offs, a task for auction research and pilots is to explore alternative mechanisms for providing information and conduction auctions that balance the gains from participation against the costs.

7.6.3 *Learning and Budget Efficiency*

Seller learning is an inherent feature of repeated auctions. Landowners are not isolated. They can learn from public announcements about auction results and from their own success or lack of success in bidding. They can share their findings with others through various networks. This learning improves bidding behavior and reduces the cost savings that can result from competition between producers. This effect has been found in the CRP (Shoemaker 1989; Khanna and Ando 2009) and in the Great Miami River (Woodward and Newburn 2012). The implication is that there may be some shelf life to auctions.

If increasing information rents is a serious problem for budgetary efficiency, there are possible remedies. First, the conservation agency should not report the average or maximum accepted bids or the distribution of bids in the preceding auctions. This information provides data for learning and adjusting bids. Second, the conservation agency could amend the rules of auctions in each bidding round. This would make the system less predictable (Lacatz-Lohmann and Schilizi 2005).

One alternative is an adaptive approach that has been suggested that involves the use of auctions as discovery mechanisms to inform the implementation of alternative instruments for spatial targeting (Latacz-Lohmann and Schilizi 2005). To illustrate, suppose the submitted practices can be nicely grouped into some classes in terms of their nutrient loads so that the classes differ from each other, but practices within the class produce relatively similar load reductions and the auction results are intended to inform a subsequent flat-rate payment scheme for suites of BMPs. Using these classes, the size of flat-rate payments could be differentiated so that the environmentally highest class receives the highest flat-rate payment.

7.6.4 The Best Case for Conservation Auctions

As with other policy instruments, spatial heterogeneity of the costs of water protection practices and water quality responses to their implementation are key factors to consider in the design and performance of water quality auctions. Table 7.3 illustrates the suitability of auctions and flat-rate policies based on these dimensions.

A flat-rate payment works best as a pay-the-polluter approach when there is no spatial heterogeneity in costs or water quality responses. In this case, uniform payments will make the most of the budget. When costs are heterogeneous, but water quality responses are not, differentiated payments are needed to maximize the water quality gains that can be

Table 7.3 Suitability of economic instruments: conservation cost and environmental quality

Environmental quality	Conservation costs	
	Homogenous	*Heterogeneous*
Homogenous	Flat-rate policy	Differentiated payments
Heterogeneous	Auction, bargaining solution	Auction

Source: Authors' creation

purchased with a given budget. They might also be considered fairer in the sense that differentiated payments reduce information rents accruing especially to low-cost projects relative to the flat-rate policy. In this case, auctions will maximize water quality gains from the budget, but if the agency knows costs well, a differentiated payment system is easier to administer. If costs are not known, then auctions have merit based on the theory of conventional auctions.

When both costs and water quality responses are heterogeneous the merits of auctions increase. In this case selecting bids based on the required payment relative to water quality performance indicator maximize the water quality improvement for the given budget. Auctions are not advantageous with homogenous costs and heterogeneous environmental responses. In this case the environmental agency can simply gather information on the costs and use either a flat-rate policy adjusted by the performance indicator or directly negotiate with the producers.

Out of these cases, auctions perform best when heterogeneity prevails for both costs and the water quality responses. This is the case for most environmental goods, and certainly for programs aiming at water quality improvement from agriculture.

7.7 Conclusions

Water quality auctions are an appealing option to improve cost-effectiveness and environmental targeting of water quality policies in a pay-for-performance policy paradigm. They can be applied in multiple regional levels and, when environmental indicators are well designed, facilitate spatial targeting to critical source areas. As with any policy instrument, details and context affect the size of the gains. Research experiences support the proposition that well-designed auctions allocate budgets more efficiently than flat-rate policies.

There are aspects of auctions where research and experience can help improve their performance. We need a better understanding on the optimal length of contract periods. While capital-intensive practices require longer periods to cover the cost of adjusting capital, uncertainty concerning prices and producers' risk-bearing behavior tends toward shorter periods. Administrative and other transaction costs relating to running the program and helping producers making bids can be reduced by simpler rules and modern IT technologies. Finally, performance indicators can always be improved for better targeting. Furthermore, like water quality trading, the likely scenario for auctions at present is not one in which they replace other subsidy schemes but instead augment them. In this case, understanding the effects of program interactions on participation rates and bidding behavior is important to design and coordination. There is little research on this topic.

References

Asker, J., and E. Cantillon. 2008. Properties of scoring auctions. *Rand Journal of Economics* 39 (1): 69–85.

Breetz, H.L., K. Fisher-Vanden, H. Jacobs, and C. Schary. 2005. Trust and communication: Mechanisms for increasing producers' participation in water quality trading. *Land Economics* 81 (2): 170–190.

Brown, L.K., E. Troutt, C. Edwards, B. Gray, and W. Hu. 2011. A uniform price auction for conservation easements in the Canadian prairies. *Environmental and Resource Economics* 50 (1): 49–60.

Cason, T., and L. Gangadharan. 2004. Auction design for voluntary conservation programs. *American Journal of Agricultural Economics* 86: 1211–1217.

Connor, J.D., J.R. Ward, and B. Bryan. 2008. Exploring the cost effectiveness of land conservation auctions and payment policies. *The Australian Journal of Agricultural and Resource Economics* 51: 303–319.

Ekholm, P., E. Turtola, J. Grönroos, P. Seuri, and K. Ylivainio. 2005. Phosphorus loss from different farming systems estimated from soil surface phosphorus balance. *Agriculture, Ecosystems and Environment* 110: 266–278.

Ekholm, P., P. Valkama, E. Jaakkola, M. Kiirikki, K. Lahti, and L. Pietola. 2012. Gypsum amendment of soils reduces phosphorus losses in an agricultural catchment. *Agricultural and Food Science* 21: 279–291.

Groth, M. 2008. *An empirical examination of repeated auctions for biodiversity conservation contracts*, Working Paper Series in Economics, No. 78. Lüneburg: Leuphana Universität Lüneburg, Institut für Volkswirtschaftslehre.

Hajkowicz, S., A. Higgins, K. Williams, D.P. Faith, and M. Burton. 2007. Optimisation and the selection of conservation contracts. *The Australian Journal of Agricultural and Resource Economics* 51: 39–56.

Hamilton, J. 2010. *Conserving data in the conservation reserve*. Washington, DC: RFF Press.

Hansen, L. 2007. Conservation reserve program: Environmental benefits update. *Agricultural and Resource Economics Review* 36: 267–280.

Hellerstein, D. 2017. The US Conservation Reserve Program: The evolution of an enrollment mechanism. *Land Use Policy* 63: 601–610.

Hellerstein, D., and N. Higgins. 2010. The effective use of limited information: Do bid maximums reduce procurement cost in asymmetric auction. *Agricultural and Resource Economics Review* 39: 288–304.

Hellerstein, D., N. Higgins, and M. Roberts. 2015. Options for improving conservation programs: Insights from auction theory and economic experiments (No. ERR-181). *US Department of Agriculture, Economic Research Service*, January.

Iho, A., J. Lankoski, M. Ollikainen, M. Puustinen, and J. Lehtimäki. 2014. Agri-environmental auctions for phosphorus load reduction: Experiences from a Finnish pilot. *Australian Journal of Agricultural and Resource Economics* 58: 205–222.

Juutinen, A., and M. Ollikainen. 2010. Trading in nature values as a means of biodiversity conservation in boreal forests: Theory and experience from Finland. *Forest Science* 56: 201–211.

Khanna, M., and A. Ando. 2009. Science, economics and the design of agricultural conservation programmes in the US. *Journal of Environmental Planning and Management* 52 (5): 575–592.

Kirwan, B., R.N. Lubowski, and M.J. Roberts. 2005. How cost-effective are land retirement auctions? Estimating the difference between payments and willingness to accept in the Conservation Reserve Program. *American Journal of Agricultural Economics* 87: 1239–1247.

Klemperer, P. 2002. What really matters in auction design. *Journal of Economic Perspectives* 16: 169–189.

Latacz-Lohmann, U., and S. Schilizzi. 2005. Auctions for conservation contracts: A review of the theoretical and empirical literature. *Report to the Scottish Executive Environmental and Rural Affairs Department.*

Latacz-Lohmann, U., and C. van der Hamsvoort. 1997. Auctioning conservation contracts: A theoretical analysis and an application. *American Journal of Agricultural Economics* 79: 407–418.

———. 1998. Auctions as a means of creating a market for public goods from agriculture. *Journal of Agricultural Economics* 49: 334–345.

McAfee, R.P., and J. McMillan. 1987. Auctions and bidding. *Journal of Economic Literature* 25: 699–738.

Myerson, R.B. 1981. Optimal auction design. *Mathematics and Operations Research* 6: 58–73.

Newburn, D.A., and R.T. Woodward. 2012. An ex post evaluation of Ohio's great Miami water quality trading program. *Journal of the American Water Resources Association* 48: 156–169.

Ollikainen, M., B. Hasler, K. Elofsson, A. Iho, H.-A. Andersen, M. Czajkovski, and K. Peterson. 2019. Toward a Baltic Sea socioeconomic action plan. *Ambio* 48: 1377–1388.

Ribaudo, M.O., L.H. Dana, M.E. Smith, and H. Ralph. 2001. Environmental indices and the politics of the conservation reserve program. *Ecological Indicators* 1: 11–20.

Rolfe, J., S. Whitten, and J. Windle. 2017. The Australian experience in using tenders for conservation. *Land Use Policy* 63: 611–620.

Schilizzi, S., and U. Latacz-Lohmann. 2007. Assessing the performance of conservation auctions: An experimental study. *Land Economics* 83: 497–515.

Shoemaker, R. 1989. Agricultural land values and rents under conservation reserve program. *Land economics* 65 (2): 131–135.

Shortle, J. 2013. Economics and environmental markets: Lessons from water-quality trading economics and environmental markets: Lessons from water-quality trading. *Agricultural and Resource Economics Review* 42: 57–74.

Stoneham, G., V. Chaudri, A. Ha, and L. Strappazzon. 2003. Auctions for conservation contracts: An empirical examination of Victoria's BushTender trial. *The Australian Journal of Agricultural and Resource Economics* 47: 477–500.

USDA Farm Serving Agency. 2015.

Woodward, R.T., and D.A. Newburn. 2012. An ex post evaluation of Ohio's Great Miami water quality trading program. *Journal of the American Water Resources Association* 48 (1): 156–169.

Woodward, R.T., D.A. Newburn, and M. Mezzatesta. 2016. Additionality and reverse crowding out for pollution offsets in water quality trading. *Ecological Economics* 128: 224–231.

CHAPTER 8

Credit Stacking

8.1 Introduction

Environmental credit stacking refers to selling credits for two or more distinct environmental benefits that are produced jointly at a location by pollution control or other resource conservation practices (Gillenwater 2012a, b, c; Marshall and Selman 2011; Robertson et al. 2014). Nitrogen credits, carbon credits, and biodiversity credits are often cited as possible credit types for stacking. To illustrate the concept, consider a farmer who establishes a riparian buffer to sell nitrogen credits. Given that the buffer is planted in grasses or trees, it may also sequester carbon and provide wildlife habitat. If markets exist for nitrogen, carbon, and habitat credits, then stacking enables the farmer to sell carbon and habitat credits, in addition to nitrogen credits. In this case, the farmer would receive a larger compensation for establishing the buffer than would be the case with only the nitrogen credit if credit prices are not affected by the stacking option.

The concept of stacking is relatively new and has emerged with the interest and development of markets for selling various types of environmental credits. Nutrient credits were discussed at length in Chap. 6. Other types include biodiversity conservation or enhancement, carbon sequestration, stream restoration, and wetlands credits (Robertson et al. 2014). Stacking is appealing for agri-environmental policy within voluntary compliance

J. Shortle et al., *Water Quality and Agriculture*, Palgrave Studies in Agricultural Economics and Food Policy,
https://doi.org/10.1007/978-3-030-47087-6_8

paradigm as it offers a means to increase financial resources available for purchasing environmental benefits produced on private lands (Cooley and Olander 2011). This potential leads to significant interest in stacking in nitrogen credit markets that include agricultural sources. Increasing the resources available to fund pollution control practices in agriculture beyond what is available from government subsidies is a key driver of interest and initiatives for water quality trading programs. Proponents see credit stacking as means to further enhance revenue generation.

Stacking has far more potential for improving agri-environmental policy than the promise of greater funding for environmental investments. It offers a mechanism for integrating the pursuit of multiple complementary environmental goals in a way that improves the overall economic efficiency of agri-environmental policies. Recent research concludes that multi-pollutant management implemented via integrated market mechanisms is a key facet of next-generation pollution control (Reeling et al. 2020). From this perspective, stacking can provide a step toward the desired direction. In Chap. 5 we emphasized that we need to get prices right for agriculture. Given multiple pollutants, and positive as well as negative externalities, multiple prices are required. Stacking offers a policy architecture for generating multiple prices for complementary environmental benefits.

To a greater degree than other innovative mechanisms introduced in this book, stacking to increase resources for water pollution control in agriculture is more concept than practice. While examples of stacking can be found in other domains, it has yet to find much use in public policy for agri-environmental management. In this chapter we examine the economics of stacking and the conditions under which the method can increase resources for voluntary adoption of pollution control practices in agriculture. While the concept is simple, linking multiple market makes the outcomes economically complex. A key lesson is that complex inter-market responses can either enhance or diminish the potential water quality benefits of stacking.

8.2 The Economics of Stacking

In this section we present a model of "full stacking" in which market rules allow farmers to generate multiple credit types from the adoption of water pollution control practices and sell credits in the corresponding markets. For simplicity we consider two credit types, nitrogen and carbon, and that credits are earned from reduced nitrogen fertilization on cropland and conversion of some proportion of cropland to an alternative use that eliminates nitrogen fertilizer applications, filters nitrogen from runoff, and sequesters carbon. Following Chap. 3, we use the term green infrastructure to refer to the alternative land use.[1] Examples are riparian buffers, grassed waterways, constructed wetlands, and retiring land for long-term fallow. Farms located near waterways can reduce both nitrogen and carbon and participate in both nitrogen and carbon markets. Farms further from waterways may have relatively little or no impact on water quality and therefore be ineligible for producing and selling nitrogen credits but are still eligible for participating in carbon markets. Active markets are assumed for both credit types.[2] Furthermore, we assume that both markets are regional so that prices are determined by regional credits supplies and demands.

For simplicity, farms in the region produce a single crop with nitrogen as the variable input. Land quality is homogenous within farms but varies across farms. As in Chap. 5, nitrogen input is denoted on a farm n, and the crop yield, denoted y, is given by the crop response function $y = f(n)$. The crop price per unit is p and the price per unit of fertilizer is c. In the absence of regulation, farmers will choose the nitrogen rate so that the marginal revenue from the fertilizer ($p * MP$) is equal to fertilizer price where MP is as previously the marginal product. The privately optimal application rate, denoted $n^0 = n^0(p, c)$, depends on the crop and fertilizer price, which are the same for all, while land quality varies from farm to farm. Without regulation or incentives, farmland is not allocated to green infrastructure to produce environmental credits, as these credits would provide no economic benefit for the farmer.

8.2.1 Baselines for Credit Calculations

Suppose now that baseline-and-credit markets for nitrogen and carbon are established and that the managing agencies' rules allow credit stacking. Credits can be generated by reducing fertilizer applications and/or establishing green infrastructure on farmland. Reducing nitrogen fertilization reduces nitrogen loads to surface and groundwaters. In addition, reduced nitrogen fertilization lowers emissions of nitrous oxide, a greenhouse gas,

[1] The model is based on Lankoski et al. (2015) and Lankoski and Ollikainen (2020).

[2] Livestock is also a source of greenhouse gas emissions, methane, and nitrous dioxide. Currently, the means of reducing livestock emissions are very limited beyond just reducing animal numbers. For simplicity we focus merely on arable fields and crop production.

for which carbon credits may be earned. Green infrastructure reduces nitrogen loads and sequesters carbon. Credits are calculated according to rules established by the managing agencies linking management actions with reductions in nitrogen and greenhouse gas emissions and carbon sequestration. For simplicity, we ignore all uncertainty about the relationship between farm practices and emissions. The reductions are calculated relative to baseline levels defined by the managing agencies' rules. Farmer participation in the markets is profitable if the revenues from selling credits net of costs from management actions results in higher profits than nonparticipation.

A critical design choice for baseline-and-credit markets introduced in Chap. 6 is the determination of the baselines for calculating credits as it affects participation decisions and credit supplies. The baseline concept suggests that the regulatory rules defining the baseline call for an estimate of nitrogen and carbon emissions from farms based on their practices,[3] prior to the establishment of the markets. Credits generated relative to this baseline would measure the actual changes in nitrogen and carbon relative to the pre-market situation. This baseline would vary from farm to farm. As discussed in Chap. 6, baseline choices in water quality trading programs indicate alternative approaches to baseline definition, reflecting, for example, minimal standards of good practice, regulatory principles that producers are expected to meet before being eligible to profit from selling credits, or concerns for fairness in the treatment of farmers. Regulatory practice suggests that baselines based on minimal standards of good practice or regulatory principles would be applied uniformly across market participants.

We denote the nitrogen and carbon credits by O^N and O^C, respectively. Both are defined as the difference between the baseline emissions and the emissions resulting from a farmer's credit supply decisions based on the agency's rules.

Emission reductions from fields will depend on fertilization rates (n) and the land area in green infrastructure (b). We can therefore use the fertilizer-buffer choice model presented in Chap. 5 to examine farm decisions for supply credits in nitrogen and carbon credits. We write the

[3] Given year to year change in production, a baseline based on actual practices would likely be an average over some period. This is analogous to the concept of grandparenting in the distribution of initial allowance allocations in cap-and-trade.

computed credits given the fertilization rate and field-edge area as $O^N(n, b)$ for nitrogen and $O^C(n, b)$ for carbon.

8.2.2 Credit Supply Decisions with Stacking

With the ability to sell credits, farm revenue is crop sales plus credit sales if the farmer participates in the market. Farm costs are fertilizer expenditures plus, if the farm participates in the credit markets and to generate credits by establishes green infrastructure, its costs. The cost of the green infrastructure is the expenditures on establishment plus the opportunity cost of land removed from production. Let P_N denote the market price of nitrogen credits and P_C the market price of carbon credits. The net revenue to a participating farm is:

$$\Pi = (1-b)\pi^0(n) - kb + P_N O^N(n,b) + P_c O^C(n,b). \tag{8.1}$$

This equation is a modification of the expression (Eq. 5.1) for net returns presented in Chap. 5. The difference is the addition of revenues from nitrogen and carbon credit sales, $P_N O^N(n, b) + P_c O^C(n, b)$. The term $(1 - b)\pi^0(n)$ is the net revenue from crop production. The cost of establishing the green infrastructure is kb where k is the per hectare expenditure to establish the green infrastructure. Note that to keep the formula simple we have omitted transaction costs relating to finding buyers and making contracts. We return to this issue later in the chapter.

Following the principles of marginal analysis, the net revenue-maximizing fertilizer and green infrastructure choices are:

$$P_N O_n^N + P_c O_n^C = pMP - c(\text{fertilizer}) \tag{8.2a}$$

$$P_N O_b^N + P_c O_b^C = \pi^0 + k(\text{green infrastructure}) \tag{8.2b}$$

Equation (8.2a) is the marginal condition for fertilizer. Equation (8.2b) is the marginal condition for the area of green infrastructure. Consider first farms located close to waterways. The left-hand side of Eq. (8.2a) is the marginal revenue from nitrogen and carbon credits sales obtained from a marginal reduction in fertilizer use. The right-hand side is the reduction in net revenue from crop sales at the margin due to the reduction in fertilizer use. This reflects the loss of revenues from the reduced yields but also a reduction in fertilizer expenditure. The fertilizer condition requires equalizing the farmer's marginal benefits and marginal costs from fertilizer reductions.

Equation (8.2a) can also be rearranged as:

$$pMP = c + \left(P_N O_n^N + P_c O_n^C\right)$$

In this form, the marginal costs are on the left side and the marginal benefits on the right side. Essentially, the option to sell credits creates an opportunity cost for fertilizer use in the same manner as an increase in the price of fertilizer. The fertilization choice is illustrated in Fig. 8.1.

The left-hand side of the green infrastructure area marginal condition (Eq. 8.2b) is the marginal revenue from nitrogen and carbon credits sales generated by a marginal increase in the area of green infrastructure. The right-hand side is the marginal opportunity cost of taking land out of crop production plus the marginal establishment cost. The condition equates the producer's marginal benefits from selling credits obtained by increasing the size of the green infrastructure with marginal costs of doing so. Recall from Chap. 5 that green infrastructure exhibits diminishing returns to nitrogen load reduction. This means that the marginal return from the buffer, $P_N O_b^N$, becomes smaller as the buffer area is increased. Analogous to the fertilization decision, diminishing marginal returns to nitrogen load reductions will limit the extent to which buffer establishment is economically beneficial to the producer so that only a part of the field is allocated to it. The second term, $P_c O_b^C$, indicates the ability of green infrastructure to sequester carbon is increasing in its size.

Farmers' choice architecture and consequently their decisions for fields that are too distant from waterways to be eligible to participate in the nitrogen credit market is different. In such cases, farmers still have an incentive to reduce nitrogen fertilization to produce carbon credits according to Eq. (8.2a) which in this case reduces to $P_c O_n^C = pMP - c$. Correspondingly, the green infrastructure Eq. (8.2b) reduces to $P_c O_b^C = \pi^0 + k$. If all parts of the field sequester carbon similarly, and provided that the price of credits exceeds the costs (lost revenue plus establishment cost), the whole field is allocated to green infrastructure. This choice is illustrated in Fig. 8.2 for credit prices with and without stacking. Higher credit prices increase the returns from green infrastructure and, correspondingly, its area.

8.2.3 *Credit Supply Without Stacking*

If stacking is not permitted, farmers can participate in the nitrogen market or the carbon market but not both. A profit-seeking farmer will choose the

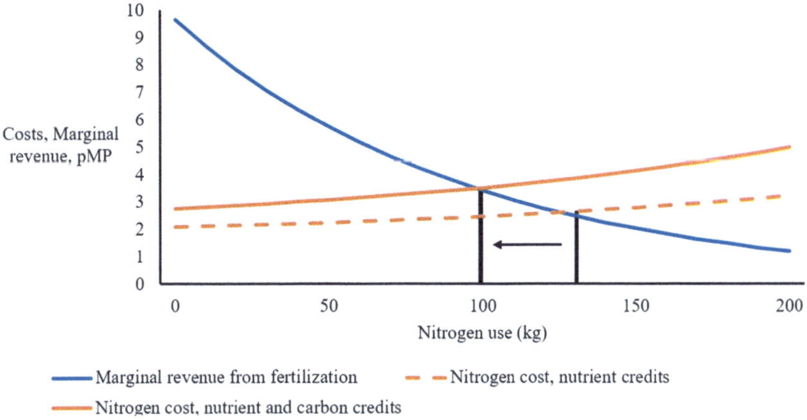

Fig. 8.1 Optimal choice of fertilizer application under credit stacking. (Source: Authors' creation)

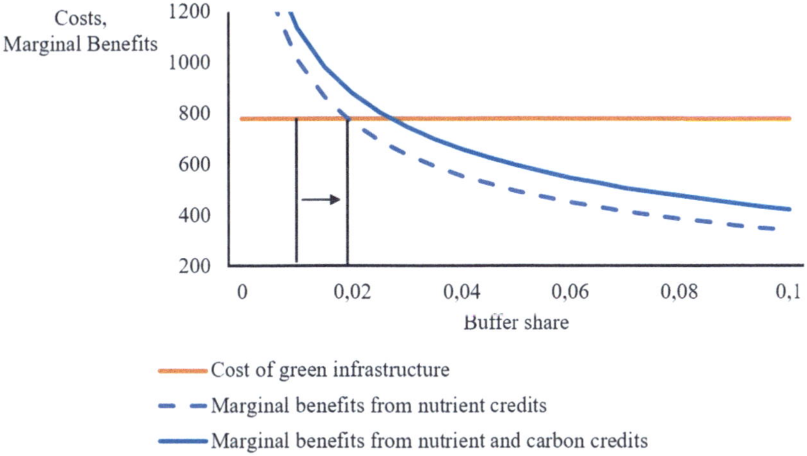

Fig. 8.2 Optimal choice of green infrastructure under credit stacking. (Source: Authors' creation)

option with the greatest net returns. Removing the option to stack credits reduces the potential returns to farmers from reducing fertilizer use or establishing green infrastructure if prices in the credit markets are the same with as without stacking. This is true no matter which credit market, carbon or nitrogen, provides a farmer with greater net returns.

Stacking increases the incentives for water pollution control and for greenhouse gas emission reductions if the credit prices are unaffected by stacking. However, as stacking affects the optimality conditions of individual farmers it will also affect the total supply of credits and therefore the market prices.

8.3 Stacking with Endogenous Prices

To examine market interactions, we need to move from the farm level to the nitrogen and carbon credit markets. The demand side of the market consist of firms or other institutions (e.g., municipal wastewater treatment plants) that are regulated for water quality protection or climate mitigation purposes. As described in Chap. 6, in the presence of high-abatement costs, regulated firms may demand credits as a means for regulatory compliance when credits offer a cheaper option than reducing their own emissions. Some businesses and institutions may also seek to purchase credits as part of their social responsibility programs. Suppliers are agricultural landowners. We assume initially that demanders can offset their emissions using credits at a ratio of one to one.

8.3.1 Credit Supply Functions

The market supply of credits describes the credits that will be offered in the market by farmers at various credit prices. The supply will depend on how many farmers choose to participate in the market and on the credit supply decisions of those who participate.

A key determinant of credit supply is land productivity as it affects the opportunity cost of decreasing fertilizer applications and allocating land from crop production to green infrastructure. The opportunity costs are higher for high-productivity soils than for low-productivity soils. In consequence, farmers with low-productivity soils are more likely to participate at any set of prices than farmers with high-productivity soils. Higher prices will provide greater incentives for high-productivity soils to supply credits.

Farmers' choices result in supply functions of nitrogen and carbon credits that depend on both credit prices. We express the aggregate carbon and nitrogen credit supply curves as: $S^N(P_N, P_C)$ and $S^C(P_C, P_N)$. For any set of nitrogen and carbon credit prices, these functions describe the credits supplied.

Figure 8.3 illustrates the nitrogen credit supply curve. The supply curve for carbon credits would be similar.

In the absence of stacking, the supply of each credit type increases with the credit price due to increased supplies from those who are participating and an increased number of participants. For those already participating, higher credit prices increase the opportunity cost of fertilizer use, leading to reduced application rates, and increase the returns to allocating land to green infrastructure. At the extensive margin, higher prices make it profitable for farmers with more productive lands to participate in credit markets. The "own-price effect" on credit supply is therefore positive.

When stacking is allowed, the supply curve for credits for each credit type is to the right of the supply curve when stacking is not allowed. The shift occurs because the quantity supplied at any price in each market will be greater. Farmers receive compensation for supplying credits in the other market. This reasoning also implies that the supply of each credit type would increase with an increase in the price of the other credit type. With regard to the supply curves, this means the supply of nitrogen (carbon) credits shifts to the right with an increase in the price of carbon (nitrogen) credits. We therefore say that both the *own-price* and *cross-price effects* are positive: higher prices of either credit type increase credit supplies. Nitrogen and carbon credits are therefore referred to as being complementary goods in supply.

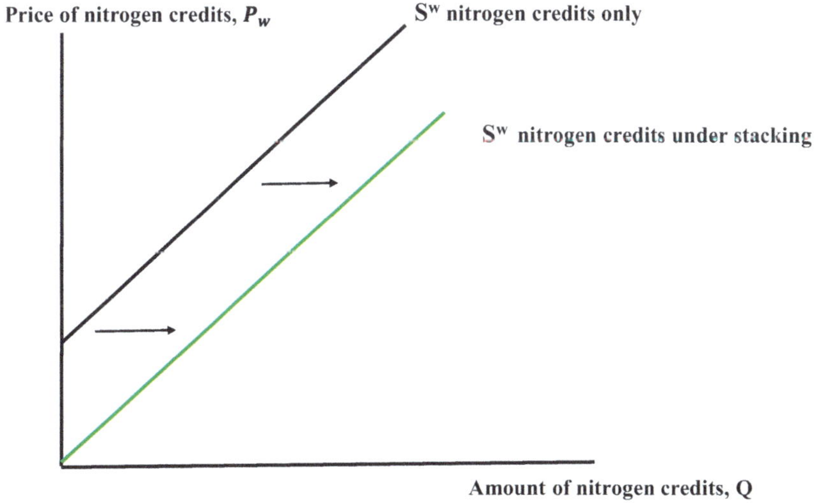

Fig. 8.3 Nitrogen credit supply: role of stacking. (Source: Authors' creation)

8.3.2 *Supply, Demand, and Credit Market Equilibrium with Stacking*

With the supply functions in hand, the next step is credit price determination in the two markets. As illustrated in Chap. 2 a market equilibrium is a price-quantity combination in which supply and demand are equal. In this case, we must consider equilibria in two connected markets. For one to be in equilibrium, the other must also be.

The aggregate demand for credits depends on the prices of credits and the stringency of the regulatory limits on carbon and nitrogen emissions. Lower prices will increase the quantity of credits demanded as firms can meet regulatory requirements at lower cost by purchasing credits. Increased regulatory limits will increase the demand for credits as polluters are pushed to increase abatement and seek lower cost options. The aggregate demands for credits are written as:

$$D^N = D^N\left(P_N\right) \text{ and } D^c = D^c\left(P_C\right). \tag{8.3}$$

Carbon and nitrogen credit prices appear in the carbon and nitrogen credit supply functions because the reductions are jointly produced. In general, we do not expect carbon and nitrogen credits to be jointly demanded. Therefore, the demand for nitrogen credits is assumed to be independent of the price of carbon credits and vice versa. Credits are therefore complementary in supply but not in demand.

In Fig. 8.4, the intersection of the nitrogen credit demand and supply curves determines the market equilibrium. A similar figure would describe equilibrium in the carbon credit market. The vertical axis describes the price of credits and the horizontal axis describes the quantity of credits.

The market equilibrium $\left\{P_N^*, Q^*\right\}$ gives the actual exchange of nitrogen emissions between credit buyers and sellers.

It is important to recognize that stacking nitrogen credits with a complementary good makes the level of nitrogen emissions in the market equilibrium depend on conditions of the market for the complementary good. A question then is how changes in the complementary good credit market affect the equilibrium in the nitrogen credit market? Consider a carbon price increase due to an increase in demand for carbon credits. The impacts on the market equilibria are shown in Fig. 8.5.

Given that the credits are complementary goods, the increase in the price of carbon credits shifts the supply function of nitrogen credits to the

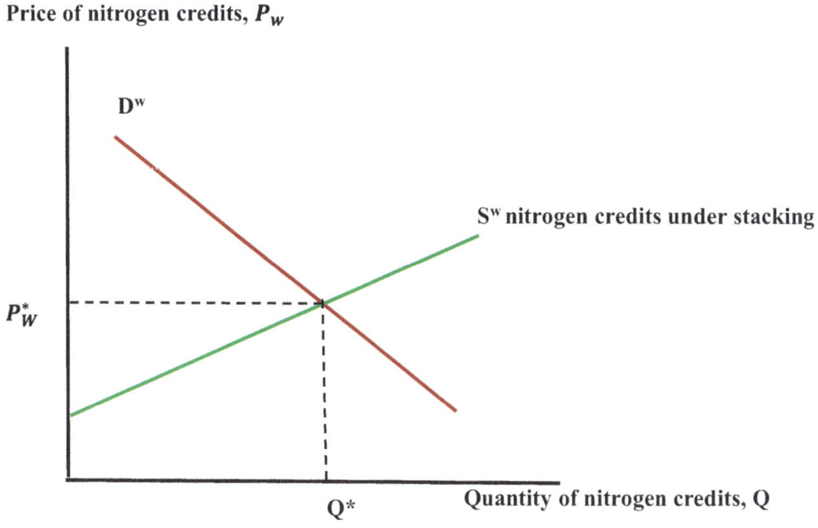

Fig. 8.4 Credit stacking: nitrogen credit market equilibrium. (Source: Authors' creation)

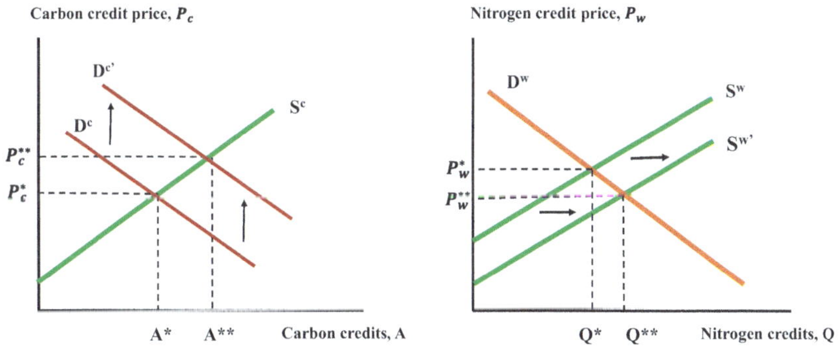

Fig. 8.5 Complementarity of carbon and nitrogen credit markets under stacking. (Source: Authors' creation)

right. This reflects reduced fertilization rates and more land to green infrastructure that result in reduced nitrogen loads and increased quantity of credits. How much the supply function on nitrogen credits shifts is an important question for the functioning of nitrogen credit markets. Figure 8.5 is drawn as a case where the change in the nitrogen credit price is small. At the other extreme, the increase in the nitrogen credit supply could be large, leading to a large drop in the nitrogen credit price and threatening the functioning of the whole market.

Thus, stacking raises a policy question of how sensitive the price of nitrogen credits is to the changes in carbon market price. Much depends on the price elasticities of demand and supply functions and the size of both markets. The elasticity of supply depends on the nature of credits produced.

A final comment concerning the role of transaction costs. Reeling et al. (2020) examined how the presence of multi-pollutant policy, stacking in our case, could promote farmers' market participation. They consider the transaction costs of buyers seeking to purchase credits, consisting of search costs and contracting costs. Search costs are incurred in seeking trading partners to meet the buyers' demand. Contracting costs are those of negotiating an agreement to deliver credits. Transaction costs reduce the demand for credits and increase the costs of providing credits. In the context of the model presented here, transaction costs shift the supply function upwards. Provided that the presence of stacking reduces the average per-market contracting cost, Reeling et al. (2020) show that the supply function of credits shifts outwards and reduces credit prices relative to the case of no-stacking when transaction costs are present under both stacking and no-stacking.

8.3.3 Supply, Demand, and Prices Without Stacking

Alternatively, suppose stacking is not allowed. Farmers may participate in one but not both credit markets. When this is the case, farmers will choose to participate in the market that provides the highest profit. That decision will depend on the relative prices of nitrogen and carbon credits and the possibilities and costs of farmers to reduce nitrogen loads and greenhouse gas emissions.

The supply functions also change. Most importantly, the cross-price effects switch signs, meaning that an increase in the carbon credit price decreases the supply of nitrogen credits. This is because it becomes more

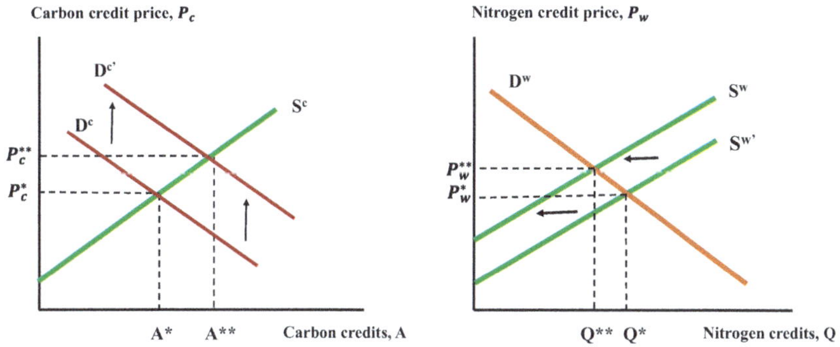

Fig. 8.6 Substitution between carbon and nitrogen credits: no-stacking. (Source: Authors' creation)

profitable for farmers to participate in the carbon credit markets instead of nitrogen credit markets. We illustrate the impact in Fig. 8.6. We consider again the effect of an increase in carbon credit prices due to an increase in the demand for carbon. The increased demand shifts the demand curve to the right. Therefore, carbon price increases as does the quantity of carbon credits sold. This increase in supply takes place via increased participation and increased abatement and sequestration measures by the participating farmers. What happens to nitrogen credits? Participation in this market decreases and the supply function of nitrogen credits shifts inwards. The use of nitrogen credits decreases in the equilibrium.

When stacking is not allowed, the interplay between markets changes, as a higher price in the carbon credit markets increases the price of nitrogen credits via reduced nitrogen credit supply. Thus, the risk of deterioration of nitrogen credit markets is absent – albeit at the cost of reduced incentives to produce nitrogen credits. The risk to the nitrogen market is instead a reduced demand for nitrogen credits due to higher carbon credit prices.

8.4 Market Design, Performance, and Environmental Integrity

As discussed in Chap. 6, environmental integrity is an issue often raised in the context of offsetting, credit trading, and stacking. One set of concerns has to do with the environmental validity of credits in general. The key issue is whether the environmental benefits provided by the credits are a valid substitute for those that are lost by the offset actions. In a typical water quality trading context, regulations require a wastewater treatment plant to increase its nutrient abatement. With trading, the plant has the option of meeting the regulation by installing new equipment to reduce its own emissions or purchasing credits. For the trade to be environmentally sound, the credits bought must be truly additional and provide an environmentally equivalent outcome to what would result from the emissions reduction from the plant.

The issues of additionality, slippage, baseline selection, and equivalence discussed in connection with nutrient credit trading in Chap. 6 apply equally to stacking (see especially Box 6.3). Additionality is not an issue in the model we have presented, as the changes in farming practices are costly and would not normally be undertaken without the incentives. The simplicity of the model presented here (e.g., single-crop, no-rotation, no-till practices) also rules out slippage and other induced economic effects that may be present in real-world settings. However, it should be recognized that these will be concerns in real-world settings when trading models are of the hybrid baseline-and-credit type presented in Chap. 6. Indeed, these issues become more complex due to the trading of multiple environmental goods. The only methods for addressing these issues is whole-farm environmental accounting and limiting practices eligible for credit generation to those with little if any net private benefit.

In this section we take up two issues connected to environmental integrity in detail. One is double dipping, which pertains to landowners receiving multiple payments for the same activity (Robertson et al. 2014; Woodward 2011). Double dipping becomes an issue in nitrogen trading when farmers can sell credits generated from implementing Best Management Practices (BMPs) that were subsidized by other programs. Some trading programs prohibit generating credits from practices that were subsidized, some do not. Double dipping has taken a large role in discussions of the environmental integrity of credit stacking and therefore deserve special attention in the chapter. A second is baseline selection for credit generation. This topic was discussed at some length in Chap. 6 but in this chapter, we take a deeper look at how baseline selection can affect the economic and environmental performance of markets.

8.4.1 Double Dipping vs. Credit Stacking

Some critics of stacking conclude that it does not lead to additional environmental protection but simply increases payments to landowners. For example, suppose a landowner receives a subsidy to install a buffer to reduce nitrogen loads to a stream and then sells a nitrogen reduction credit from this buffer. The landowner is double dipping. While seemingly an unproductive use of funds for purchasing the environmental benefits, double dipping can be economically and environmentally useful if the additional payment can be used to influence farmers' pollution control decisions (see Box 8.1).

Credit stacking in contrast refers to receiving differentiated payments for jointly produced environmental benefits (Robertson et al. 2014). In the case of the green infrastructure, if credit markets exist for nitrogen, carbon, and habitat, separate prices apply to reach revenue stream. Economic rents may result but there are no double payments for the individual benefits.

Stacking is a policy issue because environmental credit markets are typically the result of regulations that create a demand for credits for regulatory compliance and define the terms under which they can be bought and sold (Liu and Swallow 2016; Robertson et al. 2014). Double dipping and reduced effectiveness of public spending are a natural concern in that context. However, as is evident from the analysis we have presented, offering multiple payment streams can be a productive use of public funds when the result is to increase the level of conservation activity.

Various private goods are routinely supplied from joint production processes. Examples are oil and meal from soybeans, meat and eggs from poultry, and hides and meat from cattle. Market prices for jointly produced components are routine, reflect the demands for the components, and define the overall incentives for providing the goods and the proportions in which they are provided (Liu and Swallow 2016). The same can be true in properly constructed stacked markets.

From our analysis it is apparent that credit trading and stacking lead to changes in farming practices that produce environmental benefits if the goods are complementary in supply and if the procedures used to compute credits are valid. A useful dimension of nitrogen and carbon trading and stacking in this regard is that they are quantified by weight (pounds or kilograms) (Gardener and Fox 2013). Water quality and carbon credits are well-defined private (exclusive and divisible) goods that, while produced jointly on a property, can be unbundled and sold separately. Selling the same good to different buyers is not possible.

Box 8.1 Using Double Dipping to Enhance Outcomes
We have described previously how major agri-environmental programs are limited in their effectiveness and efficiency for water quality protection by weak targeting and spatial differentiation. Taking these programs as given, the overall efficiency and effectiveness of water quality protection can be improved by augmenting these programs by targeted water quality programs such as the water quality trading programs described in Chap. 6 or the water quality auctions described in Chap. 7. A study of the potential for using water quality trading to improve the overall results of agri-environmental programs in the Chesapeake Bay watershed found that double dipping can be used to improve the economic and environmental performance of coordinated program designs (Horan et al. 2004). Essentially, in the coordinated design the water trading program is used to fine-tune farmers' incentives. Double dipping is essential for this outcome.

8.4.2 *Baseline Choice*

We now consider the economics of baseline choice using the model developed in this Chapter. Consider a farmer whose fertilization rate without market participation is n^0. This is the true baseline level. The baseline set by market rules is denoted by $\bar{n}$. The regulatory baseline may be equal to, greater than, or less than the true baseline. Under the true baseline, the amount of nitrogen credits produced is the difference of nitrogen loads (denoted by z) in the baseline and loads under re-optimized use of inputs: $O^N = z(n^0) - z(n^*, b^*)$. Similarly, carbon credits, defined in CO_2-equivalents over greenhouse gases are given by the difference of N_2O emissions from fertilizer application in the baseline and re-optimized use of nitrogen and carbon sequestering land area (denoted by $\mu(b)$): $O^C = \varepsilon n^0 - [\varepsilon n^* + \mu(b^*)]$, where ε is a coefficient transferring nitrogen applied to fields into nitrous oxide emissions. Suppose first that the regulatory baseline is lower than the true baseline ($\bar{n} < n^0$). In this case, the credits certified for sale by the market authority will be less than the carbon and nitrogen reductions actually produced. The certified credits are fractions of the actual pollution reductions, as $\bar{O}^N = z(\bar{n}) - z(n^*, b^*)$, so that $\bar{O}^N < O^N$. The effect is analogous to a tax on the pollution reductions. The costs of producing a certified credit exceed the cost of producing actual pollution reductions. The reverse would be true if the baseline as defined by the market rules is higher than the true baseline ($\bar{n} > n^0$). In this case the credits certified for sale would exceed the actual pollution reductions. A farmer would be selling "hot air" to carbon credit markets and "hot water" to the nitrogen credit markets. Farmers would be overcompensated and for any given trade ratio, determining how many credits are required to offset one unit of emissions, buyers of credits would be unable to offset their emissions. The credit market system would be environmentally questionable, as demanders may not actually offset all their emissions. This conclusion holds true for a 1:1 trade ratio where one credit offsets one unit of emissions. As emphasized in Chap. 6, for agriculture the trade ratio is often set to 2:1 or greater, indicating the market rules require overcompensation. Note, however, that such a high trade ratio reduces gains of trade and the actual reduction of nitrogen loads in agriculture due to lowered demand for nitrogen credits.

The true baseline is economically optimal in that it will maximize the economic gains from trading (Ghosh et al. 2011; Ribaudo et al. 2014). It is also environmentally desirable for accurate environmental accounting, as credits generated against this baseline reflect actual environmental performance. However, determination of the true baseline requires information on farmers' choices prior to establishment of the market. As demonstrated by baseline choice in water quality trading programs, environmental agencies may choose an alternative baseline rule for this or other reasons (Willamette Partnership 2015). Whatever the reason for the difference between the adopted and true baseline, the fact that they differ means that credited pollution reductions will differ from actual pollution reductions. This matters to the environmental and economic performance of market.

Figure 8.7 illustrates the implications of true and alternative baseline on nitrogen credit supply for the case where the alternative is "tight." The alternative baseline may be tight as a policy choice or as a result of imperfect information about the true baseline. In this case the credits certified for sale are less than the true pollution reductions and the farmers will not receive compensation for all actual reductions in pollution they have produced. The supply curve with the tight baseline lies above the supply curve with true baselines and is steeper, reflecting the increase in the marginal cost of producing certified credits that result from the tight baseline. The demand curve in the figure intersects both supply functions producing two alternative market equilibria. (The story for carbon supply curves would be analogous.)

In addition to supply and demand curves, the market equilibrium will depend on the trading rules, especially the trade ratio if there is one. Figure 8.7 shows demand for a 1:1 trade ratio. In this case, to offset one unit of pollution the demander must buy one unit of nitrogen credits.

Under a true regulatory baseline and 1:1 trading, the market equilibrium is $\{P_W^{**}, Q^{**}\}$ and demanders' credit purchases offset their emissions.[4] If the regulatory baseline is tighter, the credit supply is reduced and the resulting equilibrium is $\{P_W^{*}, Q^{*}\}$. The price is higher and the quantity of credits is smaller than with the true baseline. The actual reduction in nitrogen loads at the market equilibrium is greater than what the equilibrium Q^* would indicate but smaller than Q^{**}. This is because of the steeper slope of the supply function.

Figure 8.5 also illustrates the welfare implications of the tight baseline. First, there is an economic loss because the quantity of credits traded falls short of the quantity actually produced. The size of this economic loss is the triangle C which is the sum of the changes in consumers' plus producers' surpluses due to the tight vs. true baseline. Credit buyers lose because of the higher price. The loss is the reduction in consumers' surplus, which is given by the area A. Despite the lower number of credits sold relative to the market with true baseline, sellers gain. They sell Q* credits at a price of P_W^*, even though they would be willing to sell this amount at a much lower price (the lowest dotted line). The increase in the producers' surplus is the area A+B. Thus, while the tight baseline seems at a first glance to result in reduced revenues to farmers, the increase in price counters the effect of the baseline.

Both equilibria in Fig. 8.7 are those in which certified credits are equal to or less than the actual pollution reductions. In these cases the trading equilibria will produce environmental improvements. But if the estimated baseline is lax, that is, based on a high level of nitrogen application, the upper supply functions would be located below the supply function with the correct baseline. Thus, if the agency sets an exogenous baseline or seeks to use the true baselines, there is the Skylla of the lack of environmental integrity and Charybdis of reduced participation and higher social costs.

[4] Note that if the trade ratio is set to 2:1, the supply function of tradable units would shift upwards and the resulting equilibria would entail higher price and lower quantity and reduced gains of trade.

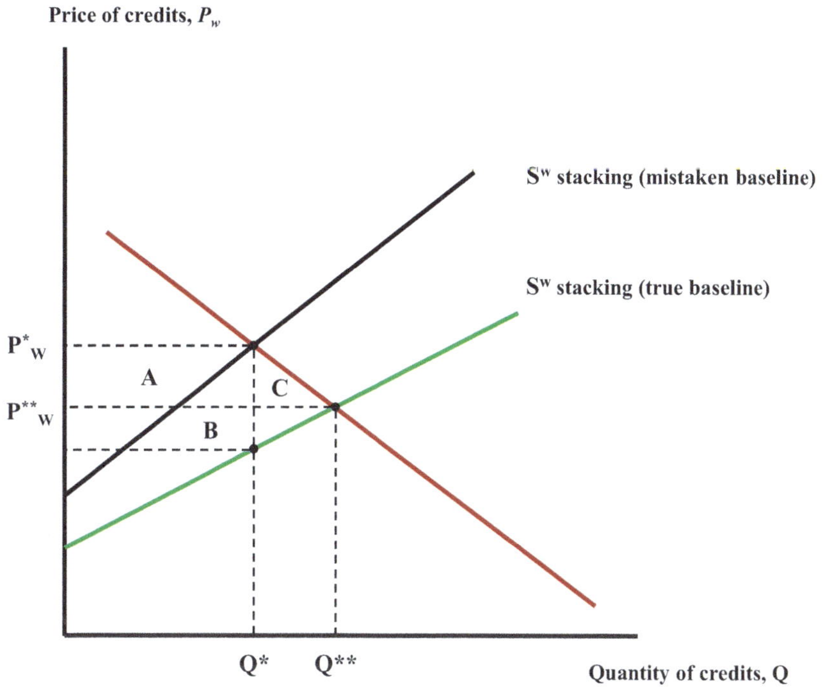

Fig. 8.7 Economic impacts of a mistaken baseline under stacking. (Source: Authors' creation)

If the trading ratio is 1:1, there are at least two ways of ensuring that the market equilibrium entails an environmental outcome with no more and possibly less pollution than would occur otherwise. First, the agency could use stringent baselines, but this approach would bring with it increased social pollution control costs and decreased market participation. An alternative is to pursue environmental integrity at the program level. An example is the use of credit reserves in water quality trading programs in which a fraction of each credit generated is taken by the environmental authority to provide a margin of safety (e.g., see Box 6.4). Since trade ratios influence the level of trading activity, environmental integrity can also be managed using the trade ratio. A ratio greater than 1:1 ratio can be used to offset the distortion that results from too lax a baseline (Horan and Shortle 2005).

In sum, this analysis indicates a trade-off between environmental integrity and economic efficiency in the choice of baseline. Tight baselines ensure that environmental conditions are made no worse by the market and may result in environmental improvements. But the environmental assurance comes at a cost of reduced economic efficiency. Throughout this book we have been very attentive with respect to environmental integrity, additionality, and slippage. In the same vein we should ask, how great is the loss of the long-term benefits from credit production caused by inefficient economic incentives? The strength of economic incentives is not only cost-efficiency but especially long-term technological and other emissions-reducing developments. This indicates that distorting markets to fully avoid environmental risk may over time lead to worse environmental outcomes than setting the prices and market right. This is an empirical question of high relevance.

8.5 Stacking in Practice

This discussion of stacking has been fairly abstract. This is because stacking is still rare and almost nonexistent for water quality (Fox et al. 2011; Gardner and Fox 2013; Kane 2008; Robertson et al. 2014). Most cases of stacking generally are found in the US and emerge largely from regulations related to land development requiring actions affecting wetlands or endangered species habitats to be offset by ecological compensations. Stacking in the US is regulated under federal and state laws. Variations are found across the states.

There are very few instances of water quality credit stacking. The instances are analogous to the offset trading described in Chap. 6 in that they are typically one-off events rather than routine activity in well-functioning markets (Morgan and Wolverton 2008). This is not surprising given the absence of robust markets for trading nitrogen credits or of markets that incorporate agriculture in carbon trading. A case cited as an early example was a sale of wetland credits from a wetlands bank in North Carolina in 2000 and a subsequent sale of nitrogen credits generated from the same wetlands in 2008. This case was subject to criticism for double dipping, which led to a moratorium on certifying nitrogen credit from land originally used to produce wetland credits in North Carolina (Gardner and Fox 2013; Kane 2008; Robertson et al. 2014). Stacking water quality credits with carbon credits or endangered species credits is receiving interest in California, Ohio, and Oregon but the outlook is largely prospective (Gardner and Fox 2013; Kane 2008; Robertson et al. 2014; Vegh and Murray 2020). Testing stacking is an objective of the Ohio River Basin Trading Project (Gardner and Fox 2013).

8.6 Summary

Environmental credit stacking is often said to offer another mechanism for increasing the resources available for purchasing pollution control from farms. Our discussion suggests that credit stacking has a much larger content and potential. Stacking provides framework for integrating markets to address the challenge of multi-pollutant management in agriculture. The mechanism has a significant appeal in theory as stacking goods like carbon that are complimentary in supply with nitrogen can be expected to increase pollution control efforts compared to water quality markets alone. This boost emerges from the fact that stacking promotes setting prices right for multiple types of emissions. This market integration makes stacking unique and promising.

Like water quality trading markets without stacking, the gains from trade in stacked markets depend significantly on design features like baseline requirements and trade ratios, in this case, in the interconnected markets. Recent work (Reeling et al. 2020) and our discussion on cap-and-trade models for agriculture in Chap. 6 suggests that stacking could be best rooted in full cap-and-trade policy rather than the current hybrid baseline-and-credit trading models as it would help avoid problems relating to additionality, slippage, and overall demand.

An important feature of stacked markets that is not present for unstacked water quality markets is the potential sensitivity of the market to events in the interconnected market. The presence of high transaction costs promotes thin markets with only a few sellers. Such markets are more volatile than markets with multiple agents. Stacking as such reduces transaction costs and increases market participation. The prerequisite for stacking to be significant for water quality protection are robust markets for water quality and the complementary good. These do not exist now. Without these, stacking is likely beneficial for specific cases but not consequential for addressing the agricultural water quality problem.

References

Cooley, D. and L. Olander. 2011. Stacking ecosystem services payments – risks and solutions. Nicholas Institute Working Paper 11-04, Duke University.

Fox, J., R.C. Gardner, and T. Maki. 2011. Stacking opportunities and risks in environmental credit markets. *Environmental Law Reporter* 2.

Gardner, R.C. and J. Fox 2013. The legal status of environmental credit stacking. *Ecology Law Quarterly* 40 (4): 713–757.

Gillenwater, M. 2012a. *What is additionality? Part 1: A long standing problem*. GHG Management Institute, Discussion Paper No. 001.

———. 2012b. *What is additionality? Part 2: A framework for more precise definitions and standardised approaches*. GHG Management Institute, Discussion Paper No. 002.

———. 2012c. *What is additionality? Part 3: Implications for stacking and unbundling*. GHG Management Institute, Discussion Paper No. 003.

Ghosh, G., M. Ribaudo, and J.S. Shortle. 2011. Baseline requirements can hinder trades in water quality trading programs: Evidence from the Conestoga watershed. *Journal of Environmental Management* 92 (8): 2076–2084.

Horan, R.D., J.S. Shortle, and D.G. Abler. 2004. The coordination and design of point-nonpoint trading programs and agri-environmental policies. *Agricultural and Resource Economics Review* 33 (1): 61–78.

Horan, R.D., and J.S. Shortle. 2005. When two wrongs make a right: Second-best point-nonpoint trading ratios. *American Journal of Agricultural Economics* 87 (2): 340–352.

Kane, D. 2008. EBX is paid twice for wetlands work. *Charlotte News and Observer*, December 8. Charlotte. www.newsobserver.com/2009/12/08/230607/ebx-is-paid-twice-for-wetlands.html

Lankoski, J., M. Ollikainen, E. Marshal, and M. Aillery. 2015. *Environmental co-benefits and stacking in environmental markets*, OECD food, agriculture and fisheries papers, No. 72. Paris: OECD Publishing.

Lankoski, J., and M. Ollikainen. 2020. *Stacking in environmental credit markets*. A submitted manuscript.

Liu, P., and S.K. Swallow. 2016. Integrating cobenefits produced with water quality BMPs into credits markets: Conceptualization and experimental illustration for EPRI's Ohio River Basin Trading. *Water Resources Research* 52: 3387–3407.

Marshall, E., and M. Selman. 2011. *Markets for ecosystem services: Principles, objectives, designs, and dilemmas*. Washington, DC: World Resources Institute.

Morgan, C., and A. Wolverton. 2008. Water quality trading in the United States: Trading programs and one-time offset agreements. *Water Policy* 10 (1): 73–93.

Reeling, C., Horan R., and Garnache, C. 2020. When the Levee Breaks: Can Multi-Pollutant Markets Break the Dam on Point–Nonpoint Market Participation? *American Journal of Agricultural Economics* 102 (2): 625–640.

Ribaudo, M., J. Savage, and J. Talberth. 2014. Encouraging reductions in nonpoint source pollution through point-nonpoint trading: The roles of baseline choice and practice subsidies. *Applied Economic Perspectives and Policy* 36 (3): 560–576.

Robertson, M., T.K. BenDor, R. Lave, A. Riggsbee, J.B. Ruhl, and M. Doyle. 2014. Stacking ecosystem services. *Frontiers in Ecology and the Environment* 12 (3): 186–193.

Vegh, T., and B.C. Murray. 2020. Incentivizing the reduction of pollution at U.S. dairies: Addressing additionality when multiple environmental credit payments are combined. *Journal of Agriculture, Food Systems, and Community Development* 9 (2): 123–139.

Willamette Partnership, World Resources Institute, and the National Network on Water Quality Trading. 2015. *Building a water quality trading program: Options and considerations.* http://willamettepartnership.org/wp-content/uploads/2015/06/BuildingaWQTProgram-NNWQT.pdf

Woodward, R.T. 2011. Double-dipping in environmental markets. *Journal of Environmental Economics and Management* 61 (2): 153–169.

CHAPTER 9

The Way Forward

9.1 The Challenge

We began this book with the observation that high-income countries of the world can show no systematic success in reducing agricultural non-point source pollution. This is not because the problem has gone unrecognized or unaddressed. To the contrary, agriculture has been known to be a major cause of water quality problems and a target of water quality programs for decades. The problem persists because the programs implemented generally fall short of what is required to meet water quality goals. Developing policies that are both effective and efficient is urgent. Pressures on land and waters affected by agricultural production are increasing as population and income growth drive increase agricultural production. At the same time, climate change will increase challenges to both agricultural production and to water quality protection from agricultural lands. These pressures and evolving and potentially transformative technological changes emphasize the need for policies to steer agriculture towards efficient, flexible, and adaptive production systems.

Throughout the book, we have identified possible reasons for the limited success. Some stem from the inherent challenges posed by the agricultural pollution control problem. The diffuse pathways and weather

J. Shortle et al., *Water Quality and Agriculture*, Palgrave Studies in Agricultural Economics and Food Policy,
https://doi.org/10.1007/978-3-030-47087-6_9

dependence of agricultural loads, the spatial scope of agriculture, the economic and technological complexity of pollution abatement in agriculture, and the heterogeneity of climate, land resources, and farming systems do not facilitate the kinds of approaches that water pollution authorities have used effectively for the control of point sources and know best. The uniform technology standards and/or effluent standards of conventional regulation are not suited to agricultural systems. Policy innovation is needed. The regulatory response has generally not been policy innovation but instead to draft established agricultural institutions and voluntary compliance instruments to the cause of water quality protection. The leading exception is the regulation of pesticides, which is prompted by an array of risks to human health and the environment. These regulations are effective in reducing external costs when enforced but research shows that the social net benefits would be increased by greater use of economic incentives to better align private and social benefits (Sexton et al. 2007; Waterfield and Zilberman 2012; Zilberman and Millock 1997; Zilberman et al. 1991).

The default approach of drafting established agricultural institutions and instruments for water pollution control would make great sense if they could be molded to do the task effectively and reasonably efficiently. The approach has been demonstrated to pay dividends in some cases with positive net benefits from water quality improvements and other environmental gains. But after decades of application, it is apparent that the approach has not been adequate, leaving agriculture as the leading cause of major water quality problems. Although there are variations between and within countries, common problems are that traditional policies are not designed to address specific water quality problems, do not prioritize the allocation of management effort and resources to places that are high priorities for water quality protection, focus on means rather than ends, and give producers too much discretion in their environmental performance. That these are the results should not be entirely surprising given the historical objectives, capacities, and constituencies of agricultural institutions (Harvey 2004).

9.2 Political Ambition

The question is how to move forward. Perhaps the elephant in the room is political ambition. Current policies have significant support from an array of interest groups. Voluntary compliance policies provide benefits to producers and landowners either by lowering the costs of adopting farming practices or by paying them to convert cropland to lower-intensity uses. Rural interests support governments investing in the agricultural economy and rural amenities. Environmental groups support governments allocating resources to provision of agri-environmental goods. Large agricultural bureaucracies exist to deliver agri-environmental programs.

The constellation of interests supporting conventional agri-environmental programs is a significant barrier to policy reform, especially to reforms that would reduce farm income via polluter-pays policies. But even within the pay-the-polluter paradigm, agricultural interests may resist reforms that improve effectiveness and efficiency. Spatial targeting, for example, would provide resources to some farmers but not others. Although obviously beneficial for making the most of public spending for water quality improvements, spatial targeting has not been a political priority as it would remove eligibility for subsidies from some and possibly many within the agricultural sector. Similarly, auctions do not find great political favor within the agricultural sector because they do not distribute payments broadly but instead favor some farmers over others based on performance criteria. Furthermore, fine-tuning the objectives of agri-environmental programs to prioritize specific water quality problems in specific places and allocating resources accordingly is not favored, as it would concentrate rather than expand the number of program beneficiaries and constituents.

Missing ambition shows up in multiple ways. Effective political ambition requires clearly stated goals and accountability. Yet, governments rarely set specific enforceable targets for agricultural load reductions. Most agri-environmental programs are intended to serve multiple environmental objectives, which can attract interest group support, but lack prioritization across objectives needed for effectiveness and efficiency in program design and accountability for outcomes. Additionally, programs are intended to provide income support for farms and promote rural development. Programs almost always focus on land use or farming practices rather than predicted water quality outcomes. Little use is made of water quality science to connect means to explicit ends. The result for water quality and other environmental ends is generally a muddle, lacking clear purpose, tight connection between means and ends, and clear accountability.

Famously, necessity is the mother of invention. Political ambition is essential for pollution policy innovation for agriculture. Places where there is significant political interest in policy innovation with the potential to transform policy are places where significant political pressures have emerged. A prime example is the Chesapeake Bay where the Chesapeake Bay TMDL set explicit regulatory load limits for agriculture and established explicit accountability measures for regulatory agencies.

Instilling political ambition is beyond the scope of this book. Our primary goal is to provide guidance for policy design where there is a prerequisite political support for change. However, important for effective political ambition is knowledge that better results can likely be achieved with fewer public resources, which may itself be an essential ingredient for bringing about change. Also famously, ideas have consequences. An example often cited in the economic research on environmental policy is the uptake of environmental markets. Research making a strong case for environmental markets as we describe in Chap. 6 began in the latter half of the 1960s. The concept and merits had been mainstreamed in economics by the 1980s when policy dilemmas created openings for the idea to move from academia to environmental management (Anthoff and Hahn 2010). The research and thinking presented in this book do not provide specific blueprints for efficient and effective policy. The book does introduce concepts and results that help identify and explain defects in current policy and essential elements of better policies.

Also potentially useful to fostering political ambition is an understanding that some policies that appeal for cost-effective pollution control can be structured to varying degrees to address issues of political acceptability and fairness. Environmental markets are again an example. Anthoff and Hahn (2010) argue that the political appeal of markets for air pollution in the US is due in part to the ability of policy makers to control winners and losers among key interest groups.

9.3 Essential Elements of "Good" Policy

Some principles of good policy we describe below are standard management principles. Others are derived from economic and other research on the problem.

9.3.1 *Explicit Water Quality Policy Goals*

A fundamental principle of good water quality management is explicit water quality goals. Ideally, goal-setting establishes for specific water resources (aquifers, stream reaches, lakes, segments of lakes, estuaries, etc.) an ambient water quality goal, required changes in the determinants of ambient water quality conditions (pollution loads, conditions of wetlands, stream banks) needed to achieve the goal, and a timetable. Goals may also set limits on sectoral loads, and if so, ideally following principles of cost-effectiveness. Stakeholder engagement can contribute to the scientific quality of plans, acceptance, and contribute to political ambition (OECD 2017; Royer et al. 2020). These are standard elements of water quality management templates (e.g., Shortle et al. 2020; OECD 2018). The US TMDL approach described in the previous chapters is, for example, consistent with this principle. If multiple goals are set, they should be accompanied by priorities and procedures for evaluating trade-offs (NRC 2012).

Underlying these choices is ideally scientific understanding of water quality conditions, the stressors that determine water quality conditions and their sources, the changes in stressors needed to achieve the ambient water quality goal, and, for the timetable to be realistic, an understanding of what is technically and economically feasible. Clearly, the size and complexity of the problem, the state of scientific understanding, data availability, the technical capacity of planning agencies, and resources for technical support matter greatly to the quality of the plan (e.g., NRC 2001, 2012). As described in Chap. 5, goals should consider uncertainty. They should also allow for adaptation in response to new information and learning (NRC 2001).

When setting goals, we strongly suggest that explicit consideration be given to the instruments that will be used for plan implementation in the planning stage. In the standard water quality management paradigm, policy makers first choose environmental goals and then select instruments to attain them. Economic trade-offs may or may not be considered in the goal setting. An assumption in this paradigm is that policy instruments that can achieve the goal are or will be available. Progress toward goals set under this assumption will be unrealistic if this assumption is false. For example, a plan based on the assumption of adequately funded pay-for-practice or pay-for-performance programs may be doomed if budgets turn out to be insufficient. The history of agricultural nonpoint pollution control illustrates the relevance of this consideration. Assessments of realistic policy options in the goal-setting process should facilitate realistic goals and identify needed policy development when existing institutions are limiting. A case in point is the Chesapeake Bay TMDL described in various chapters of this book. The TMDL development process used high-quality environmental science but did not adequately integrate social, economic, behavioral, and decision sciences to inform fundamental policy questions on how to bring about needed changes in the agricultural economy of the watershed (NRC 2012). Agricultural reductions have not met established milestones due to limitations in the policy architecture of the kind we have described (Stephenson et al. 2020).

9.3.2 *Accountability Frameworks*

A second principle of good water quality management is accountability. The absence of explicit goals and timetables for water pollution reductions for agriculture eliminates an essential element of accountability for government agencies administering programs. Explicit goals and timetable are a first step. Explicit assignment of responsibilities and reporting requirements to agencies is also needed, as is effective monitoring and enforcement of compliance with contracts in voluntary compliance programs and standards or mandatory incentives in polluter-pays programs (Chap. 5).

9.3.3 *Spatial Targeting of Source Areas*

A fundamental question for pollution control is who to regulate. Regulating a source that contributes little or nothing to a water quality problem imposes social costs without water quality benefits. There is ample research that demonstrates that agricultural lands vary in their propensity to generate pollution due to characteristics of soils, topography, and location relative to impaired waters. As we have described extensively in the book, spatial targeting of sources is an essential element of cost-effective control and budgetary efficiency. This principle is well recognized in water quality management science and planning and facilitated by evolving tools to assist in identifying areas for targeting. Identification of high-priority locations is essential to good policy design. The policy failure is the disconnect between targeting in planning and targeting in policy implementation. As described in the previous chapters, high-priority areas for nutrient control measures are recognized, for example, in planning to reduce eutrophication problems in the Baltic Sea, the Chesapeake Bay, and the Gulf of Mexico dead zone. What is missing are corresponding implementation policies.

When choosing who to regulate, farm structure should also be considered. The presence of soil or other conditions that might make a farm a likely pollution source does not imply that it is an actual source. What is done on the land matters. Given that monitoring and enforcement are expensive, additional criteria for selecting who to regulate based on size, types of farm operations, or other factors might be considered. For instruments to induce the right practices in the right places, policies must be informed by both water science and economics.

9.3.4 Policy Coherence

Agricultural pollution policy is not made in a vacuum but in the presence of multiple policies affecting decision making in agriculture. These policies are intended to serve various societal goals related to agriculture: farm income, rural economic activity, international trade, local land use and rural amenities, and environmental services. They are often implemented through interventions in agricultural input and output markets affecting input costs and output prices. They may either promote or hinder water policy targets.

Good governance requires coherence between policy goals and policy instruments. Perfect coherence is a suite of instruments that are effective and efficient in achieving their individual goals without diminishing the effectiveness and efficiency of others. As described in Chap. 2, a challenge for water pollution policy for agriculture is agricultural interventions that encourage expansion or maintain agricultural production in environmentally sensitive locations, pollution-intensive commodity production, and over production. Policy coherence is suboptimal.

As with traditional water quality protection policies, all agricultural policies have political interest groups that they serve and bureaucracies to administer them. This means that improving policy coherence is inherently a political challenge. There are, however, useful principles and practices. One is decoupling, in which farm income support is independent of producers' production choices. Policies that seek to increase farm incomes by increasing crop prices or reducing input costs, encourage production, intensification, and generation of agricultural externalities. Decoupling has long been advocated by economists as an approach to supporting farm incomes without distorting agricultural markets (Beard and Swinbank 2001; OECD 2010). The 1992 EU Common Agricultural Policy reform that eliminated farm price supports and shifted income support to area-based payments is a significant example of the approach. Decoupling is required to varying degrees of countries that are members of the World Trade Organization. Another useful principle is environmental cross-compliance, which requires producers to meet environmentally based conditions to be eligible for income support payments. Cross-compliance requirements in various forms are found in several OECD countries (OECD 2010). In the US, cross-compliance requirements apply to production on highly erodible cropland and conversion of wetlands into arable lands. In the EU, all recipients of single-farm payments are required to meet cross-compliance requirements.

9.4 Instrument Choice and Design

The principles described above apply regardless of the policy instrument or mix of instruments. We now turn to issues of instrument choice and design. In general, the task of instruments is to impact producers' behavior and induce them to choose crops, input intensities, and land use so that the water quality goals will be achieved.

In Chap. 5 we presented a taxonomy of instruments based on two domains of choice. One is the compliance base or bases of the instrument which we define as an action or outcome that is under the control of the polluter that is targeted by the policy instrument. Compliance bases were broadly distinguished as practice based or performance based. The second is the incentive or regulatory mechanism applied to the compliance base. This mechanism is the lever used to induce change in the selected compliance base or bases. Mechanisms were broadly distinguished as regulatory standards (i.e., command-and-control) or incentives. Incentives were further subdivided according to implicit property rights (pay-the-polluter vs. polluter-pays) and whether the incentive mechanism involved prices set by the regulator or by the market. Beyond whom to regulate (spatial targeting), other basic questions are what to regulate (the choice of compliance base) and how to regulate (the choice of mechanism).

9.4.1 *Single-Purpose vs. Multipurpose Instruments*

The taxonomy of pollution instruments presented in Chap. 5 assumes that the instruments have a single objective. Introduced previously, the Tinbergen principle of policy design requires that for each objective there must be at least one instrument. A defect in contemporary incentives for agricultural water pollution policies is that they are nested in a suite of other objectives that are not perfect complements. The multi-objective approach mutes the effect of the incentives for water quality protection and can have perverse outcomes. Where water quality protection is a high priority, policy goals and instruments should reflect the priority. This is not to suggest that co-benefits should be disregarded in policy design, but it remains that for optimal policy design for each purpose there should be at least an instrument.

9.4.2 *Standards Vs. Incentives*

One essential choice is between regulatory standards and economic incentives. Regulatory standards in the form of technology standards or emissions standards are the conventional tool for managing point sources of water pollution. Except for pesticides and large confined animal operations, there is very limited use of standards for water pollution control in agriculture. This is not to be lamented. Economic research concludes that the complexity and heterogeneity of agricultural nonpoint pollution make cost-effective regulation via standards an impossibility. Incentives are generally to be preferred, especially when attached to valid indicators of pollution control performance, both for cost-effective control and to promote innovation and technology adoption.

While we strongly favor incentives, limited use of context-specific regulatory standards may sometimes be useful in a mix with incentives. This is most obviously the case when a practice is either unacceptable or essential in a certain setting. When this is true, using incentives to induce utilization of the good or eliminate the bad is unnecessary. Some candidates for quantity controls for reducing nutrients and pathogens in water are prohibitions on discharging barnyard wastes directly into streams and limiting livestock access to streams for drinking water. Some candidates for quantity controls for reducing sediment and nutrient pollution include prohibitions on certain forms of tillage on steep slopes, conducting intensive farming adjacent to streams or lakes, and spreading manure on frozen ground.

9.4.3 Practices Vs. Performance

Economic research demonstrates that polluting emissions, in the case of uniformly mixed pollutants, or contributions to ambient loads, in the case of nonuniformly mixed pollutants, are the optimal compliance bases when emissions can be metered and are nonstochastic. The diffuse nature of agriculture eliminates this option as emissions cannot be routinely monitored by source.

With actual emissions eliminated from the set of feasible compliances bases, instruments must use something that is observable at reasonable cost and significantly connected to water quality outcomes. One set of choices for compliance bases are input or practice based. A second set is estimated emissions, environmental benefit indexes, or performance proxies. This set uses formulas or models to construct farm-specific indicators of pollution control performance that integrate over multiple variables affecting pollution loads. A third option is ambient-based group incentives (see Box 5.3).

The economically preferred choice for cost-effective control is a performance base, provided that a valid and reliable performance indicator is available and can be implemented at reasonable cost. Performance-based instruments allow farmers maximum flexibility to use their specialized knowledge about the land they farm and their farming operations to improve their performance as measured by the indicator and to do so at minimum cost.

Performance-based instruments also require fewer design choices from policy makers than input- or practice-based policies. For example, suppose that the practices observed on a farm are the tillage practices, nutrient and pest management practices, and land allocated to grassed waterways or buffers. A practice-based policy would require a price for each practice. Because the crop or crop rotation affects pollution runoff, the prices would be contingent on the crop. If the practice-based policy is spatially differentiated as required for cost-effectiveness, a farm may have different prices for each practice for each field contingent on the crop. That is a lot of prices to calculate, and to implement would require many decisions by the policy maker. The likelihood of pricing mistakes that diminish cost-effectiveness and possibly have unintended consequences would not be trivial. Transaction costs would be high. A performance indicator that uses the same information to compute a single metric for farm performance eliminates the need for multiple prices. A price is needed only for the performance indicator.

One lesson that has been learned is that the performance is optimally measured at the whole-farm level. This is necessary to limit slippage or other economic spillovers within a farm that offset the effects of pollution control practices.

9.4.4 Public Vs. Private Funding

Another fundamental problem with the voluntary compliance approach is the reliance on public funding. The effectiveness of the voluntary compliance approach ultimately depends on the public funds allocated to incentivize practice adoption. The budgets of even the richest countries are under stress. Tying environmental outcomes to increasingly stressed public budgets is fundamentally unwise. Shifting from a pay-the-polluter to a polluter-pays paradigm would align agricultural policy with generally accepted principles for pollution policy and diminish the budgetary limits that exist and are likely to become increasingly tight. This is of course, a significant political hurdle.

9.4.5 Innovations in Incentive Design

The instruments that we find most interesting are water quality trading and water quality auctions. Both mechanisms are inherently single-objective, performance-based instruments that facilitate the integration of water science in policy design. Both mechanisms can facilitate designs that are spatially targeted and spatially differentiated as described in Chap. 5. Both mechanisms incentivize cost minimization by sources, incentivize innovation, and address adverse selection. In the case of water quality trading, an additional virtue is the potential to integrate the management of point and nonpoint sources to improve the overall cost-effectiveness of water pollution control.

Several experiments suggest that the water quality trading mechanism has potential as an element of better water quality policy (Shortle 2013). The Greater Miami River, Lake Taupo, and South Nation River cases provide especially encouraging experiences for the potential of water quality trading programs that include agriculture. Unfortunately, the overall experience from point-nonpoint source trading, which is almost exclusively that of US programs, does not yet make a compelling case for routine use of the mechanism. The partially capped baseline-and-credit trading model that has become the norm is cumbersome and invites problems from additionality and leakage. In application, participation requirements are often steep, deterring participation, and high point-nonpoint trade ratios intended to address uncertainty about nonpoint emissions reduction are a disincentive to trading and may have perverse effects on emissions variability. But the most significant limitation for trading activity has been limited demand for agricultural credits. For institutional and economic reasons, point sources have generally shown little interest in purchasing agricultural credits. This is lethal for the partially capped baseline-and-credit trading model which by design relies on regulated point sources purchasing pollution reductions from agricultural sources. Would economic and environmental outcomes be better if the point source demand were greater? Economically yes, as trading could then provide the point source cost saving that the partially capped model is designed to provide. But environmentally, probably not much. There is little room to produce large reductions in agricultural loads where agriculture is by far the dominant pollution source. For trading markets to realize their potential, agricultural sources must be placed under an emissions cap as in the Lake Taupo model.

Water quality auctions are rare, but the basic auction mechanism has substantial demonstration. Auctions can be used to facilitate spatial targeting to hot spot regions and watersheds. Unlike trading, government conservation budgets implement an effective demand for purchasing water quality improvements if conservation budgets remain adequately funded. Thus, water quality auctions are an appealing option to improve cost-effectiveness and environmental targeting of water quality policies in a pay-for-performance policy paradigm. They can be applied in multiple regional levels and when environmental indicators are well designed, facilitate spatial targeting to critical source areas.

Credit stacking involves a significant expansion of the use of markets for agri-environmental management. In our context, credit stacking would allow activities done to generate credits for sale in a water quality market to also generate credits for sale in other extant markets, such as a carbon market. Stacking has significant appeal in theory and has recently received much attention as a potential mechanism for agri-environmental management because it would increase resources available for purchasing pollution reductions from farms. As a mechanism, stacking goods like carbon that are complementary in supply with nutrients can be expected to increase pollution control efforts compared to water quality markets alone, because it entails using multiple market-driven prices to target multiple pollutants stemming from a given arable land area. Experience from credit stacking is still lacking and awaits the development of the underlying markets for water quality, carbon, or other environmental goods. Research suggests that the gains from trade in stacked markets depend significantly on design features like baseline requirements and trade ratios in the interconnected markets, and the potential sensitivity of the market to events in the interconnected market.

Crucial to all three of the innovative mechanisms are methods for determining changes in environmental outcomes due to policy changes in farming practices. Are policies that have this requirement plausible? We believe the answer is yes for nutrients from a technical perspective based on policy developments and the evolving science of managing nutrients. Emerging water quality trading programs use performance indicators of varying complexities. Several provide examples of science-based policy architectures that are promising for economically and ecologically beneficial nutrient trading. The Pennsylvania trading program, for instance, has well-developed rules for computing credits using relatively simple spreadsheets. Underlying the rules are physical relationships based on agronomic principles and pollution load models that have been developed and are used for regulatory purposes. The program design clearly benefits from the prior development of the Chesapeake Bay Watershed Model by the Chesapeake Bay Program.[1] The major limitation of the program as a pollution control mechanism is that instead of capping whole-farm emissions, it is a partially capped baseline and credit program. The cap-and-trade program for Lake Taupo offers both science-based trading rules and a cap needed to drive the market.

This is a domain in which perfect should not be the enemy of the good. As discussed in Chaps. 3 and 6, there are known unknowns about the effectiveness of Best Management Practices (BMPs), optimal watershed

[1] A strong argument for the capacity of water quality managers to use performance-based approaches at fine spatial scales using high-resolution spatial data along with watershed modeling tools developed for the Chesapeake Bay has been made by Rich Batuik, former associate director for science, analysis and implementation of USEPA Chesapeake Bay Program Office, and Joel Dunn, president and CEO of the Chesapeake Conservancy (Batuik and Dunn 2020).

management, and the validity and reliability of pollution load models. These can be diminished by continued research. The highest priority for research to support effective nonpoint management is the development and improvement of models for estimating performance at farm and watershed scales. This is the necessary foundation for scientifically valid policy of any kind. Also important is adaptive learning by doing. With multiple applications, lessons can be learned about what works and what does not and under what circumstances in the application of performance-based policies. Policy experiments are essential to learning how to get prices right.

9.4.6 *Markets Vs. Charges*

In the domain of performance-based economic incentives which is to be preferred, charges or markets based on the estimated emissions? In Chap. 6 we described cap-and-trade as having an advantage when the water quality goal is achieved by a specific quantitative limit on polluting emissions. When emissions can be controlled deterministically, the cap in cap-and-trade places an upper limit on emissions. Changes in the demand to pollute will cause variations in the price of allowances but total emissions will not vary. An emissions tax fixes the price of emissions but allows emissions to fluctuate with changes in the demand to pollute. Furthermore, given uncertainty about the demand for pollution, a tax cannot be designed to meet a specific quantitative limit. Accordingly, if the policy goal is a quantitative limit, cap-and-trade has the advantage when emissions are deterministic.

Variability in pollution loads due to stochastic weather will always be present in nonpoint pollution. But if the goal is to limit steady state annual average loads, the appeal of cap-and-trade remains. If policy is not committed to a quantitative limit, charges may have some economic appeal. Charges may offer an advantage of lower transaction costs compared to markets (Stavins 1995). Furthermore, research demonstrates that depending on the characteristics of the benefits and costs of pollution, there are conditions in which a cap will provide higher social net benefits on average than a charge and vice versa (Weitzman 1974). The conditions are not fully understood (e.g., Horan and Shortle 2021) but empirical research has demonstrated the ambiguity (Borisova et al. 2005). The implication is that the instrument that maximizes net benefits may be either charges or quantity controls, and which is the case is an empirical question. That said, charges find very little use in water pollution control. Quantitative limits on pollution and markets seem the primary path of innovation at this time.

9.5 Fitting the Policy to the Problem

Agricultural nonpoint pollution is pervasive, but its significance and the severity of policy challenges vary from place to place. In some places, it is a leading cause of water quality problems. In some cases, the problems are small and localized, and perhaps of little or no consequence. Essential to efficient and effective control is fitting the mix of policy choices to the problem.

Consider the simple case in which one or a small number of farms damage a small local resource but have no water quality impact beyond that. For example, runoff from a dairy barnyard or a few dairy barnyards is transported a short distance to a small lake. Damages might be pathogens if the lake is used for drinking water, seasonal algae blooms, or perhaps periodic fish kills with large manure spills. With small numbers and basic water science, the pollution sources are known. Technologies for reducing or eliminating manure are well known. Water quality damages are well known.

Small, localized problems of this kind are generally better suited to decentralized than centralized solutions (Weersink et al. 1998). One reason is that cost-effective public policy interventions are costly to develop, implement, and enforce, especially when problems are heterogeneous. Small problems may not provide sufficient social net benefits to warrant the costs of public intervention. A second reason, as discussed in Chap. 2, is that decentralized solutions can produce social net benefits for small-scale externalities. If property rights pertaining pollutant discharges are well defined, the decentralized solution might take the form of Coasian bargaining between the producer(s) and the owner(s) of the lake. Alternatively, if polluting events are infrequent (e.g., breached manure storage facilities, accidental spills), the decentralized solution might take the form of a liability rule under which the responsible producer(s) must pay damages when they occur. The role of the state in these cases is to provide the legal frameworks for these mechanisms.[2]

For problems involving larger numbers of producers and multiple affected resources, and those that result from routine farming practices rather than infrequent events, active and continuous policy interventions are required to address water quality externalities. But severity may still vary from relatively low to very high. Public costs of intervention should be measured against likely benefits. Low-intensity interventions will be appropriate in cases where agriculture is somewhat of a problem but not enough of one to require a significant public response. Low-intensity interventions include nudges (Chap. 5) to encourage producers to adopt pollution control practices and perhaps provision of technical assistance (Chaps. 4 and 5) for farm planning and practice implementation. Nudges are generally low-cost information programs. Technical assistance requires the time of experts and will be correspondingly expensive. A somewhat more intensive intervention would be to add to these low-intensity measures planning requirements for producers to meet minimal standards for sediment and erosion control, and nutrient, manure, and pest management.[3] A step-up in intensity would be to require certain types of producers based on size, location, and farming operations to actually implement plans and be subject to monitoring and enforcement actions. Another step-up in intensity would be to issue "threats" (Chap. 5) of future intensive regulation should water quality conditions in an area not meet established goals.

When significant change is considered essential, as is true for many regions, some elements of the "softer approaches" described above may be maintained in the policy mix, but significant change requires

[2] A caveat. If small, localized problems are of a common type (e.g., cattle in streams, barnyard runoff) and frequent and pervasive across an agricultural region, a central authority might consider interventions (e.g., runoff controls for barnyards, restrictions on cattle access to streams) for farms located in high-risk locations. Problems being well understood and of a common type would suggest that solutions are generally also of a common type and not highly specialized for specific locations. A centralized approach could provide social benefits if the public costs of developing and enforcing measures were less than the private cost of negotiating and enforcing private agreements, or the private costs of lawsuits to recover damages. Yet another approach would be to establish standards that are unenforced unless triggered by water quality problems. This approach would save enforcement costs but still provide farmers incentives to take actions that minimize their costs.

[3] In suggesting minimal standards, we do not mean to endorse the extensive use of intrusive regulatory standards. Rather, we have in mind truly minimal context-specific requirements we have described in previous chapters.

implementation of instruments capable of bringing it about. This will be a mix of mandatory compliance instruments, ideally implemented following the guidelines we have described above and in previous chapters.

Finally, while national policies are crucial to national successes, the importance of context to placed-based solutions means that one-size-fits-all approaches are suboptimal. Flexibility is needed to implement place-based solutions and to foster innovation. The most interesting developments in pollution policy for agriculture have been the product of regional and local initiatives to address regional and local policy dilemmas.

References

Anthoff, D., and R. Hahn. 2010. Government failure and market failure: On the inefficiency of environmental and energy policy. *Oxford Review of Economic Policy* 26 (2): 197–224.

Batuik, R., and J. Dunn. 2020. Let's start rewarding Bay cleanup performance. *Bay Journal.* https://www.bayjournal.com/opinion/forum/lets-start-rewarding-bay-cleanup-performance/article_15d38ece-0f1d-11eb-bcc1-e3e3099509ca.html. Accessed 16 Nov 2020.

Beard, N., and A. Swinbank. 2001. Decoupled payments to facilitate CAP reform. *Food Policy* 26 (2): 121–145.

Borisova, T., J. Shortle, R.D. Horan, and D. Abler. 2005. Value of information for water quality management. *Water Resources Research* 41 (6).

Harvey, D.R. 2004. Policy dependency and reform: Economic gains versus political pains. *Agricultural Economics* 31 (2-3): 265–275.

Horan, R.D., and J.S. Shortle. 2021. The song remains not the same: prices vs. quantities with correlated uncertainty. *Journal of the Association of Environmental and Resource Economists.* Forthcoming.

National Research Council (NRC). 2001. *Assessing the TMDL approach to water quality management.* National Academies Press.

———. 2012. *Science for environmental protection: The road ahead.* National Academies Press.

Organization for Economic Cooperation and Development (OECD). 2010. *Environmental cross compliance in agriculture. OECD studies on water.* OECD Publishing.

———. 2017. *Diffuse pollution, degraded waters: Emerging Policy Solutions.* OECD Publishing.

———. 2018. *Implementing the OECD principles on water governance: Indicator framework and evolving practices, OECD studies on water.* OECD Publishing.

Royer, M.B., R.P. Brooks, J.S. Shortle, and S. Yetter. 2020. Shared discovery: A process to coproduce knowledge among scientists, policy makers, and stakeholders for solving nutrient pollution problems. *Journal of Environmental Quality.*

Sexton, S.E., Z. Lei, and D. Zilberman. 2007. The economics of pesticides and pest control. *International Review of Environmental and Resource Economics* 1 (3): 271–326.

Shortle, J. 2013. Economics and environmental markets: lessons from water-quality trading. *Agricultural and Resource Economics Review* 42 (1203-2016-95493): 57–74.

Shortle, J.S., J.R. Mihelcic, Q. Zhang, and M. Arabi. 2020. Nutrient control in water bodies: A systems approach. *Journal of Environmental Quality* 49 (3): 517–533.

Stavins, R.N. 1995. Transaction costs and tradeable permits. *Journal of Environmental Economics and Management* 29 (2): 133–148.

Stephenson, K., Z. Easton, L. Shabman, and J.S. Shortle. 2020. Confronting the agricultural nonpoint policy problem. *Bay Journal.* Forthcoming.

Waterfield, G., and D. Zilberman. 2012. Pest management in food systems: An economic perspective. *Annual Review of Environment and Resources* 37: 223–245.

Weersink, A., J. Livernois, J.F. Shogren, and J.S. Shortle. 1998. Economic instruments and environmental policy in agriculture. *Canadian Public Policy/Analyse de Politiques*: 309–327.

Weitzman, M.L. 1974. Prices vs. quantities. *The Review of Economic Studies* 41 (4): 477–491.

Zilberman, D., and K. Millock. 1997. Pesticide use and regulation: Making economic sense out of an externality and regulation nightmare. *Journal of Agricultural and Resource Economics*: 321–332.

Zilberman, D., A. Schmitz, G. Casterline, E. Lichtenberg, and J.B. Siebert. 1991. The economics of pesticide use and regulation. *Science* 253 (5019): 518–522.

Index[1]

[1] Note: Page numbers followed by ‘n’ refer to notes.

J. Shortle et al., *Water Quality and Agriculture*, Palgrave Studies in Agricultural Economics and Food Policy,
https://doi.org/10.1007/978-3-030-47087-6

The manufacturer's authorised representative in the EU is Springer Nature Customer Service Centre GmbH, Europaplatz 3, 69115 Heidelberg, Germany. If you have any concerns regarding our products, please contact ProductSafety@springernature.com

Printed and bound by CPI Group (UK) Ltd, Croydon, CR0 4YY
30/09/2026
02225646-0008